Method	Type of change	Time pattern	Pros	Cons
Time series	Movement or change in character	Trend cycle before and after	Strong visual impact if change is substantial; shows conditions at each data/time	Readers have to visually compare maps to see where, and how much change occurred
Tracking map	Movement	Trend cycle before and after	Easier to see movement and rate of change than with time series, especially if change is subtle	Can be difficult to read if more than a few features
Measuring change	Change in character	Trend before and after	Shows actual difference in amounts or values	Doesn't show actual conditions at each time; change is calculated between two times only

meaningful difference between numbers, w/o real origin - Interval

establish identity - Nominal

establish rank between features w/o values is - Ordinal

identifies meaningful differences between numbers w/real origin - Ratio

relates to ruggedness - fractal

A sensor's dynamic range defines the range of - brightness that can be accureately recolded.

FUNDAMENTALS OF GEOGRAPHIC INFORMATION SYSTEMS

THIRD EDITION

Michael N. DeMers

New Mexico State University

WILEY

JOHN WILEY & SONS, INC.

ACQUISITIONS EDITOR   Jerry Correa

ASSOCIATE EDITOR   Denise Powell

SENIOR PRODUCTION EDITOR   Valerie A. Vargas

EXECUTIVE MARKETING MANAGER   Clay Stone

SENIOR DESIGN MANAGER   Harry Nolan

NEW MEDIA EDITOR   Thomas Kulesa

PRODUCTION SERVICES   Hermitage Publishing Services

COVER DESIGN   Harry Nolan

COVER PHOTO   © Dr. Sigrid Hehl-Lange, Dr. Eckart Lange, ETH Zurich, Swiss Federal Institute of Technology. Used with permission of Dr. Sigrid Hehl-Lange, Dr. Eckart Lange & Dr. Rodney Hoinkes, Center for Landscape Research, University of Toronto.

This book was set in 10/12 Cheltenham by Hermitage Publishing Services and printed and bound by Hamilton. Phoenix Color, Corp. printed the cover.

This book is printed on acid free paper. ∞

Library of Congress Cataloging-in-Publication Data

DeMers, Michael N.
  Fundamentals of geographic information systems / Michael N. DeMers.—3rd ed.
  p. cm.
  Includes bibliographical references.
  ISBN 0-471-20491-9 (cloth : alk. paper) — ISBN 0-471-45149-5 (WIE : alk. paper)
  1. Geographic information systems. I. Title: Geographic information systems. II. Title.

G70.212.D46 2005
910'.285—dc22   2004042236

Printed in the United States of America

10 9 8 7 6 5 4 3 2 1

GIS technology continues to proliferate at an alarming rate, putting increasing pressure on community colleges, four year colleges, and universities to produce graduates who are knowledgeable in the concepts and ideas that underlie the technology. At the same time, these graduates need to be capable of employing the technology in real-world problem solving environments regardless of which software package is being employed. While software vendors are increasingly providing tutorials and other hands-on materials for teaching their own software, they often leave large portions of conceptual content out of these materials. This requires that colleges and universities become responsible for placing much of the current concept level educational resources in the hands of our future GIS technicians, applications specialists, domain experts, and programmers.

When I wrote the first edition of this text there were no formal tutorial books available and very few conceptual level resources outside of the National Center for Geographic Information and Analysis (NCGIA) curriculum materials. This left the instructors with limited guidance as to what topics should be covered, at what levels, and in what sequence. Today there are many tutorials (including on-line tutorials), several textbooks, and a wide range of approaches. Each has its own sequence, its own focus, and often its own audience. Some audiences are narrow, while others are broad. Some texts are focused on the technology and its immediate use while others attempt to explain the underlying geographic and analytical concepts of GIS. I have always maintained that it is a university's fundamental mission to teach, not to train—to provide the background, theory, and conceptual material that will allow anyone to use and adapt to any software package for nearly any task. While university instructors, myself among them, continue to get pressure from the student to have "less ideas" and more "software," I feel it is imperative that we not succumb to this pressure. While using software is often more entertaining, and I am fully aware of the need for trained technicians, please understand that without the conceptual background the software is of limited use. The more you know about the concepts of GIS, the more you will be able to demand of the software and perhaps even of yourself.

In this third edition I have attempted to maintain the original look and feel of the first edition, but have also incorporated some of the robust ideas that have been generated and formalized over the last six years by a wide range of GIS specialists and educators. The Geographic Information Science and Technology strawman document available for download at (www.ucgis.org) outlines, in some detail, the body of knowledge in geographic information science and technology. This document, despite being somewhat unfinished, even larval, provides an enormous amount of content, concepts, learning objectives, and resources. While it awaits public comment and input, I have taken the liberty of borrowing liberally from it in the preparation of this book.

With this particular incarnation of *Fundamentals of Geographic Information Systems* I ask that you, the reader, and/or your instructor take the time to visit the University Consortium for Geographic Information Science (UCGIS) website, download the straw document, and provide the authors of that report with your ideas. Feel free to compare what you see there with what is contained in this text. If you are a student, please provide feedback to your instructor about this textbook and ask them to pass it along to the editorial staff at John Wiley & Sons. I assure you I will take the time to incorporate, to the best of my ability, all your helpful comments.

ACKNOWLEDGMENTS

As with the previous incarnations of this text, I am deeply grateful to the many students, faculty, and reviewers who have suggested changes for the third edition. I would like to acknowledge the following reviewers for their help: Rajendra Kumar Aggarwala—University of Michigan; John Wilson—University of Southern California; Wei Luo—Northern Illinois University; Patrick Colgan—Northeastern University; and Michael Emch—Portland State University. I am particularly indebted to Dr. Elizabeth Wentz whose students at Arizona State continue to keep me on my toes. I wish to thank the editorial staff at Wiley for their hard work, talent, and professionalism. As always I thank Dr. William A. Dando for his continuous support of my efforts with regard to this text. Finally, I want to thank all the members of the model curriculum task force (particularly Dr. Duane Marble for initiating it) for their hard work, for their many and diverse insights, and for their friendship over the years that we worked on the straw document together.

CONTENTS

Chapter 12   Comparing Variables Among Maps   303

Chapter 13   Cartographic Modeling   328

Chapter 16   GIS Applications   403

Chapter 17   GIS Research   418

Appendix A   Software and Data Sources   433

Appendix B   Using the Wiley World Wide Web to Find
Data and GIS Examples   439

Glossary   441

Index   461

Photo Credits   468

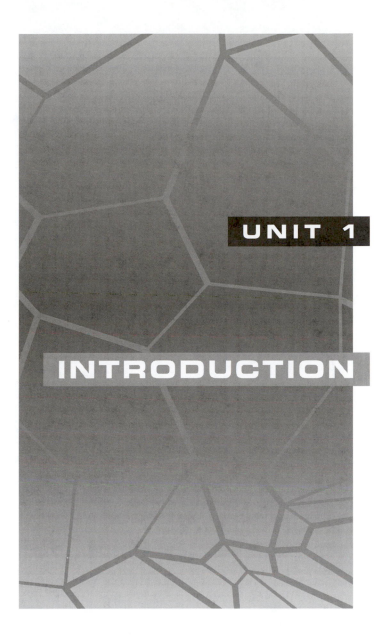

UNIT 1

INTRODUCTION

CHAPTER 1

Introduction to Digital Geography

This is a book about geography. Specifically it is about geographic information systems (GIS)—a fundamental set of digital ideas and concepts rooted in over 2500 years of exploration and geographic research (Dobson, 1995) and designed to provide answers to questions based on mapped data. GIS reflects a change from pure exploration to discovery and locating new lands, new people, and new resources through the mapping of their distributions, to explanation of these patterns. It most specifically demonstrates the change toward questioning the impacts of pattern and the use of this knowledge to predict future distributions and effects.

The GIS is a modern enabling technology for the practitioner, an integrating set of ideas for a wide range of disciplines, and a science in its own right for the theoretician. No matter what field of science or social science you pursue, the ideas you will encounter in this book will both fortify you and enhance your own understanding of your chosen discipline. You will learn to think both spatially and quantitatively. Hopefully, you will become more confident in an increasingly technological world, and this confidence will enable you to be successful and to enjoy the contributions you can make to your society and to your natural world.

LEARNING OBJECTIVES

When you are finished with this chapter you should be able to:

1. Define what a geographic information system (GIS) is.

2. Explain the parallels between the evolution of geographic thought and the GIS.

3. Explain the initial impetus for its development and describe some of the difficulties encountered during its early development.

4. Describe the relationships among a GIS, computer-assisted cartography (CAC), and computer-assisted drafting (CAD).

5. Place GIS in a hierarchical framework with other types of information systems.

6. Describe the relationship between the traditional analog map and a GIS on the basis of the four-subsystem definition.

7. Describe some basic analytical capabilities of a modern GIS.

8. Suggest possible users of a GIS and how it might benefit them.

WHY GEOGRAPHIC INFORMATION SYSTEMS?

In the early 1960s, the Department of Forestry and Rural Development of Canada decided to pursue a large-scale project to manage the resources for much of their territory (Tomlinson, 1984). Among the initial tasks was the inventory and mapping of the available forest and mineral resources, wildlife habitat requirements, and water availability and quality, to list just a few. The inventory was meant as a first step in a larger endeavor to compare the maps to each other to predict the longevity of the resource base and to develop a management strategy to ensure the long-term availability of both renewable and nonrenewable resources without damaging the environment.

The mission just outlined is a daunting one, requiring enormous amounts of data gathering, compilation, evaluation, analysis, and modeling. Yet this is often the very task that faces natural resource managers. Immediately, it becomes obvious that a map of the resources would permit viewing the extent, quality, and current rate of use at a single glance. The manual production of such very helpful maps covering an entire country, especially a large one, would call for the employment of perhaps hundreds of cartographers and considerable amounts of time and money. Many of you are painfully aware of the limited coverage and variable quality of topographic, vegetation, and soil maps for your own region or nation. Depending on the size of your area, it is entirely possible that long before such a task is completed, the resources themselves will have disappeared, leaving the environment despoiled and the local residents up in arms.

A chance meeting between the director of the Department of Forestry and Rural Development (Roger Tomlinson) and an IBM executive resulted in a suggested application of the newly emerging computer technology to the problem. What was needed was a GIS for Canada. Thus the newly instituted Regional Planning Information Systems Division, funded by the Canadian government, was assigned to produce what was to become the first geographic information system ever built—the **Canada Geographic Information System (CGIS)**. Its initial task was to classify and map the land resources of Canada.

Today there is an ever increasing recognition of the need to perform large-scale mapping and map analysis operations for a wide variety of traditionally manual tasks. Foresters wanting to keep an up-to-date inventory of their timber resources see GIS as an efficient management tool for their day-to-day operations. Fire departments have a need for GIS to enhance their routing capabilities to ensure rapid response in emergencies. The military could use GIS to determine appropriate battle plans and to organize troop movements. Cellular phone companies, wanting to provide the best service for a mobile customer base, must site their transmission towers to avoid conflicts with neighbors and still allow clear line of sight for signal transmission. Local governments employ GIS to develop growth and development plans and to modify zoning regulations to account for increasing population pressures. Businesses are using GIS to market products and even to develop mailing lists based on selected spatial criteria. Real estate companies are beginning to use GIS to isolate available housing on the basis of customer criteria such as proximity to schools, type of neighborhood, or access to highways. Police departments are currently using GIS to compile information to characterize the movements and operational settings of suspected serial killers. Academic disciplines such as geography, biology, geology, landscape architecture, range science, and wildlife management now have the capability to employ the technology to develop and test hypotheses concerning patterns of natural phenomena on the earth. The potential users of GIS are nearly limitless, and the types and numbers of users are growing at a logarithmic pace.

This growth is indicative of the nature of GIS as an empowering technology. It is not unlike the development of the printing press, the creation of the first telephone, the replacement of the horse and buggy with the automobile, or the introduction of the first computer. All these innovations had a profound impact on the way in which we communicated, the way we moved from place to place, and the way we solved problems—even on the nature of the problems we solved. The modern GIS has enhanced the utility of the map by replacing it with a large number of mapped coverages, each with an interrelated theme. These maps can be automatically analyzed, and their themes combined to give meaningful answers for decision makers. The GIS is changing the way we do things with maps, the way we think about geographic information, even the way in which geographic data are collected and compiled. Tasks that were impossible with traditional maps are now commonplace. Recent market trends for the technology indicate that it is a major growth industry, far outstripping many others, even during recession years (*U.S. News & World Report,* 1995). *Directions* Magazine estimated that there was over $1 billion in GIS software sales alone in 2001, with additional billions spent for services, hardware, and related activities (http://www.directionsmag.com/pressreleases.php?press_id=5946). As more organizations become familiar with the GIS, the need to become familiar with its basic principles will grow as well. There will also be an increase in the demand for people knowledgeable about the concepts behind the technology. We will examine these concepts here, with an aim toward understanding how spatial phenomena can be manipulated and how the technology can help us in an increasingly complex world.

WHAT ARE GEOGRAPHIC INFORMATION SYSTEMS?

Geographic information systems are tools that allow for the processing of spatial data into information, generally information tied explicitly to, and used to make decisions about, some portion of the earth. This working definition is neither comprehensive nor particularly precise. Like the field of geography itself, the term is difficult to define and represents the integration of many subject areas. As a result, there is no absolutely agreed upon definition of a GIS. The term itself is becoming hybridized and modified to conform to intellectual, cultural, economic, and even political objectives (Table 1.1). This terminology has, in fact, become extremely elastic, resulting in an increasingly confusing jargon due to new definitions that constantly creep into both the scientific and the popular literature.

This lack of accepted definition has resulted in many gross misconceptions about what a GIS is, what its capabilities are, and what such a system might be used for. It has led some people to believe, for example, that there is no difference between **computer-assisted cartography (CAC)**, **computer-aided drafting (CAD)**, and GIS. Because the graphic display from these three systems can look identical to both the casual and the trained observer, it is easy to assume that they are, with minor differences, the same thing. Anyone attempting to analyze maps will soon discover, however, that CAC systems, computer systems designed to create maps from graphical objects combined with descriptive **attributes**, are excellent for display but generally lack the analytical capabilities of a GIS. Similarly, for pure mapping purposes it is highly desirable to use a CAC system developed specifically for the input, design, and output of mappable data rather than working through the myriad analytics of the GIS to produce a simple map. CAD—a computer system developed to produce graphic images but not normally tied to external descriptive data files—is excellent software for the architect, speeding the process of producing architectural drawings and simplifying the editing process. It would not be as easy to use for producing maps as would CAC, nor would it be capable of analyzing maps—generally the primary task assigned to the GIS (Cowen, 1988).

For the experienced GIS user, there is no need for an explicit definition. The complex geographic queries that demand its use normally could not be addressed by CAC and CAD. But for those who have only heard of these tools a definition might prove useful. A preliminary definition for consideration might be that of David Rhind, who defined GIS as "a computer system for collecting,

TABLE 1.1   Examples of Synonymous Terms for *Geographic Information System* and the Source or Motivation Behind Their Derivation

Terminology	Source
Geographic information system	United States terminology
Geographical information system	European terminology
Geomatique	Canadian terminology
Georelational information system	Technology-based terminology
Natural resources information system	Discipline-based terminology
Geoscience or geological information system	Discipline-based terminology
Spatial information system	Nongeographical derivative
Spatial data analysis system	Terminology based on what system does

checking, integrating and analyzing information related to the surface of the earth" (Rhind, 1988). This definition has some highly worthwhile elements that should be examined. First, it indicates that the GIS deals with the surface of the earth. Although this is not an absolute requirement, the vast majority of GIS applications do deal with portions of the earth. Moreover, the statement that the GIS is used to collect, check, integrate, and analyze information enumerates a large number of the necessary groups of operations for any GIS.

Many additional definitions of GIS have been proposed. Some have shown the strong linkage between manual and computer-based methods of map analysis (Aronoff, 1989; Dickinson and Calkins, 1988; Star and Estes, 1990). Most other definitions have explicitly stated among GIS's primary objectives is its intent to act as a tool for analyzing data about the earth. (Some examples are Aronoff, 1989; Cowen, 1988; Dueker, 1979; Koshkariov et al., 1989; Parker, 1988; and Smith et al., 1987.) As we will see at the end of this text, one can also extend the definition to include the organizations and people involved in working with spatial data as well (Carter, 1989). Like any technology that changes as quickly as does GIS, the definitions themselves will likely change as well.

For this text I have chosen to use a definition that more closely resembles the way the GIS operates as a series of subsystems within a larger system. That definition—proposed as a standard by Marble and Peuquet (1983) and used in some form by others in their own definitions (Burrough, 1986; Ozemoy et al., 1981; Parker, 1988)—pretty much sums up what it is we do with a GIS and how we do it. It states that GIS deals with space–time data and often, but not necessarily, employs computer hardware and software. More importantly perhaps is the subsystem nature of this definition, which provides an easily understandable framework for the study of GIS. The GIS, according to this definition, has the following subsystems:

1. A data input subsystem that collects and preprocesses spatial data from various sources. This subsystem is also largely responsible for the transformation of different types of spatial data (i.e., from isoline symbols on a topographic map to point elevations inside the GIS).

2. A data storage and retrieval subsystem that organizes the spatial data in a manner that allows retrieval, updating, and editing.

3. A data manipulation and analysis subsystem that performs tasks on the data, aggregates and disaggregates, estimates parameters and constraints, and performs modeling functions.

4. A reporting subsystem that displays all or part of the database in tabular, graphic, or map form.

The subsystem definition allows for comparison between the modern digital GIS and its analog counterpart, particularly when considering the steps in the **cartographic process** (Table 1.2). The first GIS subsystem, the data input subsystem, is roughly equivalent to the first and second steps in the cartographic process—data collection and map compilation (Robinson et al., 1995) (Table 1.3). In traditional cartography the cartographer compiles or records a map made up of points, lines, and areas on a physical medium such as paper or Mylar. The data are collected from such sources as aerial photography, digital re-

TABLE 1.2 Comparison of the Cartographic Process as Applied to Traditional Cartography (Map) and Geographic Information Systems (GIS)

Map	GIS
Data collection: aerial photos, surveys, etc.	Data collection: aerial photos, surveys, etc.
Data processing: aggregation, classing, etc.; linear process	Data processing: aggregation, classing, plus analysis; circular process
Map production: final step except for reproduction and dissemination	Map production: not always final step; normally one map used to produce still more
Map reproduction	Map reproduction

mote sensing, surveying, visual descriptions, and census and statistical data. The automated counterpart uses electronic devices to record or **encode** points, lines, and areas into a computer system. Data collection sources are often the same as those used for traditional mapping but now include a wide variety of digital sources: **digital line graphs (DLGs)**, **digital elevation models (DEMs)**, **digital orthophotoquads**, and many more. Although the mechanics differ between the two technologies, the actual methods are strikingly similar.

This is also the case for the second subsystem, the storage and retrieval subsystem (Table 1.4). Although there is no actual counterpart in the cartographic method, the map itself is the storage and retrieval pool. Points, lines, and areas that have been placed on the cartographic document are stored there for retrieval by the map reader. It has been said that the map is the most compact medium for the storage of spatially related information and may be the most complex form of graphic device available. In fact, the compactness of the map and its complexity frequently hamper the map reader's ability to extract information.

TABLE 1.3 Analog Versus Digital Geographic Information System (GIS): A Comparison of Input Subsystem Functions

Map	GIS
Input recorded (compiled) on paper from a collected source	Input: "encoded" into the computer from a collected source
• Points • Lines • Areas	• Points • Lines • Areas
Sources	Sources
• Aerial photography • Digital remote sensing • Surveying • Visual descriptions • Census data • Statistical data, etc.	• Same as map data • Digital line graph (DLG) • Digital elevation models (DEMs) • Digital orthophotoquads • Other digital databases

TABLE 1.4   Analog Versus Digital Geographic Information System (GIS): A Comparison of Storage and Retrieval Subsystem Functions

Map	GIS
Points, lines, and areas are drawn on paper with symbols.	Points, lines, and areas are stored as grid cells or coordinate pairs and pointers in computer.
Retrieval is simply a matter of map reading.	Attribute tables are associated with coordinate pairs.
	Retrieval requires efficient computer search techniques.

The GIS storage and retrieval subsystem has some advantages over the graphic map in that queries can be made of the data and only the appropriate, context-specific information recalled (Table 1.4). This format places more emphasis on formulating queries and asking the appropriate questions and less on overall map interpretation. In general terms, this subsystem stores the graphic locations of point, line, and area objects **(entities)** and their associated characteristics **(attributes)**. Computer search methods are inherent in the GIS programs themselves to allow questions to be asked and appropriate answers to be given.

In the analysis subsystem, once again there is no exact cartographic method counterpart, except that the map is a fundamental tool for the analysis of spatially related data (Table 1.5). The analog map requires rulers to measure distances, compasses to find directions, and dot grids or **planimeters** to measure areas (Marble, 1990). Furthermore, the map analyst is restricted to the graphic methods used to present the data on the piece of paper or Mylar. Still, these map analysis tools have been used for a great many years because of the known utility of comparing spatially related phenomena in a quantitative manner.

The analysis subsystem is the heart of the GIS. The need to analyze maps to compare and contrast patterns of earth-related phenomena, exemplified by the long-standing tradition of doing so with traditional maps, provides an impetus to find more convenient, faster, and more powerful methods. GIS analysis uses the power of the modern digital computer to measure, compare, and describe the contents of the databases. It allows ready access to the raw data and per-

TABLE 1.5   Analog Versus Digital Geographic Information System (GIS): A Comparison of Analysis Subsystem Functions

Map	GIS
Requires rulers, planimeters, compasses, and other tools all used by the human analyst	Uses the power of the computer to measure, compare, and describe contents of the database
Restricted to the data as they are aggregated and represented on the paper map	Allows ready access to the raw data and allows aggregation and reclassification for further analysis

mits aggregation and reclassification for further analysis. Not only is it not limited in the types of data it can retrieve but it can combine selected data sets in unique and useful ways far beyond what the traditional map could provide on a single sheet (DeMers, 1991).

Of course, once an analysis has been performed, there is generally a need to report these results. In cartography, whether it be traditional analog cartography or its digital equivalent, computer-assisted cartography, the output is generally the same—a map. The most common purpose of cartography, at least from the user perspective, is to produce a map product, usually in copies for multiple recipients. In fact, production and reproduction are the final two steps in the cartographic method (Robinson et al., 1995).

A major difference between GIS and cartography, beyond the emphasis on analysis in GIS, is the method of reporting the results of analysis (Table 1.6). Although many users, perhaps even most, will still require mapped output, there are many options available in modern GIS. Some typical noncartographic output could include tables listing, for example, the anticipated crop yields per hectare by soil type or predicted changes in population densities by census district. Alternatively, either of these results also could be output as a series of histograms or line graphs. As supplements, digitally encoded photographs of selected sites could be placed on the map margins or within the tables or charts.

More advanced GIS features are available as well. Examples include output in the form of printed mailing labels for a search of a database of potential customers to facilitate the distribution of advertising. A 911 emergency system database could be connected to a police or fire department so that when a caller reported an emergency, the information could be directly routed to the nearest emergency service. This output could also be in the form of a route map showing the fastest path from the emergency branch to the site of the emergency. The types of output are often dictated more by the use for which the GIS is employed than by the software. And, like the users of maps, the outputs are many and varied.

Among the more interesting phenomena arising from the wide range of users is a new set of terms defining the system on the basis of what it does. For example, one could have a police information system, a natural resources information system, a census information system, a rangeland evaluation system, a land information system, a **cadastral** information system (one that shows landownership), and so on. Although these terms are generally descriptive of

TABLE 1.6   Analog Versus Digital Geographic Information System (GIS): A Comparison of Reporting and Output Functions

Map Output	GIS Output
Graphic device only	The map is only one type of GIS output
Many forms of maps	With minor exceptions, GIS offers same options as traditional hand-drawn maps
Modifications can include cartograms, etc.	
	Also includes tables, charts, diagrams, photographs, etc.

the use for which the GIS is employed, they do little to clarify the exact nature of the system. They generally add considerably to the confusion. Perhaps a more structured approach to classification of GIS in the form of a taxonomy would prove useful (Figure 1.1).

This taxonomy diagram clearly shows the separation between **spatial** and nonspatial information systems. The GIS appropriately fits under the spatial information systems category. Two general classes of spatial information systems are identified: geographic and nongeographic. Nongeographic information systems, although they frequently deal with some portion of geographic space, seldom have strong locational links to the earth itself. Thus, such systems as CAD and computer-aided manufacturing are nongeographic spatial information systems.

Within geographic information systems there is yet another bifurcation. GIS are divided into **land information systems (LIS)** and nonland information systems, or other GIS. Although the division is somewhat artificial, it is important because it separates the applications of GIS technology into those that are focused primarily on the land itself and those that, though being geocoded, are more focused on information that might either affect or be affected by land-related factors. These uses include census information systems whose primary focus is on populations and their housing and economic activities rather than on the land on which they reside or even on their use of the land. Another (nonland-related) GIS application might include applications surrounding political redistricting. Although political redistricting by its very nature subdivides or apportions the land into discrete portions, such activities generally have little or no direct and immediate impact on the land itself. Rather, political redistricting affects the voting patterns of those living on the land surface. A common nonland-related use of GIS is market analysis, which may include a determination of the amount of market within reasonable reach of a business **(allocation)** or might involve an analysis of existing facilities to determine the best place to put a competing or complementary facility **(location)**. Locating fire stations,

Figure 1.1   A taxonomy of information systems. The illustration shows how the geographic information system (GIS) and land information system (LIS) fit in. CAD is computer-assisted drafting; CAM is computer-aided manufacturing.

schools, and other facilities falls into this category. In general, nonland-related GIS activities tend to entail social, economic, transportation, and political types of activities.

Land-related activities provide the framework for the second and possibly the most often used type of GIS, the LIS. Such systems are based most often on the ownership, management, and analysis of portions of the earth that are most frequently of interest to humans primarily because of their condition of ownership. Land information systems are further subdivided into parcel based and nonparcel based. Nonparcel-based LIS include natural resources information systems, such as those used by national park services, the U.S. Forest Service, and land management agencies. Activities within the nonparcel-based LIS could include habitat evaluation, conservation easement procurement, wildlife evaluation, earthquake and landslide prediction, flood hazard abatement, chemical contamination evaluation, forest and range management, and scientific investigation.

Parcel-based LIS applications focus on landownership and other cadastral investigations. The defining criterion is that the land be divided into surveyed parcels having legal descriptions. Although this terminology could also apply to such portions of land as national forests, it generally assumes that the parcels are smaller than those normally encountered in national forests (National Academy of Sciences, 1980, 1983). Fundamental to applications of these types is a highly accurate **geodetic framework** on which the parcels can be precisely described. LIS applications involve traditional survey methods and are among the largest users of NAVSTAR's **global positioning system (GPS)**, a sophisticated satellite system, for acquiring this locational information. Once an accurate geodetic framework and cadastral system have been developed, many analyses of land-tenure change can be performed with the assurance of a high degree of measurement accuracy. Included in such studies are those attempting to arrive at compatible multiple land uses within selected parcels of land. Some of these studies may require the incorporation of a multipurpose cadastre—a parcelization framework that allows analysis of multiple land parcel-related phenomena.

Whether they are land related or human related, the applications of GIS technology are many and varied, offering enormous possibilities for both simple and extremely sophisticated analyses. Most of today's applications are quite limited in sophistication, however. Generally, this underuse of system capabilities seems to be related more to a lack of understanding of the existing potential of GIS than to actual software limitations. Before we can ask software to perform a particular task, we must be aware of what that task might be. Then we can see whether the software is capable of accomplishing it. People using today's GIS software are frequently heard saying, "Hey, I didn't know we could do that with the computer!" The exclamation is one of discovery, not unlike the reaction of geographers of old as they ventured into the jungle with pith helmet and machete. For the person newly introduced to GIS software, the journey into new dimensions of geographic exploration has just begun.

WHERE DO I BEGIN?

GIS is an exciting, even glamorous, field with rapidly expanding opportunities for those who are familiar with the concepts and the technology. It is a common fallacy that because GIS has become readily available and is showing up in a wide variety of organizations, it should be possible to just sit down and start using the software. GIS software, however, is not like the personal computer word processing software so familiar to us. Although most of us know some basics about writing and are perhaps skilled in computer word processing, few of us are comfortable with the analytical operations necessary to make decisions with maps. Just as word processing software assumes that you can organize your thoughts and ideas into coherent sentences and paragraphs, however, GIS assumes you are familiar with the vocabulary of maps.

When asked, most of us will say that we are fairly comfortable with maps. We use road maps routinely, and when necessary, we consult a world atlas, with its political, physical, and economic boundaries, and associated colors, graphic symbols, text, and, of course, north arrows. Most of us, however, don't think much about how much information a map contains. Nor do we give much thought to the generalization processes that take place to decide which detail gets included and which does not. Nobody wants to think about the problem of representing an essentially spherical surface onto a flat piece of paper. Because the map is such an elegantly designed document—so well thought out—we simply accept it at face value.

On occasion, however, the limitations of the cartographic art begin to show through. How often have you wondered why a road that looks straight on a map really curves all over the place? The graphic limits imposed on cartographers by available data quality, pen size, size of paper, and other conditions all require them to make conscious decisions about how much detail can and should be placed on a given map document. Much of this generalization is imposed by the map scale. The smaller the map scale (the larger the mapped area), the greater is the required generalization to produce the cartographic model.

This concept of the map as a model of reality is perhaps the most important concept that the future GIS professional must learn. Because the map has such a strong visual appeal, the viewer tends to accept it as reality. Those who work with maps, and especially those who work with the interactions of many maps, must constantly remind themselves of the limitations of the cartographic product. Here are just a few simple exercises you can do to become familiar with the cartographic model and some of its limitations.

Take a look at a number of world maps from different atlases you might find in your library. Pick out a country familiar to you. Notice how maps of it differ with respect to sizes, shapes, boundary configurations, numbers of cities, and the like on these maps. You might be surprised at the wide variation from one map to another. Consider, then, what you would have to do if you were to digitize a map of this country into a GIS. Which one would you select? Why? How does focusing on the purpose of your GIS project help you decide which map you want?

Obtain two or three adjacent topographic maps for your areas. What are the dates for each map? Are they the same? different? Now the fun begins. With clear plastic tape (preferably removable), tape the maps together so that all the lines match. Be sure to turn on some relaxing music while you do this. What do you discover? The lines don't match exactly. Imagine how you are going to input 20 or 30 of these documents to a GIS if the lines don't line up with any pair of them.

Soil data might be nice to include in your GIS. Try the last experiment with your local soil map sheets. The match between sheets is even worse. If you are using soil maps from the U.S. Department of Agriculture Soil Conservation Service, you might be quite taken with the use of the aerial photography in the background. Admittedly, this feature is nice to have. But the addition comes at a cost. If you are going to input this map with other maps inside a geographic information system, you will have to coregister it with all the others so that the features match. This requires that the locational coordinates be specified on all maps. Try to find these on the soil survey maps. How do you solve this little dilemma?

If the foregoing examples haven't convinced you of the importance of understanding the vocabulary of maps before you begin speaking GIS, perhaps this one will. You need to create a map of presettlement vegetation for your state or region. It turns out that three well-known vegetation mappers have compiled such maps for portions of your area of interest. Taking a trip to the library to obtain these maps, you discover that the first shows vegetation classified by its structural components (herbaceous, grasses, trees, shrubs, etc.), and the second map, which intersects with the first map, shows vegetation classed by floristics (based on species). You also note, to your annoyance, that the two systems seem to have only limited map areas that correspond. Hoping for help from the third map, you discover that, although it is classed on the basis of a combination of floristics and structure, its area does not overlap either of the other two maps; in fact, it is well separated from them.

Classification problems of the type just described are common ones requiring the student of GIS to become more than a student of the terminology. Before you can master the technology, you should first master its concepts. We will begin this first step in the journey in the next chapter, where we will look closely at the nature of geographic data and the methods by which they can be represented on map documents. This first step will give us a better appreciation of the fundamental building blocks of GIS and will ensure a more cautious approach when we begin implementing geographic analysis and cartographic modeling.

Terms

allocation	computer-aided drafting (CAD)	geodetic framework
attributes	digital elevation model (DEM)	global positioning system (GPS)
cadastral		land information systems (LIS)
Canada Geographic Information System (CGIS)	digital line graphs	location
cartographic process	digital orthophotoquads	planimeters
computer-assisted cartography (CAC)	encode	spatial
	entities	

Review Questions

1. What was the initial impetus for the development of the first GIS? Why was it so difficult to develop in the first place?

2. Provide a description of the change in geographic thought from pure exploration through to prediction and planning. Describe the place of GIS in this context.

3. What is a GIS? How does your definition differ from that offered by David Rhind? Why are there so many names for geographic information systems?

4. What is the relationship between the traditional map and its automated counterpart? What are the relationships among the four subsystems in GIS and the map?

5. What is the difference between GIS and CAC? between GIS and CAD?

6. Create a diagram of the hierarchy of information systems and place GIS in its proper place in that hierarchy.

7. What are the basic analytical capabilities that one would normally find in a modern GIS?

8. Who would normally use a GIS? What accounts for its popularity?

References

Aronoff, S., 1989. *Geographic Information Systems: A Management Perspective.* Ottawa, Canada: WDL Publications.

Burrough, P.A., 1986. *Principles of Geographic Information Systems for Land Resources Assessment.* Oxford: Clarendon Press.

Carter, J.R., 1989. "On Defining the Geographic Information System." In *Fundamentals of Geographic Information Systems: A Compendium,* W.J. Ripple, Ed. Falls Church, VA: AS-PRS/ACSM, pp. 3–7.

Cowen, D.J., 1988. "GIS Versus CAD Versus DBMS: What Are the Differences?" *Photogrammetric Engineering and Remote Sensing,* 54(11):1551–1554.

DeMers, M. N., 1991. "Classification and Purpose in Automated Vegetation Maps." *Geographical Review,* 81(3):267–280.

Dickinson, H., and H. Calkins, 1988. "The Economic Evaluation of Implementing a GIS." *International Journal of Geographical Information Systems,* 2:307–327.

Dobson, J.E., 1995. "Defining the University Consortium for Geographic Information Science," *GIS World,* 8(3):44–46.

Dueker, K.J., 1979. "Land Resource Information Systems: A Review of Fifteen Years Experience." *Geo-Processing,* 1:105–128.

Koshkariov, A.V., V.S. Tikunov, and A.M. Trifimov, 1989. "The Current State and the Main Trends in the Development of Geographical Information Systems in the USSR." *International Journal of Geographical Information Systems,* 3(3):257–272.

Marble, D.F., 1990. "The Potential Methodological Impact of Geographic Information Systems on the Social Sciences." In *Interpreting Space: GIS and Archaeology,* Allen, Zubrow, and Green, Eds. Washington, DC: Taylor and Francis.

Marble, D.F., and D.J. Peuquet, 1983. "Geographic Information Systems in Remote Sensing." In *Manual of Remote Sensing,* Vol. 1, 2nd ed., R.N. Colwell, Ed. Falls Church, VA: American Society of Photogrammetry, pp. 923–957.

National Academy of Sciences, 1980. Need for a Multipurpose Cadastre. Panel on a Multipurpose Cadastre, Committee on Geodesy, Assembly of Mathematical and Physical Sciences. Washington, DC: National Academy Press.

National Academy of Sciences, 1983. Procedures and Standards for a Multipurpose Cadastre. Panel on a Multipurpose Cadastre, Committee on Geodesy, Commission on Physical Sciences, Mathematics, and Resources, National Research Council. Washington, DC: National Academy Press.

Ozemoy, V.M., D.R. Smith, and A. Sicherman, 1981. "Evaluating Computerized Geographic Information Systems Using Decision Analysis." *Interfaces,* 11:92–98.

Parker, H.D., 1988. "The Unique Qualities of a Geographic Information System: A Commentary." *Photogrammetric Engineering and Remote Sensing,* 54(11):1547–1549.

Rhind, D.W., 1988. "A GIS Research Agenda." *International Journal of Geographic Information Systems,* 2:23–28.

Robinson, A.H., J.L. Morrison, P.C. Muehrcke, A.J. Kimerling, and S.C. Guptill, 1995. *Elements of Cartography,* 6th ed. New York: John Wiley & Sons.

Smith, T.R., S. Menon, S. Star, and J.L. Estes, 1987. "Requirements and Principles for the Implementation and Construction of Large-Scale Geographic Information Systems." *International Journal of Geographical Information Systems,* 1:13–31.

Star, J., and J.E. Estes, 1990. *Geographic Information Systems.* Englewood Cliffs, NJ: Prentice Hall.

Tomlinson, R.F., 1984. "Geographic Information Systems: The New Frontier." *The Operational Geographer,* 5:31–35.

U.S. News & World Report, 1995. "20 Hot Job Tracks: 1996 Career Guide," October 30, pp. 98–108.

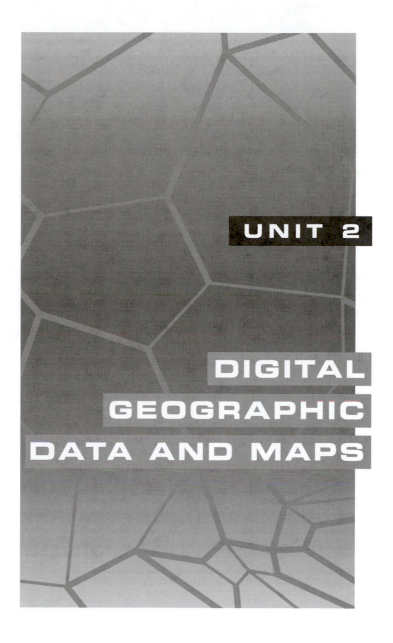

UNIT 2

DIGITAL GEOGRAPHIC DATA AND MAPS

CHAPTER 2

Basic Geographic Concepts

Unlike the real world, our digital environments are composed of cartographic objects representative of the individual components of the earth. These objects differ in size and shape, in color and pattern, and in scale of measurement and degree of importance. They can be measured directly by instruments on the ground, sensed by satellites hundreds of miles above the surface, collected by census takers, or extracted from the pages of documents and maps produced in ages past. Some will be very important for our journey, others merely useful, and still others will need to be discarded as we begin our navigations. Thus, to explore the modeled world, cartographic objects must be collected, organized, and synthesized.

The nature of the data often dictates not only how we will later represent the earth inside a GIS but also how effectively we will analyze and interpret the results of that analysis. In turn, how we view and experience our environment affects what features we note and how we will eventually represent them. The points, lines, and areas we encounter are all different. Moreover, their representation and utility depend, in large part, on our ability to recognize what features are important and to identify those that may be modified by the temporal and spatial scales at which we observe them. This information, in turn, dictates how the data are stored, retrieved, modeled, and finally output as the results of analysis.

In addition to the temporal scale and the physical sizes of objects stored in a GIS, we must consider the measurement level we will use to represent their descriptive conditions or **attributes**. At one scale, for example, points could include whole cities with no areal extent, whereas at another, objects as small as insects or even microbes will be the important point data under consideration. The cities may also include descriptive attributes, such as their names (**nominal** measurement scale); whether their viability for placing an industry would be considered major, moderate, or minor (**ordinal** measurement scale); their average annual temperature (**interval** measurement scale); or the average annual per capita income (**ratio** measurement scale). Each of these types of data represents a fundamentally different criterion, measured with a substantially different measuring device, and with a different level of data precision. The same can be said for lines, areas, and surfaces.

The first step toward better GIS skills is to begin to think spatially. We all operate in a spatial environment, but we are often oblivious to the space around us, paying no attention to how we and other objects occupy, move through, interact with, and even modify our space. Become familiar with all the possible patterns, interconnections, distances, directions, and spatial interactions in your world. As you become more sensitive to the objects themselves, you will find it an easy next step to imagine how the objects and interactions can be measured and what meter stick you will need to record them.

It is important to know these two interconnected concepts of objects and measurements well before you begin producing results from a GIS. As you progress through the rest of the book, you will begin to see how these are represented in map form, how they will be encoded and stored, and some ways in which they could be analyzed. With this firsthand experience at moving from the conceptual data model to the computer representational model, you will be well prepared to deal with the analytical models you are likely to encounter as you use GIS for your own specialty.

LEARNING OBJECTIVES

When you are finished with this chapter you should be able to:

1. Show how an improved spatial or geographic vocabulary improves your perception of the world.

2. Illustrate the impact of scale on our perceptions of the world.

3. Understand the difference between spatial and aspatial data and indicate how surrogates can be used when spatial data are lacking.

4. Understand the difference between discrete and continuous data and give examples for point, line, area, and surface features.

5. Understand the relationships among nominal, ordinal, interval, and ratio scales of data measurement and give examples for point, line, area, and surface features.

6. Understand the necessity of a structural framework or grid system for locating yourself and your relationships with regard to other earth features in both absolute and relative terms.

7. Understand the meanings of terms such as *orientation, arrangement, diffusion, pattern, dispersion, density,* and *spatial association* and be able to use them when discussing geographic phenomena.

8. Understand the use of ground sampling methods and the advantages and disadvantages of directed ground sampling, probability-based ground sampling, and remote sensing for gathering data about the earth.

DEVELOPING SPATIAL AWARENESS

We live in a complex world. To succeed, we must be aware of this complexity and be able to organize it around a framework that allows us to understand how such a seemingly disordered system continues to function. The focus on one simple underlying principle—the search for spatial order—separates geography from all the other fields of study you have encountered. Many disciplines you encountered in your studies have, on occasion, asked how people, places, critters, natural features, and the like changed or interacted spatially, but only geography continues to concentrate on spatial relationships as its principal intellectual framework. Because of this focus, geographers have developed a language that reflects the way they think about space. This spatial language, like any language, allows the geographer to think more clearly and communicate more concisely about space, examining only the structures and arrangements of the data that are absolutely necessary to explain spatial phenomena.

The spatial language, like any other language, becomes an intellectual filter through which only the necessary information passes (Witthuhn et al., 1974). It modifies the way we think, what we observe as important, and how we make decisions. When you first began to speak, you had difficulty explaining exactly what it was you wanted. Your language skills were not well developed, and the complexity of your ideas was proportionately limited. As your language grew, so did your understanding of the world and your ability to communicate it to others.

Childhood experiences did more than teach verbal communication. In large part you began to experiment with your spatial environment. The excitement of parents when their baby first notices his or her big toe is in part a recognition that the process of spatial exploration has begun (Piaget et al., 1960). Later spatial explorations include development of the concepts of movement and speed (Piaget, 1970) through crawling along the floor, discovering stairs, climbing furniture, reaching out to hot stovetops, opening cabinets, and the like—often to the dismay and concern of others. Still later we began climbing trees, exploring under bridges, wading in streams and ponds, wandering in the woods, finding the locations of new friends, walking to local parks, and so on. Many of these travels required us to learn route finding so that we could work our way back to where we began. Most often these early route-finding excursions were very ribbonlike. Once we found a route, we simply backtracked to return (Muehrcke and Muehrcke, 1992). Eventually we became more sophisticated spatially, and we began recognizing that there were shortcuts. In other words, our spatial world took on a two-dimensional structure. We began to comprehend that places and things were either near or far, straight ahead, or at an angle. And we began to take note of obstructions that had to be traveled around, of hills that slowed us down or allowed us to coast on our bicycles. In short we were beginning to think geographically.

As we experience space, we will encounter features and objects of many different types. Many disciplines have provided us with additional vocabularies to help us decipher them. Throughout geography's 2500-year history, it has borrowed extensively from numerous disciplines, synthesizing what others have learned and applying the filter of the spatial language. You will need to draw on this background to integrate this knowledge into a spatial framework. If you encounter an urban setting, knowledge of urban studies will allow you to see patterns of residences, stores, and industries. A background in biology is invaluable when it is necessary to perceive and interpret the locations, patterns, and interdependencies of plants and animals. Geology and geomorphology will provide an awareness of the differences in exposures of rock and the many landforms you encounter. Expertise in economics allows you to suspect underlying forces determining locations of businesses and industries. The list is nearly endless and far beyond the scope of this book.

We begin to exercise our geographic skills by examining the types of objects and features we encounter. Spatial objects in the real world can be thought of as occurring as four easily identifiable types: points, lines, areas, and surfaces (Figure 2.1). Collectively, these four types can represent most of the tangible natural and human phenomena that we encounter on an everyday basis. Inside the GIS, real-world objects will be represented explicitly by three of these object types. Points, lines, and areas can be represented by their respective symbols, which we will study in Chapter 3, whereas surfaces are most often represented either by point elevations or other computer structures, with which we will deal in Chapter 4. What is most important now is that in the GIS, all data are explicitly spatial. Phenomena that are by their nature aspatial (ideas, beliefs, etc.) cannot be explored directly in a GIS unless a way can be found to assign a representative spatial character to them. The process of finding **spatial surrogates** is a difficult one, and we will develop this idea in more detail in later chapters. For the time being, we restrict our discussion to the more tangible forms of geographic data.

Point features are spatial phenomena, each of which occurs at only one location in space. From your own experiences you can easily recognize such features as trees, houses, road intersections, and many more. Each feature is said to be **discrete** in that it can occupy only a given point in space at any time. For the sake of conceptual modeling, these objects are assumed to have no spatial dimension—no length or width—although each can be referenced by its locational coordinates. Points then are said to have "0" dimensionality. In reality, all point objects have some spatial dimension, however minute, or we would

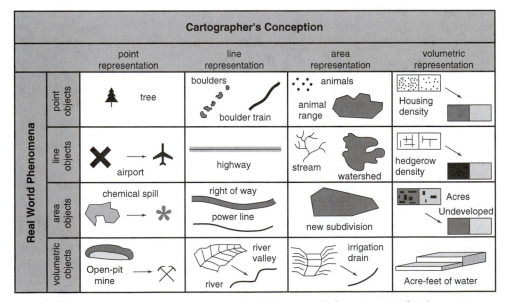

Figure 2.1   Comparison of real-world phenomena and the cartographer's conception. Point, line, area, and surface features with examples.   *Source:* P.C. Muehrcke and J.O. Muehrcke, *Map Use: Reading, Analysis and Interpretation,* 3rd ed., J.P. Publications, Madison, WI, © 1992, Figure 3.18, page 84. Used with permission.

not be able to observe them in the first place. We assume an absence of length and width, so that in measurements of air pressure, for example, which are characterized by a potentially infinite number of points, the points themselves will always occur at distinct locations with absolutely no overlap. The spatial scale at which we observe these objects or features places a framework on which we can determine whether we should think of them as points. For example, if you are looking at a house from only a few meters away, the structure seems large and occupies a substantial amount of length and width. This concept changes, however, as you move away: the house appears less like an areal object and more like a point object, the farther away you are (Figure 2.2). You choose your spatial scale on the basis of different criteria: whether you want to examine the **arrangement** of people and furniture in the house, for example, or whether you are interested in the house only in relation to other houses, perhaps for an entire city. In the latter case the house would be considered to be a point. Your observation has been filtered according to how you want to view the object.

Linear or line objects are conceptualized as occupying only a single dimension in our coordinate space. These "one-dimensional" objects may be roads, rivers, regional boundaries, fences, hedgerows, or any kind of object that is fundamentally long and very skinny. The scale at which we observe these objects once again places a fundamental limitation on our ability to conceive of them as having any width. As you are fully aware, rivers, roads, fences, and hedgerows all occupy two dimensions when observed at close range. The farther we move away from them, however, the skinnier they become. Eventually they look so skinny that it is impossible to imagine that they are anything but essentially linear objects. It also becomes impossible to represent them as more than lines because we are so far away that we can no longer measure them (Figure 2.2). Other lines, such as political boundaries, have no width dimension to be concerned about. Actually, these lines are not physical entities at all but rather a construct of political convention and agreement. Despite their lack of tangibility, however, they can be thought of as explicitly spatial because their existence separates two portions of geographic space.

Figure 2.2 Effect of scale on spatial dimensions. A house observed in a close-up aerial view appears to have length and width, but as we pull back, its length and width dimensions disappear, leaving us with the impression of a house as a point.

Linear objects, unlike point objects, allow us to measure their spatial extent by simply finding out how long they are. In addition, since they are not sited at a single location in space, we must know at least two points, a beginning and an ending point, to describe the location in space of a given linear object. The more complex the line, the more points we will need to indicate exactly where it is located. If we take a stream as an example of a linear object, the description of its many twists and turns may require a large number of points. In addition, it may be desirable to describe more than the stream's beginning and ending points. Because we have included a geometric dimension, we can also measure the shapes and **orientations** of linear objects. We will deal later with these additional spatial concepts.

Objects observed closely enough to be clearly seen to occupy both length and width are called areas. Examples of areas of "two-dimensional" objects include the area occupied by a yard, the areal extent of a city, and an area as large as a continent. To describe the locations of areas in space, we recognize that they are composed of a series of lines that begin and end at the same location. Besides merely indicating the locations of areas using lines, we are now able to envision three additional properties: as with lines, we can describe their shapes and orientations, but in addition, we can now describe the amount of territory they occupy.

Adding the dimension of height to our area features allows us to observe and record the existence of surfaces. Although we could certainly observe a house at close range and describe it in terms of its overall length and width, we often want to know whether it is a one-story or two-story structure. In this way we need to observe the house not as an area but rather as a "three-dimensional" object, having length, width, and height. Surfaces occur all around us as natural features. Hills, valleys, ridges, cliffs, and a whole host of other features can be described by citing their locations, the amount of area they occupy, how they are oriented, and now, with the addition of the third dimension, by noting their heights. As it turns out, surface features are composed of an infinite number of possible height values. We say that they are **continuous** because the possible values are distributed without interruption continuously across the surface (Figure 2.3). Because the height of a three-dimensional object varies from one

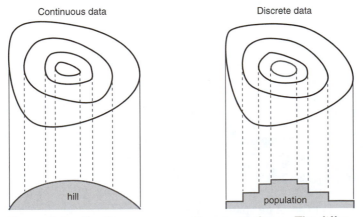

Figure 2.3 Continuous versus discrete surfaces. The difference between discrete and continuous data types.

place to another, we can also measure the amount of change in height with a change in distance from one edge to another. With this information at hand we could also determine the volume of material contained in the feature itself.

The ability to make such calculations is very useful if it is necessary to learn how much water a reservoir contains or how much surface material (overburden) lies on top of a coal seam. All these features, whether point, line, area, or surface features, occur in space. And all can be located as to exactly where they may be found. But how can we communicate the importance of a circular areal feature located in a particular area, occupying 10 hectares of space and oriented north-south? We need a way of classifying these features based on other observable properties and using terminology that others can understand.

SPATIAL MEASUREMENT LEVELS

So far we have examined the types of features and their locations in space. The objects themselves are called entities, and as we have seen, they have associated with them a set of coordinates that allow us to describe where they are located. All these spatial features or entities we observe contain information not only about how they occupy space but also about what they are and how important they are to what we are studying. A tree, for example, viewed as a point feature, might be classified, on the basis of the taxonomist's set of terminology, as a pine or an oak. We could also investigate the age of the tree by drilling a core through the trunk and counting its annual rings. With this additional information we not only know that there is a tree located at some point in space but also that it is a 35-year-old oak tree. The additional nonspatial information that helps us describe the objects we observe in space comprises the feature's attributes. As we observe our spatial environment, we begin to categorize its attributes and to note where they are. It is our nature as curious creatures to want to classify or pigeonhole the objects we find, so that we can place an additional level of organization on our space. We can now indicate that a particular feature, with a particular name, and with some measurable attributes, exists in a documented location. But before we assign these properties and attributes, we must know how to measure them. Otherwise we cannot compare the objects at one location with those at another. What meter stick do we use? How accurately can we describe the objects we encounter? What effects will these different levels of measurement have on our ability to compare them?

Fortunately, there is already a well-established measurement framework for nearly all forms of data, including geographic data. These so-called levels of **geographic data measurement** range from simply naming objects, to give ourselves something to call them, to precise measurements that allow us to directly compare the qualities of different objects. The measurement level we use will be partly determined by what we are classifying, partly by what we want to know, and partly by our ability to make measurements at our selected scale of observation. Figure 2.4 illustrates the levels of measurement in terms of three commonly used geographic features.

The first level of measurement is the nominal scale. A useful mnemonic device is to think of the word nominal as "named." These are named data. The system allows us to make statements about what to call the object, but it does

Figure 2.4   Measurement levels for cartographic objects. Point, line, area, and surface features at the nominal, ordinal, interval, and ratio levels.   *Source:* A.H. Robinson et al., *Elements of Cartography,* 6th ed., John Wiley & Sons, Inc., New York, © 1995, modified from Figures 16.1, 16.2, 16.3, pages 272, 273, 274. Used with permission.

not allow direct comparisons between one named object and another. For example, we could say that at one location we have a maple tree and at another we have an oak. Although this statement certainly separates the objects, we cannot say that one is better than the other because the two are inherently different. In geographic terms, we can no more say that a church is better than a fire station than we can make such a judgment about apples and oranges.

If we want to compare two objects, we must be more precise in our level of measurement. As always, this is determined by what it is about the objects that we want to compare. Using our tree example, if we were interested in knowing how well a maple tree, an ash tree, or a pine tree might serve in producing a comfortable setting for a picnic, we could place each one on an ordinal scale from best to worst for that particular question. Because pine trees often have low branches and tend to drop sap on the ground, we probably would class pines as the worst of the three choices for our picnic area. Although ash trees produce less debris and don't have low branches requiring us to hunker down while we eat, their leaves are small and some sunlight is likely to get through. We might want to class ash trees as moderate for our picnic setting. The maple tree produces very dense shade, however; moreover, it doesn't put a great deal of sap on the ground, and it doesn't have low branches. Therefore, we would likely assign a classification of excellent to the maple tree.

In our picnic example we produced a spectrum of values ranging from best to worst. Clearly, however, the spectrum is based entirely on the purposes for which we intended to use the information. Because we have classed maple as best, ash as moderate, and pine as worst, the classification does not extend to other uses of the trees. Neither the maple nor the ash would be a very interesting Christmas tree, for example. So our classification is based on a single spectrum representing a single set of circumstances. Clearly, then, ordinal data can give us some insights into logical comparisons of spatial objects, but comparisons are limited to the specific spectrum that serves as the basis of the questions we are asking. Put another way, the grade you receive in your GIS class cannot be compared to the one you receive from your calculus instructor. Two fundamentally different sets of criteria are involved.

If we want to be more precise in our measurements, we move to the interval level of data measurement, in which numbers are assigned to the items measured. Data measured at the interval level can be compared, as in the case of ordinal data, but they can be compared with more precise estimates of the differences. A good example of spatial data measured at the interval level consists of soil temperatures across a study area containing widely different soil types. We may find that temperatures of some very dark, organic-rich soils are much higher than those for other parts of our study area that lack organic material and are a lighter color. In fact, we can say that the soil types, measured at the same time, exhibit a difference of 8°F. The dark soil, for example, might be 85°F, while the lighter soil is only 77°F. We now have an easily measured, precisely calibrated difference between the soils at two different locations.

One limitation remains in our ability to make comparisons with data measured at the interval level. Take two additional, extremely different soils, one nearly white and the other almost black. When we measure these at the same time, we find very different temperatures: 50°F and 100°F. As before, we can determine the numerical difference between these two soils as 50 Fahrenheit degrees. The numbers 50 and 100 look appealing as we try to read more meaning into the numerical difference. Is it fair to say, for example, that the dark soil is twice as warm as the light soil? At first glance it certainly appears so. However, you should remember that the Fahrenheit temperature scale we have used has an arbitrary starting point. For us to make a ratio of the two numbers, our starting point—in this case 0°F—must represent a true starting point for measuring temperature. To make the comparison we have suggested, it would be necessary to convert all the temperatures to the Kelvin scale, where the starting point, or zero, now represents a total lack of the atomic movement commonly associated with heat. Converting our two temperatures of 50°F and 100°F to kelvins, we get 283 and 311 K, respectively. When we form a ratio from these two numbers, we see that the dark soil is not twice as warm as the light soil.

In our conversion of Fahrenheit temperatures to the Kelvin scale, we moved to the final, most useful, level of data measurement—ratio. By converting our temperatures, we were able to perform a meaningful ratio operation on the two numbers because we were operating at the ratio scale. As the name implies, this is the only level of data measurement that allows us to make a direct comparison between two spatial variables.

Another example might be useful because it uses a set of numbers with which we are frequently concerned. Suppose you are studying two sets of homes located in different parts of a city: one in an upscale neighborhood in the suburbs and the other in a low-income neighborhood in the inner city. By surveying the residents, you determine that the median annual income for the suburban neighborhood is $50,000, whereas its counterpart in the inner city is $25,000. Because $0 means no income at all, we can safely say that we are operating at the ratio level of data measurement; hence, comparisons are meaningful. By dividing one value by the other, we can now correctly say that the median family income in the suburban neighborhood is twice that of the inner-city neighborhood. Such comparisons are valid and will become quite useful for your GIS data analysis. But, as with non-GIS data, you can make them only with ratio-level data.

SPATIAL LOCATION AND REFERENCE

Thus far we have seen that we can observe a wide variety of features. We can group them, on the basis of the scale at which we observe them, according to whether they are points, lines, areas, or surfaces. And we can categorize them using measurements at four different levels of measurement—nominal, ordinal, interval, and ratio—depending on how we want to describe them and how we wish to compare them. But we can only begin to describe our world knowing simply where things are, what they are, and what measurements we can assign them. We also need to know how these spatially referenced, classed, and measured objects interact in space to produce the overall sites we observe.

Until now we have indicated that we can **locate** objects in space. Location is the first important spatial concept we need, but to locate objects means that we must have a structured mechanism to communicate the location of each object observed. The first type of location is called **absolute location** and will give us a definitive, measurable, fixed point in space. But first we must have a reference system against which to evaluate such a location. In addition, the reference system must have a fixed relationship to the earth we measure.

The earth is a roughly spherical object, with some large-scale and some small-scale deviations from that shape. If we are observing the earth as a whole, it is generally convenient to consider it to be perfectly spherical. Around that spherical shape we can use simple geometry to create a spherical **grid system** that corresponds to the rules of geometry. This grid system, known as the spherical grid system, places two sets of imaginary lines around our earth (Figure 2.5).

The first set of lines for our spherical grid starts at the middle of the earth, or equator. These lines are called **parallels** because they are parallel to each

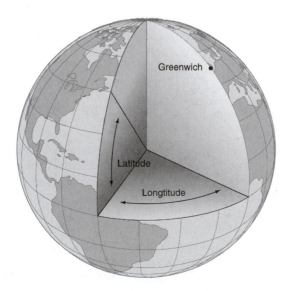

Figure 2.5 The geographic grid. Spherical grid system showing parallels and meridians. Parallels allow us to measure angular distance north and south (latitude) from the equator (0 degrees latitude) up to a maximum of 90 degrees north (North Pole) and 90 degrees south (South Pole). Meridians start at the prime meridian and allow us to measure angular distance east and west (longitude) up to a maximum of 180 degrees where they would meet at the international date line. *Source:* A.H. Robinson et al., *Elements of Cartography,* 6th ed., John Wiley & Sons, Inc., New York, © 1995, modified from Figure 4.4, page 47. Used with permission.

other, and they circle the globe from east to west. At the equator, the first of these parallel lines, we assign a starting value of 0. As we move both north and south from the equator, we draw additional parallels until we reach the poles. Because each of these lines is a given angular distance, measured between the center of the earth and the intersection of the line, each can be used to measure the angular distance from our starting point at the equator to where the last line would occur at the poles. The angular distance, called **latitude**, from the equator to either pole is equal to one-fourth of a circle, or 90 degrees. So we can measure the angular distance from the equator north or south up to a maximum value of 90 degrees.

This is only half of our spherical grid system, however. To complete the grid we need another set of lines running exactly perpendicular to the first. These lines, called **meridians**, are drawn from pole to pole. The starting point for these lines, called the **prime meridian**, runs through Greenwich, England, and then circles the globe, becoming the **international date line** on the opposite side of the earth. The prime meridian is the starting point or zero point for angular measurements east and west, called **longitude**. Longitude is then measured east 180 degrees of the prime meridian, until it reaches the international date line. Similarly, it can be measured west of the prime meridian up to 180 degrees, again until it reaches the International Date Line.

This system of angular measurements allows us to state the absolute location of any point on the earth by simply calculating the degrees of latitude north or south of the equator and the degrees of longitude east and west of the prime meridian. With this system we can then describe with relative ease the locations of any of the objects we wish to describe (Figure 2.6). In addition, these angular measurements can readily be converted to feet or miles, meters or kilometers, thus allowing us to measure short or long distances on the ground using devices designed for calculating them. Although we will need to make adjustments to this system to adapt for the development of flat maps that closely approximate the locations and arrangements of objects on the globe, what we have is a quite elegant system for determining absolute location.

The Great Circle Distance (D) on the sphere between two points A and B is calculated using the standard formula in spherical trigonometry:

$$\cos D = (\sin a \sin b) + (\cos a \cos b \cos |\delta\lambda|)$$

where a and b are the geographic latitudes of A and B, and $|\delta\lambda|$ is the absolute value of the difference in longitude between A and B. (The product of the sines will be negative if A and B are on opposite sides of the equator.

Example: Great circle distance between Nome, Alaska and Miami, Florida.

$\cos D = (\sin 63.5 \sin 25.75) + [\cos 63.5 \cos 25.75 \cos (165.33 - 80.183)]$
$\cos D = (0.895 \text{ X } 0.434) + [0.446 \text{ X } 0.900 \text{ X } \cos 85.147]$
$\cos D = (0.895 \text{ X } 0.434) + [0.446 \text{ X } 0.900 \text{ X } 0.0846]$
$\cos D = 0.38843 + 0.0339 \qquad \cos D = 0.422$

$D = \cos^{-1} 0.422 = 65.04^{\circ}$ x 69 miles per $^{\circ}$ = 4488 miles (approximately)

Miami:
25.75°N
80.183°W

Nome:
63.5°N
165.33°W

Figure 2.6   Calculating great circle distance. The illustration shows the calculation of spherical or great circle distance between two points on a sphere.   *Source:* P.C. Muehrcke and J.O. Muehrcke, *Map Use: Reading, Analysis and Interpretation,* 3rd ed., J.P. Publications, Madison, WI, © 1992, Figure 12.3, page 255. Used with permission.

But as we continue to explore our world, we quickly note that it would be very useful to be able to describe not only the absolute locations of objects but also their relationships to other objects in geographic space. In fact, this **relative location** becomes quite prominent in our GIS analyses, especially when relative location to other objects affects operation. With our absolute grid system we can determine relative locations by knowing the absolute distances between any two objects by simply subtracting the coordinates of the smaller from the larger. Relative distance is only half of what we might want to know. It would also be useful to know the difference in direction between the two. We could, for example, indicate that a landfill is located 1500 meters southeast of the center of the city. This measure of distance and direction provides us with a framework for describing the precise location of the landfill relative to the city. From the standard Pythagorean theorem (Figure 2.7a) for relating the parts of a triangle that has one 90 degree angle, we can determine the distance of the hypotenuse by the so-called distance theorem (Figure 2.7b), which is expressed as follows:

$$d = \sqrt{(X_2 - X_1)^2 + (Y_2 - Y_1)^2}$$

where

$(X_2 - X_1)$ = difference in the X direction or longitude

$(Y_2 - Y_1)$ = difference in the Y direction or latitude

d = distance between the two points

However, we could provide additional relative information that would be of use to people living in the city. If we know that the prevailing winds are from the northwest, we can say that the landfill is located 1500 meters downwind of the city, thus illustrating that the odor of the dump will not normally be a problem for city dwellers. This latter approach, though less precise in terms of measurement, has more practical utility. This also provides us with a means of interpreting what we have observed—of determining the important relationships between and among features.

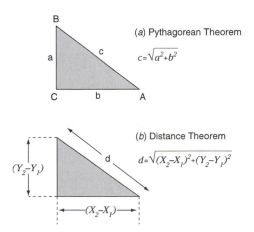

(a) Pythagorean Theorem

$c = \sqrt{a^2 + b^2}$

(b) Distance Theorem

$d = \sqrt{(X_2 - X_1)^2 + (Y_2 - Y_1)^2}$

Figure 2.7   Calculating planar distance. The calculation of distance between two points using the Pythagorean theorem *(a)* and the distance theorem *(b)*.

SPATIAL PATTERNS

In the preceding section we alluded to the importance of knowing the relationships among objects in space. A primary purpose of GIS is to analyze these relationships. That is why we put so much spatial data into the GIS database in the first place. When we begin to examine the analysis subsystem of the GIS, we will look at the many possibilities available to us. In the meantime, it will be instructive to begin a general development of our spatial language with respect to spatial comparison. This will enable us to begin thinking about what things must be collected during our excursion because we will have a broad framework for ascertaining the types of spatial relationship that can be examined and what they can tell us about our environment that we could not find out from a knowledge of features, their locations, and descriptions alone.

We begin by examining our last measure of location, in which we determined the relative location of a landfill to the nearby city. This easy calculation was a measure of proximity, the quality of being near something. Proximity can be measured, as we have already seen, by noting the distance between two features or objects. However, we can also set limits of acceptable proximity—a distance within which it is either acceptable or unacceptable for two objects to be found. In our example, if the landfill is too close to the city, even if it is downwind, some people will have to look at the unsightly feature. Beyond 2000 yards, however, we may assume that it is not visible to any city resident. We can then give a minimum acceptable proximity of 2000 yards to ensure that any landfill will be placed at least 2000 yards beyond the edge of the city.

Our landfill example showed the interaction of two spatial objects. However, many features and objects occur in far greater numbers: cities in a country, houses in cities, animals in natural areas, natural areas in states, trees in forests, roads across nations, tributaries along streams, even plants in a garden all occur as multiples. But they do not all occur uniformly located within these areas. Each set of objects exhibits a particular spacing or set of arrangements. We begin to notice that these arrangements and spacings seem to have underlying controls or processes that dictate their placement. The instructions on the tomato seedlings may tell us to place the plants 3 feet apart in the garden. By following the directions, we produce a **regular** or uniform pattern of evenly spaced plants (Figure 2.8). This uniform pattern differs considerably from the locations of trees in a forest, which seem to be scattered at **random**, with no apparent underlying design. Alternatively, when we look at the locations of cities across nations, we often see that these population centers are located near lakes, oceans, and streams. Knowing that water bodies provide sources of drinking water and recreation and are useful in commerce, we can easily see that the tendency of cities to cluster near such features is driven by these needs. The clustered distribution of cities demonstrates a type of distribution with very high **density** of features, whereas the distribution of other objects such as farmhouses, scattered about a nation's rural landscape, demonstrates a more **dispersed** pattern.

Certain features seem to be organized in yet another way. For example, the vegetation on the north side of a steep slope tends to be decidedly different from that on the south side. In other words, vegetation assemblages have a particular orientation. The property of orientation is also found in planted shelter-

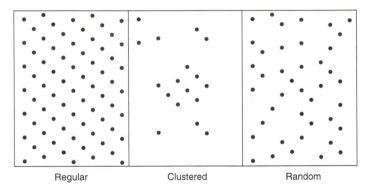

Regular   Clustered   Random

Figure 2.8   Point distribution patterns. Regular, clustered, and random distributions of spatial objects.

belts that are oriented at right angles to the direction of the wind. Some city streets are oriented along a relatively square grid, whereas others seem to be scattered about with no particular sense of direction. Some of you may have noticed that a great many American Civil War statues are oriented with the figures facing in the direction from which the foe would arrive. In some regions of the world that experienced continental glaciation, we find piles of debris called moraines oriented at right angles to the flow of ice, or lines of giant rocks called boulder trains that give evidence of the direction of retreat of the ice.

If we revisit some locations we have seen before, we find still other spatial processes that give us clues about changing geography. If you return to your hometown or city after an absence of a few years, you may see that it is larger than it was when you left. The town has experienced **diffusion** into the neighboring farmlands. Or you may see that your former downtown area is less important as a center of commerce, and instead there has been a diffusion of stores into malls on the outskirts of town. The process of diffusion can occur daily as people concentrated in the residential areas of large cities travel outward toward their workplaces elsewhere in the city. Diffusion can even be seen in the introduction of GIS concepts and technology, as the once small number of universities offering courses expands to include more and more each year.

But as we view all these spatial configurations, we also ask for the relationships that might occur between one set of features and another. We touched on this when we mentioned some causes for the development of cities near water bodies. What we discovered is that one spatial pattern may be partially or totally related to some other spatial pattern. The spatial arrangement of steep slopes was seen to have a strong **association** with the vegetation that occurs there. We can now begin to ask questions about the causes not only of single distributions but also of **spatially correlated** distributions of phenomena (Figure 2.9). This is among the more powerful capabilities of modern GIS: the ability to illustrate, describe, and quantify spatial associations, allowing us to examine clues to the mechanisms that cause these associations. This has long been a major task among geographers, as evidenced by the writings of Carl Sauer (1925). We are beginning to acquire the spatial language of geography, and now, with the use of GIS, we have the ability to automate these geographic concepts. Because we know the language, we begin to think in spatial terms—to filter our thinking

Figure 2.9   Spatial relationships between spatial phenomena. Spatial correlation between geological formations and the types of soils formed from them. *Source:* U.S. Department of Agriculture county soil survey for Gove County, Kansas.

to observe the important spatial features, to identify spatial patterns, and, finally, to ask the questions that will elicit explanations for pattern interactions.

GEOGRAPHIC DATA COLLECTION

Many data are still observed through ground survey methods not unlike those used by our predecessors. Tape measures, still fundamentally useful for measuring relatively small objects, have been replaced in some cases by relatively inexpensive range finders that send a beam of light from the observer to an intervening obstacle. Other observations, such as qualitative, or categorical data, are still acquired by means of direct visual observation or by collection of sample specimens for later identification. These observations should be evaluated carefully before they are accepted as truth, since often their quality is a function of the observer's experience. The traditional method of conducting vegetation surveys illustrates this point. Vegetation surveyors used to base landownership maps on so-called **witness trees**, trees left behind after lands had been cleared for agriculture. The surveyors relied on the assumption that these trees were representative of the commonly available forest types and generalized from these easily identifiable objects to the surrounding areas. Although some of the witness trees may have been representative of the presettlement vegetation, many were far less typical, having been left because of their prominent size or convenient position.

Absolute positions on the earth were once recorded relative to the locations of celestial bodies such as the North Star. The handheld compass is a simple device that relies on the tendency of the earth as a rotating body to create opposing poles at its north and south extremes. For more local observations, locations are also relatively easily and cheaply obtained using simple handheld

Brunton pocket transits, or the more sophisticated **plane table and alidade** survey devices. These devices are still in use today where both the amount of terrain and the budget are limited. Each device also assumes that there are some nearby points of known location against which the object locations can be compared.

From the point of known location, surveyors can measure distances and angles through a process called **dead reckoning**; they can intersect a number of line-of-sight measurements **(triangulation)**; or they can create a known baseline against which only the distances between objects and the ends of the baseline need to be measured **(trilateration)** (Figure 2.10).

Differences in elevation at different places on the landscape can be measured with the use of a device called a **dumpy level**, which has a telescopic sight much like the alidade but is placed on a taller stand and is capable of rotating up and down to examine elevation differences at a distance. A number of traditional survey devices have been updated to offer much greater accuracy, as well as digital readouts and superior ease of use. Theolodites are one example. All such improved instruments, like their older counterparts, have somewhat limited utility if it is desired to survey large portions of the earth simultaneously. The time required to survey large areas with these devices, however, is far outweighed by the value of the highly accurate positional data obtainable for applications such as locating the boundaries of property lines. Thus, their continued utility is likely to be ensured.

Technological innovations have improved the methods by which we can obtain positional information, especially for large portions of the earth (Figure 2.11). Circling the globe today are satellites whose positions are known with great accuracy. These satellites receive radio transmissions from field units on earth and return a signal that is processed by a nearby ground station and is then sent to the field unit as a set of coordinates. The coordinates give posi-

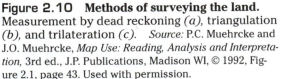

Figure 2.10   Methods of surveying the land. Measurement by dead reckoning (*a*), triangulation (*b*), and trilateration (*c*).   *Source:* P.C. Muehrcke and J.O. Muehrcke, *Map Use: Reading, Analysis and Interpretation,* 3rd ed., J.P. Publications, Madison WI, © 1992, Figure 2.1, page 43. Used with permission.

tional location as well as elevational location, providing a very useful set of data to the user. Perhaps the most promising of these devices, and now among the most widely used, was mentioned in Chapter 1: the NAVSTAR global positioning system (GPS) (Figure 2.12). GPS depends for its accuracy and precision on the number of satellites, the amount of detail or service provided by the military (currently in charge of the program), the locality of the base station providing the baseline data, the age of the field unit in its place, and the sophistication of the field and base stations. Available systems today can give locational accuracies ranging from a relatively coarse 100 meters down to at least decimeter levels. Direct line of sight is not necessary between the base station and the field unit but is required between the field unit and the satellite itself. Thus the usefulness of these devices is limited in regions with dense tree canopies that obscure the line of sight. For field surveys of animals, there is a system of devices using similar technology to track the mobile locations of deer, birds, small mammals, and even bees (if the units are small enough). Traditionally, the units are attached to the creatures, and their locations are determined by monitoring a continuously operating signal at a field unit. The field unit may be located on the ground or carried aloft with aircraft. A recent trend in this form of spatial data collection, called radiotelemetry, is to combine it with the GPS unit, allowing much greater areas to be monitored. As these two technologies continue to improve in quality, ease of use, and cost, there will be more and better data available for later input to our GIS.

When our concern is to gather information about the distribution of objects, plants, animals, or even people, rather than individual locations, we employ another form of data collection called a census. The census device depends on the nature of the data to be collected. We might, for example, use direct contact to

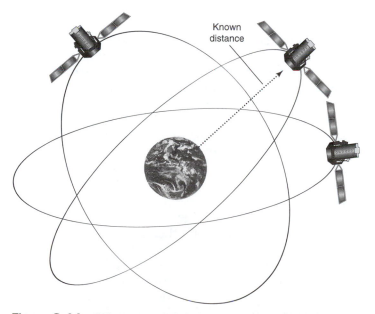

Known distance

Figure 2.11 Schematic of global positioning system (GPS) showing multiple satellites, each of which is at a precisely known distance above the earth.

Figure 2.12   Global positioning system (GPS) field unit.

physically note the exact locations and characteristics of individual shrubs in a grassy area. As implied by this example, the purpose of the method is to obtain information about an entire population of objects in space. The most common example, of course, is the census of population conducted by a government. Such survey devices attempt to gather both locational and attribute data about people so that generalizations can be made about populations in selected areas or districts. Data obtained often include marital status, income, housing, age, and gender. From these data generalizations it is possible to determine, at least approximately, which areas have the lowest per capita annual income, whether the overall population is getting older (on the basis of percentage of the total population), whether there are more people living in apartments or in houses, and so on. These results, in turn, allow governmental bodies to suggest ways of adapting to changing circumstances. In other words, governments now have a method of planning based on knowledge of the spatial distribution of people.

The potential power of the census to describe spatial relationships among a nation's people has prompted a move toward automation of the data itself. The census has become computerized, both gaining from and contributing to the move toward automated geography. We will see in Chapter 4 how the U.S. Bureau of the Census has created and modified its methods of keeping track of data to make census information accessible to most commercial GIS products.

At times, however, neither ground survey nor census is an appropriate means of gathering data and providing contextual spatial information about a large portion of the earth. This statement applies particularly to natural phenomena, but it also holds true for anthropogenic features and cultural data under selected circumstances. Obtaining data about phenomena on the regional or even the continental scale may require the use of secondary or indirect methods of data collection. These indirect methods often rely on sensing devices far removed from the observer and are therefore called collectively **remote sensing**. Although the term most often implies some form of satellite sensing of radiometric data from the surface of the earth, we will use the term more loosely to include the use of aerial photography, which is sometimes separated from the definition because of the strikingly different methods by which the data are interpreted. As a rule, remote sensing refers to the use of satellites that obtain data about the surface of large regions of the planet, most often coming in the form of computer-compatible data that can immediately be operated on by computer software.

Point remote sensing devices are normally placed at strategic locations that are meant to be indicative of the general character of the surrounding area. In addition, sensors are scattered throughout the region of interest so that data can be obtained for as much of the region as possible. The weather station is perhaps the best-known point remote sensing device. However, remote sensor networks have also been developed to collect continuous or periodic data on soil temperatures, soil moisture content, and other useful soil factors. These nets are most often connected through a telecommunications network to a base station that receives and stores the information, either as hard-copy displays in the form of paper polygraphs or as digital data inside a remote computer. Even more exotic devices have been developed to record the interception of flying insects, crawling animals, and moving sand grains or to monitor water or atmospheric pollution, seismic changes, and, of course, the prowling jewel thief. Collectively, these point-specific sources of information can provide a large regional picture of the phenomena they sense. The use of aerial photography permits the collection of data describing the continuous change in phenomena from one place to another, making it unnecessary to rely on discrete observations. This type of remote sensing most often relies on the use of aircraft to carry a photographic device designed to sense and record portions of the electromagnetic spectrum. These devices come in various sizes and can use a range of films, from traditional black and white to color to false-color infrared films, depending on the data being collected. In many cases, a particular film and filter combination is used to eliminate unwanted portions of the spectrum and enhance the visibility of regions more indicative of the features being sensed.

Aerial photography, a mainstay in a wide range of spatial analyses, has a long tradition of use for forest and range evaluation and management because the photos allow analysts to see substantial portions of areas at a glance. Soil scientists have used aerial photographs to assist in perceiving the subtle changes in soil type over large areas but also as base data on which soil maps are placed. Urban specialists have used aerial photographs to estimate populations by counting dwellings and interpolating from known averages per household. Geologists have long used aerial photography as a source of information on the spatial distribution of surface features as well as subsurface phenomena such as salt domes and fault zones. The military, of course, has used aerial photography for some time as a means of reconnaissance. Color infrared film, originally called camouflage detection film, was developed largely under military auspices. Thus the use of aerial photography as a means of examining geographic data is well established and still common for areas that are not overly large. For extremely large areas, such as states, regions, or countries, the costs in money and time generally are prohibitive. Similarly, a number of more exotic airborne devices such as **side-looking airborne radar (SLAR)**, **scanning radiometers**, **color video**, and **digital photography** tend to be ruled out for large-scale applications again, primarily because of the time and effort needed to cover major portions of the earth.

Larger areas can be surveyed by other methods, however, many of which deploy by satellite, hundreds of miles from the surfaces they are viewing, some of the same technology currently being carried in aircraft. The great distances between the sensing platform and its object allow the satellite to view large areas synoptically at each instant. In addition, because satellite remote sensing devices are orbiting the earth, they are able to sense much of the planet in a fraction of the time that would be needed by a conventional airborne sensing device. As indicated, a wide range of sensing devices is available, each with its own spectral, temporal, and spatial characteristics. Some radar satellites are designed primarily for observing weather, but others are able to interpret ground features as well. The SEASAT satellite, for example, is designed primarily for examining oceanic features such as waves and icebergs. Thermal scanners on board satellites allow geologists to evaluate the landscape for hot spots, providing data they need to help predict volcanic activity. Radiometers, such as the five-channel advanced very high resolution radiometer (AVHRR) on board the NOAA-7, -9, -11, and -14 satellites, have been used to map very large portions of the earth because these devices can sense the surface of the entire planet in a fraction of the time needed by most other satellite remote sensing systems.

Remember that remote sensing data are not gathered directly. Data from aerial photography, radar, and digital remotely sensed imagery are all more or less representative of what is on the surface. Although they do not, for example, sense types of vegetation or kinds of human activity, analysts can use the electromagnetic signals received as surrogates of what is actually on the ground. In most cases, the signals must be processed by experienced interpretation specialists before the object categories can be properly identified. Most often the classified data are used as input to a GIS database, rather than the raw data themselves. As we will see later, the ability of the human or machine interpreter to correctly identify objects in space has much to do with the utility of such data in decision making.

The field of satellite remote sensing is a vast and ever changing one, both because of the addition of new systems with new sensing devices and because remote sensing scientists are finding new and innovative ways to use them. Each system produces a set of data that provides spatial insights about what it views. Each also sees things through different eyes. Some view the land as a series of lines, whereas others produce huge checkerboard structures, building up images from the different shades and colors of each cell. Remote sensing scientists will continue to investigate the utility of these devices, and every GIS specialist should be aware of the potential capabilities of this technology for understanding our planet. As GIS scientists, our role is most often to determine how best these remotely sensed data can be incorporated into GIS databases and combined with other forms of data, to provide a more complete view of the lands we explore.

POPULATIONS AND SAMPLING SCHEMES

One problem yet remains before we are ready to explain what we see on our journey. We now have a spatial language at hand that allows us to know where to look and what to look for as we examine our environment. And we know about devices that allow us to collect large volumes of data by means of satellites. But the sheer complexity of our planet can very often become overwhelming simply because there are so many objects and so many possible factors that might be examined. And while remote sensing devices allow us to see large regions at a single glance, the resolution with which they view the earth often limits the sizes of objects that can be observed. Some features—for example, animal burrows in a prairie—are well below the ability of many remote sensing devices to explore. And because such features lie in a large territory, it is prohibitively expensive to obtain detailed aerial photography, so we must do our viewing and evaluating on the ground. This is easy because the features of interest are numerous enough to visually dominate the landscape. But there are far too many for us to count each burrow and put an absolute location for each. To determine why these features have a particular pattern requires us to know that a pattern exists in the first place, which ideally would be done by examining and recording their locations. Because we are unable to obtain a complete census of all the holes, we must sample them to estimate the entire population from a smaller, representative subset.

Sampling can be performed in a number of ways; some are harder than others, and some give us a better ability to make inferences about the larger population. Although sampling is effective for either aspatial or spatial data, we will restrict ourselves to spatial sampling because the GIS deals with explicitly spatial data. Within the spatial domain we sample data in one of two primary ways, directed and nondirected sampling. Each method will be determined both by the limitations of acquiring spatial data and by the inferences we are trying to make about the population of data using the sample.

Directed sampling, as the name implies, involves making decisions about what objects are going to be viewed and later catalogued. In other words, we direct our sampling based on our ability to make samples from and about a **target population** of objects or features. This target population occupies a given **sampled area** within which our samples are taken (McGrew and Monroe, 1993). Together, the target population and the sampled area comprise the **sampling frame**, which includes the types of object of interest bounded by specific and identifiable spatial coordinates. Within directed sampling, sometimes called purposive or judgmental samples (McGrew and Monroe, 1993), we use a combination of experience with the study area and its target population, accessibility to the objects we want to investigate, likelihood that we can obtain pertinent information about each individual (e.g., using only data from survey forms returned by people to whom they were sent), and focused (nonrandomly selected) study areas or case studies designed to demonstrate a particular phenomenon. Although this form of sampling is often discouraged as being "nonscientific," it is often necessary. Take some of the following situations.

You are planning a spatial survey of the types of vegetation that arise in an area that is cleared of vegetation, later farmed, and then abandoned. You know of a large area that has experienced these conditions, and so you plan to study that area, rather than all areas in your state or nation that also have these conditions. Your study is more focused because accessing all the areas that have been cleared and abandoned would be impossible. Upon visiting your potential study area, you find that at least half of it is surrounded by barbed wire and the owners don't want you tramping around on their property. Your study area is now further restricted. In addition, you discover that many of the areas still available for study are inaccessible by vehicle. Now you must direct your investigation to focus only on regions within a reasonable distance from roads. Thus you have had to focus on a small case study area known to have experienced the conditions of interest, and you have had to limit even this area because access to it is restricted. This type of situation is common and will limit the sampling strategy.

Another situation has already been alluded to concerning the use of surveys of populations. If, for example, you want to determine the spatial extent of cable television users and find out the programs they watch, you will use a common survey device that consists of a set of questions about which premium cable television programs the respondents watch. The questionnaires will be sent to the people in your city (the target area). However, you have to remember that some people do not have television, and many set owners do not subscribe to a cable service. You want to survey only people who have cable television (your target population), perhaps by contacting your cable companies to obtain lists of their customers who are willing to be surveyed. Once you have identified cable viewers, you send each one a survey document requesting a response to the questions and indicating how the document should be returned to you. Because most of us have received such surveys, we know that many potential respondents do not return the questionnaires. This will likely be the case with your survey as well. What you are left with is a directed survey, focusing only on cable television users and presenting results based on the responses of people who returned the survey.

Many more scenarios could be included here, but each of you will have to make decisions based on personal circumstances. Although a nondirected, probability-based sample generally is preferable because it eliminates bias, at times this approach will be impossible. Before you begin your sampling procedure, try to determine whether the data can be collected by means of a probabilistic sampling procedure. Only if this is not a possibility should you default to the directed approach. If you are able to use a probabilistic spatial sampling methodology, each object or feature you select from your sampling frame is expected to have a known chance of being selected for study. This known probability of being selected is used to design the procedure for taking the sample. Stated differently, you use the known probability to set up a sampling methodology that will allow all objects the same probability of being selected.

The methodology for probabilistic sampling can easily be divided into four general categories: random sampling, systematic sampling, stratified sampling, and homogeneous sampling (Figure 2.13). These categories could be

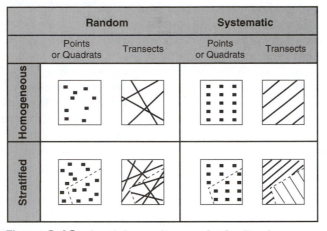

	Random		Systematic	
	Points or Quadrats	Transects	Points or Quadrats	Transects
Homogeneous				
Stratified				

Figure 2.13   Spatial sampling methods. Random, systematic, stratified, and homogeneous sampling designs. *Source:* P.C. Muehrcke and J.O. Muehrcke, *Map Use: Reading, Analysis and Interpretation,* 3rd ed., J.P. Publications, Madison, WI, © 1992, Figure 2.6, page 51. Used with permission.

combined to create a hybrid sampling design if so desired, but we will examine the basic types and let you modify them later as you see fit. *Random spatial sampling* is the most basic sampling design. Its purpose is to allow each individual point, line, area, or surface feature to be selected with the same probability as the next. If the spatial data you are sampling are discrete, such as trees, lakes, or people, your purpose is to observe something about a few of these by selecting them, at random, for examination. In such circumstances each object is assigned a unique number, say from 1 to 1000. With the use of random number generators found in nearly all computers and many hand calculators, or a set of random number tables, it is easy enough to select a fraction of these numbers, again randomly. We may, for example, select 100 of the 1000 spatially located objects' numbers for measurement. If the data occur continuously, such as with topography, barometric pressure, or soil temperature, we will randomly select points at which these properties can be measured and number them, selecting points to be examined as before. In both cases it is possible to select random points, random areas called quadrats (often used to examine the amount of above-ground biomass for grasses), or line transects for use in selecting our objects of study.

Systematic designs operate almost exactly the same as random designs, but now we decide on a repeatable pattern, rather than random numbers, to use as a basis of selection. For point data we could, for example, select every tenth tree, or trees located approximately 20 meters apart. To examine small plots or quadrats, we would select these in the same manner, by examining every nth object or an object every n meters. Similarly, if our sampling involves using line transects, a popular method for examining vegetation assemblages, we would systematically determine where each line transect would fall and take our census of the vegetation along each of these transect lines. Or, if we wanted to completely survey individual plots or quadrats, we would once again select these using a systematic, repeatable pattern for selecting each quadrat for study.

Stratified spatial sampling adds another dimension by selecting small areas within which individual spots, objects, or features are sampled. Stratifying simplifies the process of sampling by dividing the task into small regions that can, for example, be examined by one person, or in one day of sampling. Within each stratified area we can decide whether a random or a systematic design will be used to examine the objects or features. In a modification of this technique called **clustering**, we determine at the onset whether the objects to be observed are **clustered** or dispersed throughout your entire study area. Each of these groups can then be selected as a separate study subarea, much as we did when we stratified our overall study area. Again, we can use point, quadrat, or transect approaches, and we can choose a systematic or a random method of sampling within each subarea. The advantage of this approach is that homogeneous objects often imply an underlying process. Selecting these areas out for individual study is likely to give us more accurate data relating to that process than will treating it as a part of the whole, in which case we are assuming that the processes operating throughout the whole study area are generally the same. This approach, however, has the disadvantage of requiring us to make decisions about which subareas are more clustered than others that may turn out to be inaccurate.

MAKING INFERENCES FROM SAMPLES

Spatial data taken from spatial sampling yield three major types of manipulation: data at nonsample locations can be predicted from sampled locations; data within regional boundaries can be aggregated (so that a single class of data characteristics can be assigned); and data from one set of spatial units can be converted to others with different spatial configurations (Muehrcke and Muehrcke, 1992). As you continue learning about GIS, you will discover many situations involving predictions, and a general understanding of prediction problems will save you time and energy later on.

Although sampling shortens the time necessary to collect data about a region, it also leaves gaps in our knowledge about the locations not sampled. Given that most GIS software is based heavily on the idea of areas rather than points, we must be able to determine or predict the missing point values. This need commonly arises when we gather information about surfaces using point samples. For example, to gain an idea of what an entire surface looks like, we might select a number of points to measure elevation. When we view topographic maps, either as contour lines or as three-dimensional representations, the missing data are not measured but are instead spatially predicted (Muehrcke and Muehrcke, 1992). Under these circumstances, there are two general types of predictive models. Interpolation is used to predict missing values when we have values bounding, or on both sides of, the gap. If there are values on one side of the missing data, but none on the other side, we call the methodology extrapolation. **Interpolation** can be as simple as assuming a linear relationship between the known values and filling in the sequence. More sophisticated methods are based either on assumptions of a nonlinear relationship between these known values or on a weighted distance or local operator model, where nearer values are more likely to be useful in predicting missing values than points farther away. **Surface fitting models** involve fitting the observed values to an equation and then solving the equation for each missing value (Muehrcke and Muehrcke, 1992). Surface fitting models are more likely to be useful for **extrapolation** than the others because the equation can easily be extended beyond the known data values. All these methods allow us to predict missing values, but it should be remembered that predictions are not measurements, and each prediction will have its own set of problems and errors. We will look more closely at these methods later on when we study surfaces.

Extrapolation is more appropriate when we are sampling point data as estimates of areas rather than surfaces. From these areas we want to be able to make further predictions about points in other areas that have not been sampled. Let's say that we are sampling the density of trees in a number of small areas, and we want to be able to predict the densities in other surrounding areas. Such a task usually requires us to operate using three assumptions. First, we calculate the average densities, on a region-by-region basis, to prevent region size from affecting the values. Next we assign each density to a single point inside each region (usually some central point). Then, having completed these steps, we can return to the point interpolation methods to predict density values for each of our missing areas.

A final predictive problem based on sampling requires the prediction of area quantities on the basis of a set of quantities sampled from a different set of areas with different sizes and shapes. In the simplest approach, which entails a form of spatial overlay, it is assumed that the data inside each area are essentially uniform and homogeneous (that is certainly true). When these two sets of areas are overlaid, again assuming that the data are homogeneous, the amount of overlap should be directly proportional to the amount of data in each.

One other conversion needs to be considered here that might relate to sampling, although it can also apply to total population census. Suppose you are sampling discrete objects, such as the locations of animals. Once you have located the creatures, and for the sake of simplicity we will say that they don't move around much, you will want to know what portion of the earth they generally occupy. In other words, you need to know their range. This is a common problem for wildlife managers, for example, who use radiotelemetry devices to locate animals and often must operate in a GIS that is poorly adapted for working with point data but strong on area-based data. Some relatively easy computer techniques can be applied, as well as some biological approaches, and we will consider these in more detail later. For the time being it is enough that you be aware of their existence as you look for ways to explain what you observe.

A final note is in order concerning the case of area prediction using overlay techniques. To perform this nifty feat, we are going to need a device that has been mentioned only briefly but will be the major topic of the next chapter—the map. The map is the fundamental device by which we abstract our environment's space and within which the GIS will operate to analyze it. In the next chapter, then, we will expand our spatial vocabulary and improve our spatial filter by examining how we can move from spatial data in the conceptual sense to spatial data in a graphic sense. The skills we will be learning, which are called graphicacy (Balchin, 1976), will serve us well as we move toward a better understanding of automated geography inside a GIS.

Terms

absolute location	geographic data measurement	random
arrangement		ratio
association	grid system	regular
attributes	international date line	relative location
clustered		remote sensing
clustering	interpolation	sampled area
color video	interval	sampling frame
continuous	latitude	scanning radiometers
dead reckoning	locate	side-looking airborne radar (SLAR)
density	longitude	
diffusion	meridians	spatial surrogates
digital photography	nominal	spatially correlated
discrete	ordinal	surface fitting models
dispersed	orientation	target population
dumpy level	parallels	triangulation
extrapolation	plane table and alidade	trilateration
	prime meridian	witness trees

Review Questions

1. Why is it important that we understand the language of geography before we can be effective in GIS analysis? What impact does an increased spatial vocabulary have on our ability to work with spatial phenomena?

2. What impact does scale have on how we experience our world and on how we model it? Give an example, other than the one in the text, of how scale change might permit you to view an object with length and width as if it were a point.

3. Why is it important to develop a spatial vocabulary for GIS? What impact does an increased spatial vocabulary have on how we explore our environment? How does a geographic language help us in filtering the data we view in our explorations?

4. What are spatial surrogates? Can you give some examples of aspatial data that must be given explicit spatial dimensions before they can be incorporated into the explicitly spatial framework of a GIS?

5. What are discrete data? Can you give some examples of them for point, line, area, and surface features?

6. What are continuous data? Give some examples of these, especially with regard to surface data.

7. Why is it important to understand the levels of data measurement when we are observing and recording attribute data for the objects we encounter?

8. Give some concrete examples of nominal, ordinal, interval, and ratio data for each of the following features: point, line, area, and surface.

9. Why do we need a structural framework such as the latitude–longitude grid system? What does it add to the way we view our world?

10. What is the difference between absolute location and relative location? Give some examples of each.

11. Some of the terms added to our vocabulary are orientation, arrangement, diffusion, pattern, dispersion, and density. How do these terms improve the way we observe our earth?

12. What does spatial association mean? What does it have to do with GIS?

13. What is the impact of modern technology on ground sampling? Give some concrete examples.

14. What has remote sensing added to our geographic tool kit that ground sampling lacks? How will improvements in remote sensing impact on GIS?

15. Why would we choose to use directed sampling, as opposed to probability-based sampling, if directed sampling is considered to be less "scientific"? What are some of the conditions that force us to use directed sampling?

References

Balchin, W.G.V., 1976. "Graphicacy." *American Cartographer,* 3:33–38.

McGrew, J.C., Jr., and C.B. Monroe, 1993. *An Introduction to Statistical Problem Solving in Geography.* Dubuque, IA: Wm. C. Brown.

Muehrcke, P., and J. Muehrcke, 1992. *Map Use: Reading Analysis and Interpretation,* 3rd ed. Madison, WI: J.P. Publications.

Piaget, J., 1970. *The Child's Perception of Movement and Speed,* translated by G.E.T. Holloway and M.J. Mackenzie. New York: Basic Books.

Piaget, J., B. Inhelder, and A. Szeminska, 1960. *The Child's Perception of Geometry,* translated by E.A. Lunzer. New York: Basic Books.

Sauer, C.O., 1925. *Morphology of Landscapes.* Berkeley: University of California Press.

Witthuhn, B.O., D.P. Brandt, and G.J. Demko, 1974. *Discovery in Geography.* Dubuque, IA: Kendall/Hunt Publishing Company.

CHAPTER 3

Map Basics

The map is the primary language of geography. It is, therefore, the fundamental language of digital geography. This graphic form of spatial data abstraction is composed of different grid systems, projections, symbol libraries, methods of simplification and generalization, and scales. If you are comfortable with the map as a method of modeling your environment, you may be able to skip this chapter and move on to Chapter 4, especially if your background includes coursework in map use and cartography. (Even so, it might prove useful to review this chapter before moving on, as a reminder of the cartographic method and all it implies for correct analysis and interpretation.) If you have not had such courses or if your experience in mapping or map reading is minimal, you should spend some time focusing on this material. You might also find it useful to select a couple of good books on cartography or map reading to supplement the short descriptions given here. As you evaluate your map familiarity, remember that inside a GIS, you are likely to encounter a variety of maps far more abundant than you might have expected on the basis of courses in geology, surveying, or soil science. In addition to the geological, topographic, cadastral, and soil maps used in these disciplines, the thematic content of GIS map layers includes vegetation maps, transportation maps, animal distribution maps, utility maps, urban plans, zoning maps, land use maps, land cover maps, and remotely sensed imagery. These maps will be portrayed as prism maps, choropleth maps, point distribution maps, dasymetric maps, surface maps, graduated circle maps, and a host of other types. If some of these terms are new to you, then read this chapter if for no other reason than to become generally familiar with the myriad possibilities.

Remember, our examination of the earth through GIS is predicated on our ability to think spatially. Spatial thinking requires us to be able to select, observe, measure, catalog, and characterize what we encounter. But the depiction of objects in cartographic form depends on what questions are being asked, whether we are trying to display the cartographic form or analyze it in a GIS, whether we are observers or users, whether our data are collected in the field or through remote sensing, whether archived maps of previous observations are going to be used in our analysis, and many other factors. The more we know about the possible combinations of graphic elements and how they are manipulated on the cartographic document, the stronger our geographic language. And, as you have seen, the larger our spatial vocabulary, the better we are at making decisions about spatial phenomena and their distributions in space. Knowledge about cartographic methods will increase a portion of our spatial vocabulary that we have called **graphicacy**.

An improved level of graphicacy will assist us in all four subsystems of the GIS tool kit. When we consider putting existing maps into the GIS, we will be aware of the impacts of different levels of generalization, scales, projections, symbolization, and the like on what is input and how it is done. Once inside the GIS, we will be able to identify potential problems that will require editing: two adjacent input maps originally produced through different projections, for example, or large symbol sizes that permitted us to place data points in incorrect or illogical locations. As we begin to analyze our data, we will be aware of the potential for error of some map layers that were created from very small scale maps. Finally, as we produce output from our GIS, we will have a better feel for how best to display the results of analysis, because we will be familiar with the cartographic method and its design criteria. We will also learn how spatial features and their relationships can be depicted in a form that allows us to view these relationships more easily and, therefore, how we can begin to formulate methods of analysis.

LEARNING OBJECTIVES

When you are finished with this chapter you should be able to:

1. Understand the role of graphicacy and cartographic communication in improving our understanding of the world around us.

2. Understand the relationship between the analytical and holistic cartographic paradigms as they apply to GIS.

3. Understand how scale is illustrated on a map and the possible impact on the results of analysis of putting differently scaled maps into a single GIS database.

4. Show how the map legend acts to link entities and attributes on a map.

5. Know the different families of map projections, indicate how they modify the properties of the maps, and tell how this knowledge can be used in decisions regarding selection of appropriate map projections.

6. Be familiar with some basic grid systems and their operation, recognizing their advantages and disadvantages for GIS work.

7. Understand the impact of scale, class interval selection, symbolism, and simplification on the development of cartographic databases.

8. Know the difference between cartographic and geographic databases and understand the potential problems associated with producing a GIS database composed of both.

ABSTRACT NATURE OF MAPS

The map is a model of spatial phenomena. It is an abstraction; it is not a miniature version of reality that is meant to show every detail of a study area. Although this may sound rather evident, at times we have all either ignored or forgotten this simple fact. There are limits to what we can do with cartographic skills. The primary reason for our misinterpretation of the limits of maps to display reality is that they are among the most elegantly designed graphical instruments created to communicate spatial data. Maps have been around for thousands of years, and we have all become more or less accustomed to seeing them. Such familiarity combined with the compactness of information maps contain, and their powerful visual appearance, all lend a sort of authoritative infallibility that is difficult to discount. This is all the more reason for us to be aware of the basic character, level of abstraction, symbology, and methods of production that comprise the map product.

Maps come in a wide variety of forms and subject matter. The two primary types are the general reference map and the thematic map. As we have seen already, most of what we will deal with in GIS has to do with thematic maps, although at times reference maps are used as input to a GIS, primarily to permit the disassembly from the more complex map of selected (thematic) data. Although we will largely limit our discussion to thematic maps, much of what is covered in this chapter can be readily applied to the reference map.

A PARADIGM SHIFT IN CARTOGRAPHY

As we continue to move to an automated cartographic and GIS environment, we should be aware that the way maps are viewed has changed considerably over the past few decades. These changes have, in some respects, contributed to the enhanced utility of the GIS because of the way we treat the spatial data we input to the system. The traditional approach to mapping, called the **communication paradigm**, assumed that the map itself was a final product designed to communicate a spatial pattern through the use of symbols, class limit selection, and so on. This paradigm is the traditional method of viewing cartography, but it is limited because raw, preclassified data are not readily available to the map user. In other words, the user, being restricted to handling the final product only, is incapable of regrouping the data into forms more useful for changing circumstances or needs.

An alternative approach to cartography, one that maintains the raw data for later reclassification, developed at approximately the same time that map makers began to apply major advances in computer technology. This method, known as either the **analytical paradigm** (Töbler, 1959) or sometimes the **holistic paradigm**, maintains the raw attribute data inside a computer storage device and displays data based on user needs and user classifications (DeMers, 1991). An early precursor to CAD and even GIS itself, this method today is much more flexible in its application than its predecessor. The impetus for its development is the idea that a map, especially with the use of computer technology, should allow for both communication and analysis.

The analytical paradigm was originally designed for use with value-by-area mapping (covered later in this chapter), where each area would have its own unique color or shading pattern relative to the raw data it represented. These maps, called classless choropleth maps (Töbler, 1973), had the initial limitation of not being easily interpreted by the user. In this respect, they were not unlike unclassed satellite images. With the use of the computer as a storage and data classification device, however, the method expanded in concept to allow multiple classifications of the data, and each set could be readily viewed by the viewer. This enhancement prompted some to change the name from analytical to holistic because it signals the return to prominence of the map's role as a communication device. We will see this in Chapter 14, where we discuss output from the GIS, a truly holistic approach to cartography as well as to geography in general.

An example, using identical study areas, for different purposes, may be useful for understanding the different cartographic paradigms. Suppose you are creating a map for a state park area. The initial assignment is to create a map that will allow tourists to enjoy the activities and sights offered by the park. You design the map to highlight the lake, boat docks, camping areas, cabins, the park ranger station, geysers or other natural features, walking trails, fishing sites, restricted areas, food vendors, and so on, showing the roads that provide access to each feature (Figure 3.1a). The features may also be classed to illustrate, for example, the quality of the roads, the types of boat that can use the docks, and whether hookup facilities for recreational vehicles are available at the campsites. This map is provided primarily to show the spatial distribution of the phenomena of interest in a way that indicates how each one may be reached. This is illustrative of the communication paradigm.

The park rangers, however, will need information to manage the facilities and resources for which they are responsible. They may need such quantitative information as the average tree size within the forested regions, or the buildup of fire fuel in the forests, the numbers of rare or endangered species in restricted

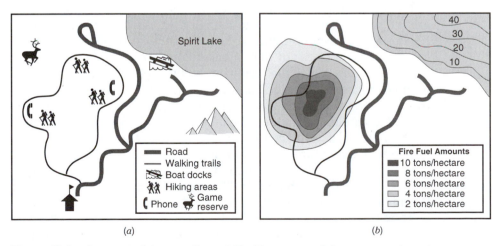

(a)   (b)

Figure 3.1 Cartographic paradigm shift. Two maps of the same study area—a state park—showing (a) the use of the communication paradigm to create a visual display of spatial relationships and (b) the added utility of the analytical paradigm for the same region.

areas, and the sizes and areal distributions of gaps in the forest that provide needed environments for browsing species. With these data, the park rangers can begin to determine whether deer populations are rising or falling, whether some trees may need to be cut to lessen the threat of fire, or where old and dying trees should be thinned to allow smaller trees to grow in. Many of these situations will call for additional attribute information about tagged species, diseased trees, or annual differences in animal or plant species composition.

Eventually, the amount of information, particularly the attribute information, becomes so dense that a single map is incapable of illustrating the necessary conditions for the park rangers to plan for future management strategies. The data must first of all be detailed for each location. If, for example, there are 200 different types of vegetation assemblage, data must be recorded for each, but displaying all these areas on a single map will obscure the information visually. Moreover, multiple maps would have to be produced for each variable at each area. The primary function of these cartographic representations is analysis, rather than simply indication of spatial distributions (Figure 3.1*b*). The communication paradigm is inadequate for this application, and it must be augmented to include more analytical capabilities such as are now available in the holistic paradigm with a GIS.

MAP SCALE

No matter which paradigm we choose, when we consider converting our spatial concept of space into a map, we must always remember that maps are **reductions** of reality. Although it might be intellectually appealing to envision a map that physically covers our entire study area, such a map would require us, once again, to explore the planet on foot. A primary purpose of any thematic map is to allow us to view important detail, for a large region at a single glance, without the distractions of inconsequential or extraneous detail. The amount of reduction is primarily a function of the level of detail we need to examine our area. If we are looking at a very small area, such as a single field (say of 20 hectares), we are not required to reduce reality as much as we would if we were looking at a study area of 1000 square kilometers.

Scale (Figure 3.2) is the term commonly applied to the amount of reduction found on maps. It can most easily be defined as the ratio of distance on the map to the same distance as it appears on the earth. For example, a **map legend** might indicate that 1 inch on the map is equal to 63,360 inches on the ground. In other words, if we placed an object 1 inch long on the ground, we would have to line up 63,359 more equivalent objects to match the amount of ground covered by a single inch on the map. A scale expressed as "1 inch equal to 63,360 inches" is called a **verbal scale**. This common method of expressing scale has the advantage of being easily understood by most map users. Another common method is the **representative fraction (RF)** method, in which both the map distance and the ground distance are given in the same units, as a fraction, thereby eliminating the need to include units of measure. Experienced map users most often prefer the RF method because it reduces confusion. The GIS practitioner is especially warned to find out which of these two methods of expressing scale is being used. Many of you will undoubtedly experience a sense

10 miles   20 miles

Graphic or Bar Scale

1" = 36 miles

Verbal Scale

1:36,500

Representative Fraction (RF)

Figure 3.2 Methods of illustrating map scale. Examples of the three most common methods of illustrating scale on a map. The verbal scale, the representative fraction (RF), and the graphic or bar scale all have advantages and disadvantages for analog or digital mapping projects.

of delight when a county extension agent or other government official offers you maps (or aerial photographs) at a scale of "1:600." You expect to see maps at a representative fraction of 1 to 600—very detailed maps. However, you are really hearing a shorthand version of the verbal scale, and your contact is trying to say that the map scale is 1 inch = 600 feet. This translates into a representative fraction of 1:7200, for a much smaller scale map covering a relatively large area.

A **graphic scale**, also shown in Figure 3.2, is another basic method of expressing scale; here measured ground distances appear directly on the map. Actual area measures may be illustrated on the map as well, but this is less common. Although the graphic scale is difficult to communicate orally to the user, it has the advantage of being easily transferred from one scale to another, which far outweighs the limitation. GIS map manipulations are likely to entail many changes in scale of output depending on the format required by the user. A graphic scale device can be placed on the map during input, and, as the map scale changes on output, so will the scale bar itself.

As you begin working with GIS, you will find that most software can accomplish changes of scale very easily. And, of course, the scale at which you input the data may differ from the scale at which you display the results. The ability of the software to convert to nearly any scale frequently imparts a feeling of confidence that can translate into problems later on. As you will see in some detail in Chapter 5, reliable analysis is greatly influenced by the quality of the data that are input to the system. This reliability is, in turn, affected to a large degree by the scale of the cartographic documents you input. Keep in mind that on a very small scale map, say of 1:100,000, a line of 1 millimeter covers over 100,000 millimeters, or 100 meters, on the ground. That's approximately the length of a football field. When the scale is changed in the GIS for later output at, say a scale of 1:1000, the line that is drawn will be 1 millimeter in width, giving the reader the impression that the line was accurately located when it was input. In fact, its location is far from accurate.

Another example might be quite useful in illustrating the problem of scale as it affects analysis. Someone not familiar with maps presents you with a map of the 48 contiguous states of the United States depicted on a piece of paper about the size of a 3.5-inch floppy diskette. The person needs to have the map blown up to a much larger size, say 1 meter on a side, so that the areas of each state can be measured and the results used to calculate population density by state. Whether this map is to be enlarged through some xerographic approach or input to a GIS to have the areas measured, the end result is the same. The values obtained through dividing the number of people for each state by the

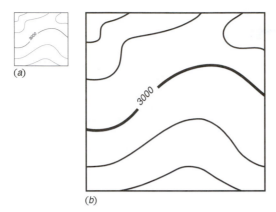

Figure 3.3   Effect of scale on accuracy. A portion of an extremely small scale map blown up to a larger scale. The lines have become thicker and show a high level of generalization, areas are less precise, and measurement and analysis are practically impossible.

highly error-filled measures of area are certain to be useless (Figure 3.3). Thus the following simple, long-held, rule of thumb: it is always better to reduce a map after analysis than to enlarge it for analysis. This rule applies as much to the automated environment as to the manual one.

MORE MAP CHARACTERISTICS

Maps, as images of the world we model with them, represent the locations of objects in space as well as their qualities or magnitudes. These two related items are called **entities** and **attributes**, respectively, and both are as necessary in the cartographic document as they were when we collected the data in the first place. But regardless of whether entities and objects represent points, lines, areas, or surfaces in the real world, they cannot be displayed as miniaturizations of reality because of the limitations of scale. Instead, we must store them in the computer's memory and then, upon output, assign a set of symbols to represent them. In turn, the symbols must have an interpretive key called a map legend to be referred to by the user. The legend effectively unites the entities with their attributes, whereupon each displayed entity can be understood to represent a real feature with measurable attributes. In this way the map reader is able to envision what was actually seen as the original data were collected, through the variety of methods we have already discussed. As you remember from Chapter 2, entities are measurable, and their attributes can be delimited using a number of different levels of data measurement. Associated with each different spatial data type and each level of data measurement is a set of symbols. We will take the spatial data types in turn and examine their potential to be mapped, the variety of symbols that can represent each one, and the ways in which the process of map abstraction limits how they can be used as input to the GIS.

MAP PROJECTIONS

Although when viewed at close range the earth appears to be relatively flat except for the surfaces we traverse, we all know that it is relatively spherical.

Maps, as we have seen, are reductions of reality, designed to represent not only its features but also its shape and spatial configurations as well. The use of globes is a traditional method of representing the earth's shape. Although globes preserve the majority of the earth's shape and illustrate the spatial configuration of continent-sized features, they are very difficult to carry in one's pocket, even at extremely small scales (often as little as 1:100,000,000). Most of the thematic maps encountered in practice, as in GIS analysis, are of considerably larger scale, say on the order of from 1:100,000 to as large as 1:1000, depending on the level of detail. A globe of this size would be difficult and expensive to produce and even more difficult to carry around or to spread out on a digitizing table for input to our GIS. As a result, cartographers have developed a set of techniques called map projections designed to depict with reasonable accuracy the spherical earth in two-dimensional media.

As the term implies, the process of creating **map projections** was originally envisioned as positioning a light source inside a transparent globe on which opaque earth features were placed, then projecting the feature outlines onto a two-dimensional surface surrounding the globe. Different ways of projecting could be produced by surrounding the globe in a cylindrical fashion, as a cone, or even as a flat piece of paper. Each of these methods as originally envisioned produces what is called a **projection family**. Thus there is a family of planar projections, a family of **cylindrical projections**, and another called **conical projections** (Figure 3.4). A fourth, referred to as **azimuthal projections**, which is based on the idea that the light is projected at a selected azimuth onto a flat material, is discussed later in this section. Today, of course, the process of projecting a spherical surface onto a flat medium is done using the mathematical principles of geometry and trigonometry that re-create the physical projection of light through the globe.

Cartographers have studied, modified, and produced new projections to better represent the earth on a flat medium. Unfortunately, many noncartographers in the geographic community have largely neglected this important aspect of spatial representation. The reasons for this are the stuff of the history and philosophy of geography. But with the advent of satellite remote sensing

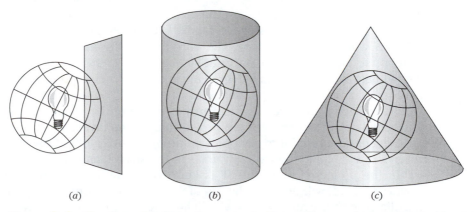

(a)   *(b)*   *(c)*

Figure 3.4 The three families of map projections. They can be represented by *(a)* flat surfaces, *(b)* cylinders, or *(c)* cones. *Source:* P.C. Muehrcke and J.O. Muehrcke, *Map Use: Reading, Analysis and Interpretation,* 3rd ed. J.P. Publications, Madison, WI © 1992, Figure C.9, page 573. Used with permission.

and geographic information systems, the subject can no longer be ignored. It is even more relevant now than in the past because of the need to minimize the distorting effects of projections on the true representation of spatial phenomena, combined with the wide range of mapped data the automated geographer will encounter every day. These matters will impact on all phases of GIS and should be understood to limit their negative influences.

Projections are not absolutely accurate representations of geographic space. Each will impose its own types and amounts of distortion on the map document. The important characteristics of our maps that must be retained for accurate analytical operations will often dictate which projections should be used. These characteristics or properties include angles (or shapes), distances, directions, and areal sizes. It is impossible to preserve all these properties at the same time when performing a map projection. We will look at each one separately, but first we must establish some useful terminology to help us understand the changes in properties as they are altered by the projection process.

The process of map projection as it is mathematically produced is twofold: first an obvious scale change converts the actual globe to a **reference globe** based on the desired scale; then the reference globe is mathematically projected onto the flat surface, where the globe's three-dimensional surface is transformed to a flat surface (Robinson et al., 1995). When we reduce the scale from the actual globe to its reference globe, we have changed the representative fraction to reflect this scale change. The representative fraction for the reference globe, called the **principal scale**, is calculated by dividing the earth's radius by the radius of the globe. We now have a representative fraction that is uniform throughout the reference globe because the shape of the latter is nearly identical to that of the earth. That is, the actual scale will be the same as the principal scale everywhere on the globe.

Before we perform the second step of projecting the reference globe to its flat map counterpart, we should note that the **scale factor (SF)**, defined as the actual scale divided by the principal scale, is by definition 1.0 at every location on the reference globe. When we move from a spherical reference globe to the two-dimensional map, however, the scale factor will necessarily change because the flat surface and the spherical surface are not completely compatible. Therefore, the scale factor will differ in different places on the map (Robinson et al., 1995). We will need to keep the idea of a changing scale factor in mind as we examine the types of distortion made through the projection process.

When working with a globe, the cardinal directions of the compass rose found on its side will always occur at 90 degrees of one another. In other words, east will always occur at a 90-degree angle to north. This property of maintaining correct angular correspondence can be preserved on a map projection as well. A projection that retains this property of **angular conformity** is called a **conformal** or **orthomorphic** map projection. As we have seen, the SF is always 1.0 in every direction for every point on the reference globe. **Conformal projections** permit us to mathematically arrange the stretch and compression so that, within the projected map, the SF is kept the same in every direction. Since, however, the SF will not then be equal to 1.0 at every point, the parallels and meridians in the resulting map will always be at 90 degrees to one another as they were on the globe, but the areas shown on the map will be distorted (Robinson et al., 1995). Keep in mind that maintaining true angles is difficult for large areas and should be attempted only for small portions of the earth.

As we have seen, the conformal type of projection results in distortions of areas, meaning that if area measures are made on the map, they will be incorrect. We can preserve areas through projections called equal area or equivalent projections, in which the product of scale factors in cardinal directions is equal to 1.0 (Robinson et al., 1995). This procedure ensures that if you find the areas of, for example, square surface features, their two dimensions multiplied together will result in an area identical to what would be calculated on the reference globe. This is because the product of these two dimensions results in the identical SF. However, once we have achieved this identity, we find that, although the amount of area is correctly measurable, the scale factor will vary in every direction about a point except the cardinal directions. In other words, by preserving area we distort angles. Thus, it is a fundamental feature of these two properties of projected maps—that areas and angles cannot be preserved at the same time.

If our purpose in projecting a map is to accurately measure distances, we must select a projection that preserves distances. Such projections, called **equidistant projections**, require us to keep the scale of the map constant; it must also be the same as the principal scale on the reference globe (Robinson et al., 1995). There are two ways in which this can be done. The first maintains a scale factor of 1.0 along one or more parallel lines called **standard parallels**. Then distances measured along these lines will be accurate representations of real distances. The second approach is to maintain the scale factor as 1.0 in all directions from either one or two points. Distance measured from those starting points will then be accurate representations of true distance. A distance measured from any other point on the map will be inaccurate. As you might guess, the choice of starting point is vital here. Usually you choose the point at which most of your measurements are going to be made.

When maps are used for navigation, the primary interest is to preserve the directions shown. The preservation of true direction on a map is limited to preservation of great circle arcs, which define the shortest distance from point to point on the earth. Normally, our purpose is to depict these great circle routes as straight lines. There are two primary methods of doing this. The first uses very limited areas to show great circles as straight lines between all points on the map. If you intersect the meridians with these great circles, however, the angular intersections will be incorrect. Both the limited area and the angular intersections between meridians and great circles severely limit the use of this projection. An alternative, called the azimuthal projection, is more commonly used for preserving directions. As with the equidistant projection, we begin by selecting one or two points of reference from which our directions will be preserved. In this case, straight lines directed from these reference points will illustrate true direction. Again, measures of direction made starting at any other points will be incorrect.

We need some rules of thumb for determining which of the many map projections we might want, depending on the types of analysis to be performed. If an analysis requires us to analyze motion or changing directions of objects—for example, the type of data obtained by a game manager using telemetry to record the location of every member of a herd of reindeer at different times—the conformal projection is best. This type of projection also is best for the production of navigational charts and when angular orientation is important, as it often is with respect to meteorological or topographic data. This group of projections includes the Mercator, transverse, **Lambert's conformal conic**, and **conformal stereographic projections**.

General reference and educational maps most often require the use of **equal area projections**, but our interest is in analysis. As the name implies, these maps are best used when calculations of area are the dominant calculations you will perform. If, for example, your interest is in the calculation of the changing percentage of a land cover type with time, or if you are trying to analyze a particular land area to determine whether it is large enough to be considered for a shopping mall, equal area projections are best. When considering the use of an equal area projection, you will need to take into account the size of the region involved as well as the distribution and amount of angular deformation. Small areas will display far less angular distortion when equal area projections are used, and this property might be relevant if both shape and size are important characteristics for your analysis. Alternatively, the larger the area you are analyzing, the more precise your area measures will be if you use an equal area projection rather than another type. Types of equal area projection for use with medium-scale maps most often encountered in GIS work are **Alber's equal area** and **Lambert's equal area projections**.

GIS projects that require a determination of shortest routes, especially for long distances, call for the use of azimuthal projections because of their ability to show great circle routes as straight lines. These projections are most often used for creating maps for airline traffic, determining ranges of radio signals, keeping track of satellites and satellite data, and mapping other celestial bodies (Robinson et al., 1995). Only recently have these projections become popular, but their use will increase as these uses for GIS increase. The most common azimuthal projections you will encounter include Lambert's equal area, stereographic, **azimuthal equidistant**, orthographic, and **gnomonic projections**. A final note about the use of these projections is that some maintain both direction and area. This property may prove useful for analyzing aerial distributions such as volcanic plumes that are likely to travel great circle routes as they disperse into the atmosphere and follow the general circulation patterns of the earth.

There are many map projections from which to choose—far more than are listed here. Some are special-purpose projections especially useful for depicting the entire earth or very large portions of it. Other projections allow for better coordination of large-area mapping programs such as the topographic mapping of whole continents, which is done in smaller portions. The list is quite large. As you will see later, the selection of projection is a fundamental process of designing a GIS. You should take the time to select a good reference on map projections, with special emphasis on the properties each best maintains. Two sources that might prove particularly useful are those of Nyerges and Jankowski (1989) and a well-known reference work by Snyder (1988).

GRID SYSTEMS FOR MAPPING

We have seen that a grid or coordinate system is necessary to reckon distance and direction on the earth. This geographical coordinate system based on latitude and longitude is useful for locating objects or features when they are confined to the spherical earth or its reference globe. Because we will most often operate with two-dimensional maps projected from this reference globe, however, we also need one or more coordinate systems that correspond to the distortions introduced through that process. These reference systems, called **rectangular coordinates** or **plane coordinates**, allow us to locate objects correctly on these flat maps.

The basic rectangular coordinate system is also the one that most of us learned when working with graphs and number lines. It consists of two lines, an abscissa and an ordinate. The abscissa is a horizontal line that contains equally spaced numbers starting from 0, called the origin, and extending as far as we wish to measure distance in either of two directions (Figure 3.5). The values are called X coordinates; they are positive if we move to the right of 0 and negative if we move to the left. The second line, the ordinate, allows us to move vertically from the same point of origin in a positive or negative Y direction. Together the X and Y coordinates allow us to locate any point or feature by combining values of X and Y. As you will see later, **digitizers**, the devices we use to input geographic coordinates into a GIS, are based on this simple **Cartesian coordinate system**. It is possible to obtain reasonably accurate results with such a system as long as the map exhibits the property of **conformality**. Maps of this type are most likely to illustrate only small portions of the earth's surface. Each of these large-scale maps must maintain its own coordinate system to ensure a degree of accuracy of measurement (Robinson et al., 1995).

By tradition, when reading maps using rectangular coordinates, we give the X value first and the Y value second. When a map is oriented with north at the top, the normal situation, the X value, is called an **easting** because it measures distances east of the origin or starting point. In like manner, the Y value is called a **northing** because it measures distance north of the origin. As you can see, there are no westings or southings. Instead, the origin is placed so that all references are positive; or, stated another way, they are placed in the northeast quadrant of our reference system. This allows us to read first right, then up

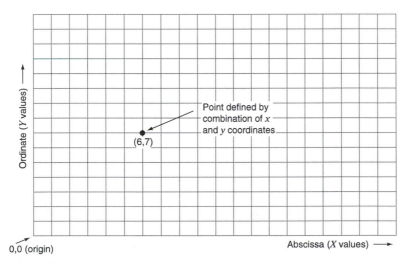

Figure 3.5   A Cartesian coordinate system. Classic rectangular grid system used to represent the number system. The abscissa (X values) is horizontal, and the ordinate (Y values) is vertical. Any point can be defined by combinations of these two.

from the origin, a process called reading right-up. In some cases, the size of the area will require us to construct a number of **false origins** to ensure that each portion of the earth will be accurately represented on a flat surface. Measures from false origins are then called **false eastings** on the abscissa and **false northings** on the ordinate line.

As indicated, plane coordinates are not normally used on small-scale maps because of the potential for distortion. For small-scale maps, adjustments must be made to compensate for the distortions introduced during projection. Despite the large number of map projections available, the vast majority of plane coordinate systems attempt to adjust for conformality by using only conformal map projections, typically the transverse Mercator, polar stereographic, and Lambert's conformal conic. This will not always be the case, however. If, for example, your study area is located in equatorial locations, a **Mercator projection** may prove more useful (Robinson et al., 1995).

In the United States, five primary coordinate systems are used, some based on properties of the map projections and others on historical land subdivision methods. When you encounter maps of other nations in your GIS work, you will need to ascertain the projections, coordinate structures, and other properties of the systems of those nations. Many countries will use one or another of the types discussed below but will require you to become familiar with their points of origin and the zones of the earth they occupy before beginning your GIS input procedures.

Perhaps the most prevalent plane grid system used in GIS operations is the **universal transverse Mercator (UTM)** grid (Figure 3.6). It has been adopted for much remote sensing work, topographic map preparation, and natural resource database development because it allows precise measurement using the metric system of measurement, which is accepted by many countries and by the scientific community at large. The basic unit of measure is the meter.

UTM divides the earth from latitude 84 degrees north and 80 degrees south latitude into 60 numbered vertical **zones** that are 6 degrees of longitude wide.

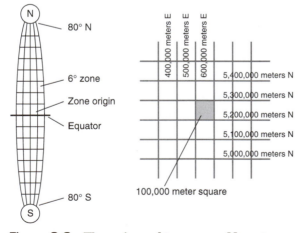

Figure 3.6   The universal transverse Mercator system (UTM).   *Source:* P.C. Muehrcke and J.O. Muehrcke, *Map Use: Reading, Analysis and Interpretation,* 3rd ed. J.P. Publications, Madison, WI © 1992, Figure 10.8, page 215. Used with permission.

To allow for all coordinate locations to be positive, the UTM has two primary ordinate starting points, one at the equator and the other 100 million meters south of the equator (located at latitude 80 degrees south). For reference, these zones are numbered starting at the 180th meridian, in an eastward direction. Each zone, in turn, is divided into rows or sections of 8 degrees of latitude each, with the exception of the northernmost section, which is 12 degrees, allowing all the land of the northern hemisphere to be covered with the system.

Because each section can be located by a combination of the number and letter (read right-up as before), relatively small sections of the globe can be isolated. With the exception of the northernmost grouping, each of these sections will occupy 100,000 meters on a side and can be designated by eastings and northings of up to five-digit accuracy. Thus measurements should be correct to within a 1-meter level of resolution.

As the name implies, the UTM system uses a transverse Mercator projection. For each of the 60 longitudinal zones, a separate Mercator projection is applied to reduce distortion. The *Y*-coordinate origin is placed at the exact center of each zone (comprising 6 degrees of latitude), and a false origin is offset 3 degrees west of that. The scale factor does not vary from its 0.99960 in the north-south direction, but it does vary in the east-west direction. Still, at its farthest point from the *Y*-coordinate origin, the scale factor is very nearly the same at 1.00158. This near equivalence illustrates the minimization of distortion possible with UTM that results in accuracies approximating 1 meter variation in every 2500 meters distance (Robinson et al., 1995).

For polar regions that extend beyond the area covered by the UTM grid system, while still maintaining the same level of accuracy, we use a **universal polar stereographic (UPS) grid**. This system divides the polar regions into concentric zones, then splits them into two equal halves east to west at longitude 0 and 180 degrees. These zone halves are assigned different designations for the North and South Poles. In the northern hemisphere the west half is designated grid zone *Y*, and the east half grid zone *Z*. In the southern hemisphere the west half is designated zone *A*, and the east half zone *B*. As in the case of the UTM grid system, measurements are made as eastings and northings of up to 2 million meters, corresponding to 180 degrees in longitude. Also as in the UTM, the zones can be divided into 100,000-meter squares, each with its own projection, resulting again in an accuracy of approximately 1 meter in 2500. Collectively, the UTM and the UPS systems provide a global coverage with only minor distortion and reasonably accurate measures.

Perhaps the premier mapping organization in the United States and among the largest contributors to GIS data, the USGS (United States Geological Survey) has found it convenient to simplify the UTM system for its topographic mapping. Because the United States falls entirely inside the northern hemisphere, you only need a zone number (1–60) and a single pair of easting and northing values to locate a point on the map. The equator is designated as the origin for northing values, and the meridian in the center of each meridian zone is given a value of 500,000 meters as a false easting. These simple modifications make the designations simpler, but the locations are identical to what would be found on the UTM. Most USGS topographic maps have this coordinate system printed in blue lettering on the margins, together with another grid system we will discuss a little later.

A system devised in the 1930s by the U.S. Coast and Geodetic Survey (now the U.S. Chart and Geodetic Survey), called the state plane coordinate (SPC) system, uses a unique set of coordinates for each of the 50 states (Figures 3.7, 3.8) (Claire, 1968; Mitchell and Simmons, 1974). This grid system uses either a transverse Mercator or Lambert's conformal conic projection tied to a national geodetic framework. Its original design was to provide a permanent record of land survey monument locations. Like the UTM system, the SPC uses two or more zones, in this case overlapping, to maintain an accuracy of one part in 10,000. Each zone has its own projection and coordinate grid that is measured in feet rather than meters, although metric equivalents have recently been included (Figure 3.7). To define a location using this system, you give the state name, the zone name, and the easting and northing values (in feet). The advantage of SPC is in its accuracy, estimated at four times that of the UTM system. For the GIS applications specialist, the disadvantage is a lack of coordination between state borders (Figure 3.8). If your study area extends across one or more state boundaries, you will have to spend time making adjustments among the disparate grid systems. As you might guess, this means that the object locations on each side of the boundary are also going to be misaligned.

The final plane or rectangular grid system in common use in the United States is the U.S. **Public Land Survey System (PLSS)**, established in 1785 as a method of land subdivision (Figure 3.9) (Pattison, 1957). As a land subdivision system it was not, and is not today, formally tied to any particular map projection, or to a reference globe at all. It was meant as a tool for recording ownership of land. Its basic unit of measure is an area of 1/640 of a square mile, or 1 acre. Each square mile of area is called a section, and sections are grouped in larger, 36-square-mile groups that collectively are called townships. The sections are numbered in a serpentine fashion starting in the upper right with section 1 and ending in the lower right with section 36. Each of the 36 sections can be divided and subdivided into halves and quarters, permitting ownership of smaller portions of land than an entire square mile (640 acres). For example, a person can own a quarter of a quarter of a quarter section, meaning that he or she owns 1/4 of 1/4 of 1/4 of 640 acres, or 10 acres. One could also own a quarter of half a section, or 1/4 of 1/2 of 640 acres, or 80 acres. By defining the location of the subdivisions as to north, south, east, or west for halves, or northeast, northwest, southeast, or southwest for quarters, individual landowners can determine which lands belonged to them.

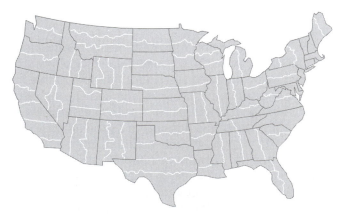

Figure 3.7   **State plane coordinate system zones for the United States.**

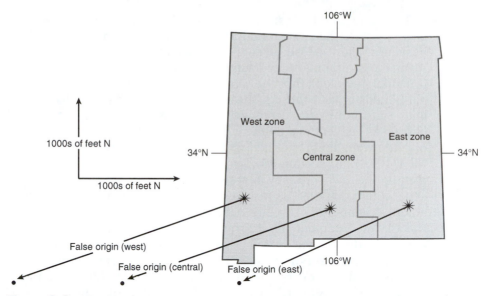

Figure 3.8   The United States state plane coordinate (SPC) system. Applied to New Mexico, this illustrates the potential for problems at cross-state boundaries when using SPC for geographic information system (GIS) work in large regions.

Figure 3.9   The Public Land Survey System (PLSS) of the United States. Township (T) and range (R) lines intersect at the center of a township. The township occupies 36-square-mile sections (S), each of which is 640 acres in area. When more than one fractional portion of a lot is given, the position in the section map is located by reading the subportions from right to left. *Source:* P.C. Muehrcke and J.O. Muehrcke, *Map Use: Reading, Analysis and Interpretation,* 3rd ed. J.P. Publications, Madison, WI, © 1992, Figure 10.20, page 226. Used with permission.

Townships and their smaller, square-mile sections are located within a larger grid of established horizontal and vertical lines. The horizontal lines, called baselines, allow measurements of township lines (not the same as townships themselves, although they bound townships), whereas the vertical lines, called principal meridians (not the same as prime meridians), measure the longitudinal bounds of land through measurements called ranges. Township and range lines run through the center of each township of land. Together, these lines, chosen as conventions by the U.S. Continental Congress, allow the designation of landownership by stating first the subportion of the numbered section (e.g., NE 1/4 N 1/2 of section 18), and then the township and range numbers that identify the township they intersect.

Although the PLSS as just described is not particularly difficult to understand, a number of problems have been encountered in its implementation in the GIS environment. First, boundaries with Canada and Mexico interrupt and often distort the system, preventing a designation of land without substantial minute subdivisions, especially along meandering stream borders. Second, past surveyors' errors are compounded when markers are lost, creating many situations in which landownership disputes occur. Finally, because the earth is round and maps are flat, and the PLSS is often portrayed on projected maps together with other rectangular grid systems, the square sections are often offset east and west to account for converging meridians as we move away from the equator. All these conditions, combined with the lack of an earthbound original geodetic framework, contribute to the limited usefulness of this system for GIS. However, you will encounter the PLSS, and you will be required to provide a projected reference system on which to map this system.

All grid systems rely on accurate measurements of the earth's size. Such measurements, in turn, are dependent on some reliable starting point from which to begin. This starting point is generally referred to as a **datum**. The datum allows geodetic scientists and surveyors to compare their earth measurements not only among themselves using a common datum but also to other coordinate systems based on other datums. Datums are based on the somewhat nonspherical shape of the earth, known as an **ellipsoid**, that resembles a flattening of the earth at the poles due to rotation (Figure 3.10).

In North America, the first datum was devised in 1866, based on reference points along the earth's surface, that characterized the earth's ellipsoid as having an equatorial axis of 6,378,206.4 meters and a polar axis of 6,356,538.8 me-

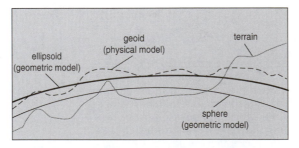

Figure 3.10 Relationships among terrain and geoid, ellipsoid, and spherical representations of the earth.

ters. These values resulted in a flatness ratio of 1/294.9787 by using the following formula:

$$f = 1 - \frac{b}{a}$$

where

a = the equatorial semimajor axis

b = the polar semimajor axis

Although a simpler flatness ratio of 1/297 was already adopted as the international standard, the original was used for the 1927 **North American datum (NAD27)** because a substantial amount of mapping in the United States had already taken place using the older standard.

Unlike the previous examples, newer datums have been established that are based on estimates of the center of the earth. In 1983, for example, a new datum was established for the United States **North American datum (NAD83)** that became the standard **geodetic reference system (GRS80)**. A modified version of GRS80, developed by the U.S. military in 1984, resulted in the **world geodetic system (WGS84)**. Many more of these datums are being used throughout the world, and one might want to refer to the National Imagery and Mapping Agency (NIMA), now called the National Geospatial Intelligence Agency (NGA), web pages to see an extensive list.

The principal primary reason for knowing the datum with which you are working is to allow for the development of a proper geodetic framework for your GIS maps. To ensure that multiple maps of the same study area are located precisely at the same locations on the earth, one must use the same datum for each. All GPS receivers refer to a particular datum and ellipsoid to obtain correct locational values. You should be aware, however, that geodetic scientists also employ another model of the earth based on compilations of all the variations from the ellipsoid. This is called a **geoid**, which, as you might guess, means "earth shaped." Elevational measurements on the earth are based variously on a reference sphere, an ellipsoid, a geoid, or the actual terrain (if measured with traditional survey equipment) (Figure 3.10).

THE CARTOGRAPHIC PROCESS

The first step of the traditional four-step cartographic process is cartographic data compilation. It has traditionally involved the selection or development of a base map on which to place the spatial data collection. Symbol sets are used to represent the points, lines, areas, and surfaces, each of which may skew exact locations to make room for their placement on the limited map surface. A preselection and sorting process is invoked to group the data by means of statistical techniques. Under the communication paradigm, grouping has most often been performed as a separate step or set of operations before mapping is begun. With the advent of the analytical or holistic paradigm, grouping can be performed within the GIS with the use of statistical or database management techniques. However, the cyclical nature of GIS operations (see Chapter 13) has obscured grouping as a separate and easily identifiable portion of the cartographic process. Many maps that will be included in the GIS database nevertheless will be input from maps created under the traditional communication paradigm. As we have already seen, this compilation process can obscure the raw data by creating categories based on the original design criteria for the map, leaving the GIS analyst without a complete set of attribute data for input.

MAP SYMBOLISM

As we have already seen, maps are not exact miniaturizations of reality but rather abstractions. We have observed that geographic objects occur as points, lines, areas, and surfaces, and we have also seen that they can be described by four different levels of measurement. When we move to cartographic abstractions, we need to represent all these objects, no matter what their measurement level, by careful selection, categorization, and symbolization so that the results will physically fit on the space provided and the reader can understand what is being presented.

A useful diagram (Figure 3.11) is an expansion of Figure 2.4, which we used earlier to show the relationships between spatial data types and their levels of measurement (Muehrcke and Muehrcke, 1992). We now have some sample sets of symbols that correspond to points, lines, areas, and surfaces at all levels of measurement. Figure 2.4 grouped the last two levels of measurement as interval/ratio, rather than keeping them separate, because the symbol set is generally the same for these measurement levels. It is important to note this because it adds another level of abstraction. If we are not aware of how the data were originally gathered, we may not know whether they are interval or ordinal. The symbols do not tell us. We may find ourselves performing a ratio-type analysis on a set of data that do not have an absolute starting point. Results from such an analysis would be meaningless.

Figure 3.11 Scale change and feature elimination. Comparison points, lines, areas, and surfaces as these features are detailed in large-scale maps and eliminated for small-scale documents. *Source:* P.C. Muehrcke and J.O. Muehrcke, *Map Use: Reading, Analysis and Interpretation,* 3rd ed. J.P. Publications, Madison, WI © 1992, Figure 3.13, page 81. Used with permission.

Note that lines can be used to illustrate surfaces. This may add some confusion as well, for two reasons. First, the novice may tend to misinterpret lines as one-dimensional features rather than as symbols of surfaces. This form of confusion is easily overcome by continued familiarity with map features and symbols as more and more maps are encountered. The second problem is that of interpreting the lines themselves as accurate representations of point elevation values (see Chapter 2). As we have seen, the lines are estimates or predictions of specific elevations made through the process we have called interpolation. Although lines may be good predictors of elevations, the GIS student should keep their predictive nature clearly in mind, especially if they are used as GIS input. The only nonpredictive elevational values are those that were actually measured, as exemplified by benchmarks on topographic maps. As we will see later, most commercial GIS software has its own interpolation capabilities. It is up to you to become familiar with both the data and the reliability of the interpolation method before deciding to input known point locations or to sample the lines themselves for input (see Chapter 5).

The use of lines to represent surfaces is one of a large number of changes of dimensionality that can occur when the data are abstracted with the use of symbols. Figure 2.1 clearly indicates the wide range of dimension changes that can be symbolized on a map. Good judgment will always be necessary in deciding whether the symbol geometry and dimensionality given are truly representative of an object. For example, if an area symbol is used to depict a point feature, you must be aware that, despite the two-dimensionality of the symbol, only a single point needs to be recorded or input for the feature.

The same can be said regarding levels of measurement. Symbols such as graduated circles, indicative of attributes at point locations, are often manipulated to achieve a particular visual response (Robinson et al., 1995). This change in symbol size is not directly proportional to the changes in the data; rather, it is designed to allow the viewer to visually perceive proportionality. Such manipulations are commonplace in cartography, and the GIS analyst should examine the map carefully before proceeding with data encoding based on symbol manipulation. As always, if the raw data are available, the attributes should be input from them rather than from their graphic depictions.

A major difference between the communication and holistic paradigms is the classification-oriented manipulation of data prior to map production. Because a single thematic map is meant to serve a single purpose, this classification process normally takes place only once and the raw data are no longer available. Consider the case of choropleth or **value-by-area mapping**. When mapping areas under the communication paradigm, we want to group the areas into meaningful as well as visually appealing aggregations. The many common methods of grouping or aggregating these areas are collectively called **class interval selection**.

Among these methods of class interval selection we find several categories that merit examination. The first category is called constant intervals. This group of methods includes having the same number of areas within each category or the same number of data points for each class, or simply dividing the range of values from start to finish into an equal number of classes. Each of these constant interval methods has its own characteristics. Some will produce well-balanced map output (see Chapter 14); others will be more convenient; and still others will ensure that all classes will have data.

The second group of classing techniques is called variable intervals. These approaches produce maps that are less intuitive but may be highly useful for isolating certain high or low values, or for highlighting variations in value. Variable interval methods can be systematic, including arithmetic, logarithmic, or other mathematical series; or they can be unsystematic, with the clustering of the data into natural groupings used to determine where the class intervals will be designated.

Both general types of class interval selection can be used for point, line, and area symbols to depict all spatial data types. Take, for example, the use of contour intervals on topographic maps. The selection of contour interval is as much a method of class interval selection as is grouping areas for choropleth maps. Similarly, creating a discrete set of point symbols to show variation in attribute variable is also a method of class interval selection because the cartographer knows that the human eye is incapable of discerning extremely subtle changes in sizes.

All these interval selection techniques create documents that to varying degrees disguise the original data and, if poorly selected, can obscure the original distributional patterns. We should keep these principles in mind when we prepare to produce GIS data output. If, on the other hand, we intend to use these classified maps as input to a GIS database, we must be very careful about the analyses we perform with the highly manipulated data.

Of course, symbolism and classification are not the only cartographic compilation methods to be noted. Among the most important compilation processes the GIS analyst will encounter is the graphical simplification that takes place on the map as these classified data and their symbols are transferred to paper. This process presents a particular problem during GIS input, but it also affects the results of subsequent measurements and other analyses. Simplification goes further than classifying the data we will find on a map or than determining the types and levels of abstraction for the symbols used. Instead, simplification eliminates some features that are not wanted, or smoothes, aggregates, or further modifies the features on the map. We saw some of this in our earlier discussion of the changes in dimensionality. Ultimately, the purpose of map simplification is to ensure the readability of the cartographic document once it is produced. Two basic methods are used: feature elimination and feature smoothing. We will discuss them separately.

In feature elimination we perform a function that is very much like the process of spatial data gathering itself (Figure 3.11) (Muehrcke and Muehrcke, 1992). When we observe a portion of the earth, we use our geographic filter to make decisions about which features we will note and which we will ignore. The importance of the features during the data gathering process is determined before we begin, and it is largely controlled by our reasons for gathering the data in the first place. The selection of objects for investigation will, by default, act as a passive process of feature elimination on our maps, because only the features selected will be placed in the database or map document. In some cases, passive feature elimination occurs because we are unable to view objects in the field with the instrumentation at hand. Sensitivity to features is also a function of the current state of scientific knowledge. For example, changes in plant and animal species affect what is recorded during census activities; mammal locations that previously could not be pinpointed specifically now can be registered by **radiotelemetry**; prior to initiation of a national population census, we were unable to collect large-scale information on population changes and socioeconomic variables. Sensitivity to environmental factors and their interactions affects how we conceptualize them, which, in turn, affects what we select for investigation and later mapping and what is eliminated.

Unlike passive feature elimination, active feature elimination can be used in data collection and in mapping and **cartographic database** development as well. When we select certain electromagnetic radiation bands for remote sensing, we are actively eliminating certain portions from our data set. The sampling schemes we discussed earlier also actively promote selectivity by eliminating large portions of the objects that could be collected. We also perform active feature elimination on the map or digital cartographic database itself, prior to final map production. Here the map, rather than the person collecting the data, acts as the spatial filter. Very small towns often do not show up on maps of highly populated states or nations, whereas towns of the same size may appear on maps of areas with low population density. Similarly, we may eliminate some smaller or less important river tributaries, lakes, or islands during the mapping process because of lack of display space. In all these cases, a set of rules is formulated to determine which are selected and which are not. The rules may be as simple as eliminating a certain proportion of the objects or perhaps every other one; or they may include a set of decision rules (e.g., eliminate towns below a certain population, eliminate the smallest tributaries of stream networks). Whichever set of techniques is applied, the result is a less detailed output. If the product of feature elimination is to be input to a GIS, you create a cartographic database with some data missing.

Another useful method of simplification is called smoothing (Figure 3.12). This process abstracts detailed geometric objects into objects reduced in detail. Much as in caricatures of famous people, important geometric features are retained by representing a given detail as a simplified geometric shape. On maps showing coastal regions, boundaries, sinuous streams, or islands, we may generalize the lines that represent these irregular features so that their existence is recorded, but their spatial detail is minimized to fit on the document. GIS input

Figure 3.12 Scale change and smoothing. Use of smoothing, including the use of change in dimensionality. Note how features are simplified to only the most representative caricatures. *Source:* P.C. Muehrcke and J.O. Muehrcke, *Map Use: Reading, Analysis and Interpretation,* 3rd ed. J.P. Publications, Madison, WI, © 1992, Figure 3.14, page 81. Used with permission.

from these maps will result in less than satisfactory measures of length, shape, area, or other geometric property we may wish to calculate. Because GIS analysis frequently results in cartographic output, however, we may find these two forms of simplification useful when we produce the final results of our analysis.

MAP ABSTRACTION AND CARTOGRAPHIC DATABASES

As we have already seen, map scale restricts the amount of data that can be contained in a single map document. Base maps used for GIS data input have had to be adjusted during compilation to account for these limitations. It is largely on the basis of scale that features must be generalized, spatially displaced, or even abstracted further to allow for map readability (Robinson et al., 1995). When these graphic entities and their associated attributes, sometimes collectively called a cartographic database, are collected from existing cartographic documents rather than from field or remotely sensed data, they must somehow be digitally captured to a computerized database. Cartographic databases move from a higher level of abstraction to an apparent lower level of abstraction. This is a result of the software suggesting a higher level of accuracy than what actually exists. Their counterpart **geographic database** data move from an actual lower level of abstraction—and thus higher level of accuracy—to a higher level of abstraction (lower accuracy) (Figure 3.13).

When data are obtained from map documents, the symbols representing points, lines, areas, and sometimes surfaces are highly abstract representations of reality. Their size and placement are themselves graphic abstractions that are less accurate than the ability of computerized input devices to define them. Therefore, when we move from map document to digital database, we must decide exactly which part of a point, line, or area symbol should be recorded. If the placement of a point symbol, for example, is physically offset from its actual location to make room for another feature, our GIS input will indicate an accuracy of location that does not exist. In addition, the symbol itself takes up space, and we attempt to locate an exact point for GIS input for this zero-dimensional object; then we must decide whether the center of the symbol is its most accurate location—hence the "apparent" lower level of abstraction.

Alternatively, if we are creating a geographic database from, say, a survey instrument, we are now faced with the opposite problem. The survey instrument may be accurate to within centimeters or even millimeters of some absolute location. However, the input device is often not capable of reproducing this information with the same level of accuracy.

Figure 3.13   Abstraction for cartographic and geographic databases. Diagram showing how cartographic databases move from higher levels of abstraction to apparent lower levels and geographic databases move from lower abstraction levels to higher.

Although each of these issues needs to be addressed individually as we create our GIS database, some of the more difficult problems arise when these two sets of compilation rules interact. For example, a wildlife scientist once attempted to create a database indicating the location of roads relative to a species of bird that inhabits regions in or around the roads. The expedience of creating a road network with available data, in this case from existing 1:24,000 USGS topographic maps, made the decision to create a cartographic database for this GIS theme a straightforward one. The birds, in this case burrowing owls, create individual burrows on the ground that do not appear on maps, so a geographic database was created by using a GPS unit with 1-meter accuracy and then entering the data into the GIS. But when the two digital maps were displayed, the locations of burrows that were known to lie well within 10 meters of the roads appeared to be offset by at least 100 meters from the roads. This incompatibility of different compilation rules resulted in failure of the GIS to analyze the interactions between roads and burrowing owls.

The same problem can arise when creating a cartographic database from maps of different scale. As we have already seen, maps at very small scales will require symbols that occupy much larger proportions of space on the ground than will very large-scale maps with the same symbol sizes. Thus the scales of input for a cartographic database should be as nearly identical as possible.

Terms

Alber's equal area projections
analytical paradigm
angular conformity
attributes
azimuthal equidistant projections
azimuthal projections
Cartesian coordinate system
cartographic database
class interval selection
communication paradigm
conformal
conformality
conformal stereographic projections
conical projections
cylindrical projections
datum
digitizers
easting
ellipsoid
entities
equal area projections

equidistant projections
false eastings
false northings
false origins
geodetic reference system (GRS80)
geographic database
geoid
gnomonic projections
graphicacy
graphic scale
holistic paradigm
Lambert's conformal conic projections
Lambert's equal area projections
map legend
map projections
Mercator projections
North American datum (NAD27)
North American datum (NAD83)
northing
orthomorphic

plane coordinates
principal scale
projection family
Public Land Survey System (PLSS)
radiotelemetry
rectangular coordinates
reductions
reference globe
representative fraction (RF)
scale
scale factor (SF)
standard parallels
universal polar stereographic (UPS) grid
universal transverse Mercator (UTM)
value-by-area mapping
verbal scale
world geodetic system (WGS84)
zones

Review Questions

1. What is graphicacy? What impact does improved graphicacy have on our ability to function as GIS specialists?

2. What is the communication paradigm? What is its primary purpose? What impact does it have on GIS? Give an example of the use of the communication paradigm for traditional maps.

3. What is the analytical or holistic paradigm? How does it differ from the communication paradigm? Give an example of its use in cartography. What impact does this paradigm have on GIS?

4. What are the basic methods of illustrating scale on a map? Describe them. What are the relative merits of each type of scale as used in GIS?

5. What are some potential problems of putting multiple scales of maps into a GIS database? How could they affect analysis and measurements?

6. What is the purpose of the legend on a map? What impact does a map legend have on the relationships between entities and attributes?

7. What are map projections? What is their purpose? What are the three basic families of map projections?

8. What basic properties of the spherical earth are affected by using map projection? Which types of projection are best for preserving each of these properties?

9. On the basis of your answer to question 8, suggest some basic decisions you will need to make in deciding which projections to use for a variety of different GIS analyses.

10. Describe the UTM grid system. What are its advantages and disadvantages for GIS use?

11. What is the state plane coordinate system? What impacts, positive or negative, might the use of this system have on GIS analysis along state boundaries?

12. Describe the U.S. Public Land Survey System. What is a section? a township? What are the problems associated with using the PLSS for GIS work?

13. What is a datum? Why do we need to know which datum we are using?

14. What impact does class interval selection have on GIS input and later analysis?

15. What impact does the physical size of cartographic symbols have on map accuracy? What about the interaction of two or more map symbols?

16. What is the difference between a cartographic and a geographic database? What potential problem exists in attempting to produce a GIS database from both these sources?

References

Claire, C.N., 1968. "State Plane Coordinates by Automatic Data Processing." Coast and Geodetic Survey, Publication 62–4, U.S. Department of Commerce. Washington, DC: U.S. Government Printing Office.

DeMers, M.N., 1991. "Classification and Purpose in Automated Vegetation Maps." *Geographical Review,* 81(3):267–280.

Mitchell, H.C., and L.G. Simmons, 1974. *The State Coordinate Systems.* U.S. Coast and Geodetic Survey, Special Publication No. 235. Washington, DC: U.S. Government Printing Office (originally published in 1945).

Muehrcke, P.C., and J.O. Muehrcke, 1992. *Map Use: Reading, Analysis, and Interpretation.* Madison, WI: J.P. Publications.

Nyerges, T.L., and P. Jankowski, 1989. "A Knowledge Base for Map Projection Selection." *American Cartographer,* 16:29–38.

Pattison, W.D., 1957. "Beginnings of the American Rectangular Land Survey System, 1784–1800." Chicago: University of Chicago, Department of Geography, Research Paper 50.

Robinson, A.H., J.L. Morrison, P.C. Muehrcke, A.J. Kimerling, and S.C. Guptill, 1995. *Elements of Cartography,* 6th ed. New York: John Wiley & Sons.

Snyder, J.P., 1988. *Map Projections—A Working Manual.* U.S. Geological Survey, Professional Paper 1935. Washington, DC: U.S. Government Printing Office.

Töbler, W., 1959. "Automation and Cartography." *Geographical Review,* 49:526–534.

Töbler, W., 1973. "Choropleth Maps Without Class Intervals." *Geographical Analysis,* 5:262–265.

CHAPTER 4

GIS Data Models

A basic understanding of the structural makeup of the GIS, whether in isolation or in tandem with other software, is essential to enable us to perform our work efficiently. Each system has its own unique structures, methods of representation, and ways of analyzing spatial data. Fortunately, these can be grouped into a relatively small number of basic data. Because you will likely move from system to system throughout your career, you should become familiar with all the basic types. You will have plenty of opportunity to focus on the system you use most often as you work with it. The representation of spatial data is another formalism not unlike the ones we have already examined when moving from actual earth features to a limited set of definable objects called points, lines, areas, and surfaces. Nor is it unlike the cartographic abstractions of these objects and their measurement levels into mappable objects. The difference is merely in how we represent the data inside a digital computer in such a way that they can be edited, measured, analyzed, and output in some useful form. In this chapter we will examine some basic **computer file structures**. Then we will move to **database structures** that enable large amounts of data to be organized, searched, and analyzed. We will look at the basic concepts involved in the representation of space and its objects by graphic data structures. Then we will develop more comprehensive data models that allow multiple sets of cartographic data and their attributes to be linked to form a complete GIS database system.

LEARNING OBJECTIVES

When you are finished with this chapter you should be able to:

1. Understand the difference between simple graphics (analog or digital) and a map.

2. Know the different types of file structure and indicate the advantages and disadvantages of each for computer search.

3. Be able to identify the differences among network, hierarchical, and relational database structures and know the advantages and disadvantages of each.

4. Be familiar with the terminology of relational database management systems, including such terms as the *primary key, tuple, relation, foreign key, relational join,* and *normal forms.*

5. Describe the basic methods of representing graphic entities on a map using raster, vector, and quadtree systems, and understand their advantages and disadvantages.

6. Describe the three basic raster data models used for multiple maps in a GIS and understand their advantages or disadvantages.

7. Describe the methods of compact input and storage of raster data and know how they work.

8. Understand the difference between simple and topological vector data models and know their advantages and disadvantages.

9. Describe how the vector data model can be compacted to save computer space.

10. Describe how the triangulated irregular network (TIN) model works to store surface data.

11. Understand the basic differences between hybrid and integrated GIS systems, especially with regard to how the data are stored and accessed.

12. Have a basic understanding of how object-oriented GIS systems organize data and analytical techniques.

A QUICK REVIEW OF THE MAP AS AN ABSTRACTION OF SPACE

Before we move into another level of abstraction, one that allows the computer to operate on spatial data, it is best to review how we have moved from the real earth to more abstract views that we can manage intellectually.

We begin the process of abstraction by conceptualizing what we encounter as a group of points, lines, areas, and surfaces. This process, as you remember, is filtered by the questions we want to ask and how we intend to answer them. We make decisions about which objects to take note of and which to ignore. Then we decide on a method of data collection, whether it is a complete census or a sampling procedure. Some objects are assigned names; others are measured at higher levels of measurement (i.e., ordinal, interval, and ratio).

Once we have obtained our data, we proceed to make decisions about representing them in graphic form. We collate and group our data, decide on which projection we will need (if any), what grid system would be best, and so on. In some cases, especially in the absence of GIS, we produce a map directly from the data, only later entering these data into a cartographic database for use inside the GIS. More often today we do enter the data into the GIS directly, creating a geographic database based on direct observations.

The processes just summarized involve examining the environment, deciding what a necessary conception of that reality should be, and abstracting it further, either as a map or in computer-compatible form for direct entry into

GIS (Figure 4.1). However, the computer forces us to modify how we envision our data. Computers do not operate directly on visual or graphical objects as we would draw on a piece of drafting paper. Instead, the computer must be addressed by means of a programming language.

When we develop our concept of space and spatial relationships, we are able to organize our data with a view to making sense of it. The map has long been the graphic language we have used to visualize space and its objects. But our graphic language has a distinctly different structure from what is available inside the computer. We must create an explicit, rule-based language that allows the computer to use its digital (0 and 1) view of the world to identify the spatial extent of each object, to locate it in a system of coordinates, to separate adjacent objects from one another, and to be able to identify and sort objects by orientation, size, location, and so on. Such matters that require analog thought for their accomplishment are very simple for us but very difficult for the computer.

Fortunately, we as students of GIS do not have to begin from scratch to explain all the detail of representing and operating on geographic objects in a spatial context. But to be good at GIS, we should be aware of some of the basic ways others have found to allow the computer to do so. We will begin at an elementary level of computer structures but not at the machine level. Instead we will begin by examining traditional computer file structures that allow the storing, ordering, and searching of pieces of data. Then we will move to a higher level of organization in the computer called database structures—composed of combinations of file structures to allow more complex methods of managing data. Next we will examine the way in which the geographic space can be represented explicitly into a form of graphic data structure, and finally we will extend this to include multiple graphic data layers and their databases into what we call GIS.

Figure 4.1   Transforming geographic space into a geographic information system (GIS) database. Flow diagram showing change in abstraction level from the real world to the cartographer's conception to cartographic abstraction, and finally to computer abstraction.

SOME BASIC COMPUTER FILE STRUCTURES

If you have had a basic computer course or a computer language course, among the first things you learned was simple computer file structures. Files are nothing more than a simple accounting system that allows the machine to keep track of the records of data you give it and retrieve these records in any order you wish. For GIS this is no different. Much of what we do in GIS consists of storing entity and attribute data in a way that permits us to retrieve (for display, for example) any combination of these objects. This requires the computer, using a representational file structure, to be able to store, locate, retrieve, and cross-reference records. In other words, each graphical entity must be stored explicitly, along with its attributes, so that we can select the correct combinations of entities and attributes in a reasonable amount of time. This is identical to having a cross-referenced list of names, addresses, and phone numbers. The list items will, of course, need to be sorted—for example, by alphabetically organizing the names.

Simple Lists

The most basic file structure is called a **simple list**. In our names and addresses example, this is much like creating a separate index card for each name in a file like a Rolodex. Rather than organizing the names in any formal order, however, you place the cards in the Rolodex in the order in which they are entered (Figure 4.2). The only advantage of such a file structure is that to add a new record, you simply place it behind all the rest. Clearly, all the cards are there, and an individual name can be located by examining the cards, but the lack of structure makes searching very inefficient. Suppose your database contains 200,000 records. If your basic file structure is a simple, unstructured, unordered list

Figure 4.2   Simple list. File structure illustrated as an unordered Rolodex.

structure, you may have to search 200,000 cards to find what you are looking for. If it takes, for example, 1 second to perform each search, searching will require you to perform as many as $(n + 1)/2$ search operations (Burrough, 1983). This translates into a maximum of $(200,000 + 1)/2$ seconds, or nearly 28 hours to search for one point. Obviously, you need to get organized.

Ordered Sequential Files

As you know, most Rolodexes, like telephone directories, are ordered according to the sequence of alphabetic characters (Figure 4.3). This method of ordering allows each record to be compared to each record before or after it to determine whether its alphabetic character is higher or lower in sequence. **Ordered sequential files**, as these are called, can use alphabetic characters, as in our Rolodex example, or numbers that also occur in recognizable sequences against which individuals can be compared. The normal search strategy here is a sort of divide-and-conquer approach. A search is begun by dividing the file in half and looking first at the item in the middle. If it exactly matches the target combination of numbers or letters, the search is done. If not, the item of interest is compared to each of its neighbors to determine whether the alphanumeric combination is lower or higher. If it is lower, the half containing higher numbers or letters is searched in the same way. If it is higher, the half containing lower numbers or letters is searched by the divide-and-conquer method. In this way, the program avoids having to search large portions of the file. The number of search operations using this strategy is defined as $\log_2(n + 1)$ operations. In the case of our graphic example we reduce the maximum amount of time to just over 2 hours rather than the nearly 28 hours that would have been needed to search the entire list if each operation took a second to perform. Of course, any computer that takes a full second to perform this operation is not likely to be very useful for GIS.

Figure 4.3   Ordered sequential lists. File structure illustrated as an ordered Rolodex. In this case, the ordering is produced using alphabetical listings of names.

But the proportional savings in time is made quite obvious by this example. The search speed does come at a cost, however, because now every new record must be entered in the correct sequence or your file will quickly resemble the simple list structure, defeating the binary or divide-and-conquer search strategy.

Indexed Files

In both preceding examples, the records were retrieved by examining and comparing a key attribute, say a number sequence or an alphabetic sequence. The search strategy was based on the key attributes themselves. In GIS, as in many other situations, the primary items you want to search are points, lines, and areas. However, you will seldom search for a particular point, line, or area based on its code numbers. Instead, each entity will often have assigned to it a number of descriptive attributes, just as we have seen before. Typically, a search will consist of finding the entities that match a selected set of attribute criteria. Thus you might ask the GIS to find all study plots in excellent condition for subsequent display or analysis. Or you might want to examine or analyze all study areas in poor condition that have slopes less than 25 percent. Because of the possibly large numbers of attributes linked to each entity, a more efficient method of search will be necessary if you are to find specific entities with associated, cross-referenced attributes. Otherwise your search method will rapidly deteriorate into an exhaustive search of all attributes associated with all entities—the same tedious process employed with the simple list file structure (Burrough, 1983). In short, you need an index to your directory (Figure 4.4), much like the Yellow Pages you would use to find a particular type of store.

Figure 4.4   Indexed data structure. Files illustrated as an index to a book. The index shows how to find information in the larger file by selecting key features that can be searched for.

Indexed files begin with a set of records and item keys (Figure 4.5*a*) and can be indexed either as **direct files** (Figure 4.5*b*) or **inverted files** (Figure 4.5*c*). In direct indexed files, the records (properties) themselves are used to provide access to other pertinent information. For example, if you search for, say excellent conditions (properties) of a set of quadrats, the computer will compare each quadrat with its recorded condition and isolate each record that contains the excellent condition. The program will then compile a list or index of all quadrats that contain excellent conditions. In this way it is now able to relate this to the other properties such as number of species, slope, aspect, and percent bare. By creating this index, the program is able to ignore all other records that do not satisfy this search criterion.

Further improvements in search speed can be obtained if a formal permanent index is created for a selected attribute to be searched. From our previous example we could create an index of condition properties and have it specifically associated with specific entities (e.g., quadrats) and their computer locations (Figure 4.5*c*). Because the index is based on anticipated search criteria, rather than on the entities themselves, the information is literally inverted, in that the attributes are the primary search criteria and the entities rely on them for selection. For this reason we call this an inverted index file structure.

To create an inverted file structure requires an initial sequential search of the data to order the entities on the basis of the attributes of interest. This search,

Item key	**Index** Record #	Items
A	1	A(1)
B	$n + 1$	A(2)
C	$(n + 1) + 1$	.
.	.	.
.	.	B(1)
.	.	.

(a)

Quadrat #	**Properties** # species	Slope (deg)	Aspect (deg)	% Bare	Condition
1	15	8	N-NW	14	Good
2	3	27	N	35	Poor
3	21	5	NE	<5	Excellent
4	11	10	S	20	Fair
5	6	18	SW	15	Poor
6	18	7	NW	10	Good

(b)

Properties	Quadrats
Poor	2, 5
Fair	4
Good	1, 6
Excellent	3

(c)

Figure 4.5 Direct and inverted files. Comparison of *(a)* direct indexed files and *(b)* inverted index files. Note the improvement made by ordering the file to select the index items in the form of an inverted file for the single category of property "condition" *(c)*.

in turn, has three requirements. First, it requires you to know beforehand the criteria you are likely to be searching. Second, any additions of data will require you to recalculate the index as well as the original data. Finally, if you fail to indicate a particular criterion needed for your search—for example, if you forget to include accident rates as one of your search criteria—you will be required to use sequential search methods to obtain the information you want.

COMPUTER DATABASE STRUCTURES FOR MANAGING DATA

We have seen three basic types of file structure for the storage, retrieval, and organization of data. However, we seldom store a single file; rather, we compile and work with multiple files. A collection of multiple files is called a **database**. The complexity of working with multiple files in a database requires a more elaborate structure for management, called a database structure or a database management system. Although new forms of database structure are being created all the time, there are three basic types with which you should be familiar in the early stage of your GIS education. These are **hierarchical data structures**, **network systems**, and **relational database structures**. Each will be examined separately.

Hierarchical Data Structures

In many cases there is a relationship among data called a one-to-many or parent–child relationship (Burrough, 1983). The parent-to-child relationship implies that each data element has a direct relationship to a number of symbolic children, and, of course, each "child" is also capable of having an association with his or her "offspring," and so on. As the name implies, the parents and children are directly linked, making access to data both simple and straightforward (Figure 4.6). This type of system is perhaps best exemplified by the hierarchical

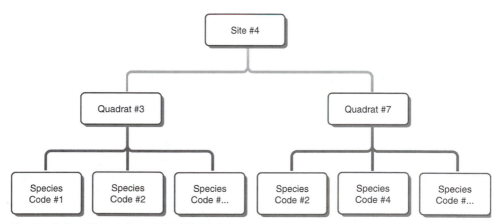

Figure 4.6   Hierarchical database structure. This illustration shows parent-to-child branching based on key attributes.   *Source:* Modified from R.G. Healey, "Database Management Systems," Chapter 18 in *Geographical Information Systems, Principles and Applications,* D.J. Maguire, M.F. Goodchild, and D.W. Rhind, Eds., Longman Scientific and Technical, Essex, England, © 1991, Figure 18.3, page 256. Used with permission.

system of classifying plants and animals found in taxonomic literature. Animals, for example, are separated into vertebrate and invertebrate forms—respectively, those with a backbone and those without. In turn, the vertebrates have a group called mammals, creatures that nurse their young. Mammals, of course, can be further broken down into other groups. The structure then begins to resemble a family tree, and in fact taxonomists use a nearly identical graphical structure to illustrate the relationships among organisms. A major feature of hierarchical systems exemplified by the taxonomic hierarchy is a direct correlation between one branch and another. The branches are based on formal criteria or key descriptors that act as decision rules for moving from one branch to another through the structure.

In a hierarchical database structure, if your criterion or key descriptor information is incomplete, your ability to navigate through the net is severely hampered. The nature of the hierarchical system requires that each relationship be explicitly defined before the structure and its decision rules are developed. A major advantage of such a system is that it is easy to search because the structure is so well defined, and it is relatively easy to expand by adding new branches and formulating new decision rules. If your initial description of the structure is incomplete, however, or if you want to examine the structure on the basis of valid criteria that are not included in the structure, a search becomes impossible. Knowledge of all possible questions that might be asked is absolutely necessary to develop a hierarchical system because questions are used as the basis for developing the decision rules or keys. If the types of search are identified, it is probably safe to assume that the system has a well-established set of linkages based on long-term use under nonautomated conditions, or that the number of questions is relatively limited.

You may have used a modern computerized bibliographic search system. This type of system was developed to simulate the ways in which people looked for books or articles before computers came into use. We might search for subject, or author, or title, or even a range of catalog numbers placing us in a focused portion of the library holdings. Systems that include the ability to add multiple search criteria or Boolean operations exhibit both requirements for easy hierarchical development: well-defined criteria and a limited number of search types. Imagine, however, that you know there was a great book on GIS located in the northwest corner of the fifth floor of the university's main library. You don't recall the catalog number, but you would like to find other books located near the one you remember seeing. You would not consider asking the bibliographic system to locate all the books in the northwest corner of the fifth floor of the main library. The technique might have worked in the long run, but as a practical matter, you need to know more about the books in your target area.

Situations like the extremely vague search just described occur more often than not when you are working with information inside a GIS. Among the most difficult things to do is to anticipate all the possible searches you might perform. After all, the GIS database normally contains widely varying types of data as points, lines, or areas, and there are many different thematic maps in the database. One of the most enjoyable features of a GIS is that you can try out searches or test out relationships you had not envisioned before you began. Unfortunately, the hierarchical structure is not very good at this because of its rigid key structure.

Burrough (1983) cited a classic example of how a system based on a hierarchical structure fails to comply with a user's request because the information needed is not even included in the system (for additional details, see Beyer, 1984). Not unlike our fifth-floor library search, the director of the Royal Botanical Gardens in Kew wanted to query the institution's botanical database system for all the available plants native to Mexico so that he could examine them before an upcoming field visit. This particular search criterion had not been anticipated, however, and the geographic locations of the plants in the system were not recorded. Thus the program was incapable of selecting plants based on location. Clearly, the rigid structure of the hierarchical system also would make it difficult to restructure to allow further searches of this nature. Beyond this severe limitation, the hierarchical structure creates large, often cumbersome index files, frequently requiring repeated entries of important keys to perform its searches. This adds substantially to the amount of memory needed to store the data and sometimes contributes to slow access times as well.

Network Systems

Searches performed in hierarchical data structures are restricted to the branching network itself. In graphic databases, users frequently need to jump to different portions of the database to acquire entity information based on queries of attributes. Since attribute and entity data may very well be located in different locations, the creation of a hierarchical structure requires that direct links be made between search criteria and the graphic devices used to illustrate the locations in space. Even though this can be done, the potential numbers of hierarchical branchings and associated keys can get very unwieldy. Such awkwardness is experienced principally because the hierarchical data structure is most useful when there is a one-to-one or a many-to-one relationship among the variables.

In addition to one-to-one and many-to-one relationships, many GIS databases have many-to-many relationships in which a single entity may have many attributes, and each attribute is linked explicitly to many entities. (An area representing a research quadrat for a study site, for example, will have numerous point locations, with multiple plant or animal species associated with each. In addition, each species might be found in more than one quadrat.) To accommodate these relationships, each piece of data can have associated with it an explicit computer structure called a pointer that directs it to all of the other pieces of data to which it relates (Figure 4.7). Each individual piece of data can be linked directly anywhere in the database, without the existence of a parent–child relationship. Pointers are common features in computer languages such as C and C++, and a basic knowledge of these languages will help you understand exactly how the devices are used. For our purposes, a graphic visualization should suffice. Figure 4.7 illustrates two quadrats (#3 and #7) for study site #4. Notice how the pointers are used to relate individual point locations to their representative species identified for each. Pointers also look back from the species to the locations and finally back to the quadrats in which they are located.

Network systems are an improvement over hierarchical structures for GIS work because they are less rigid and can handle many-to-many relationships.

Figure 4.7 **Network database structure.** This structure allows users to move from data item to data item through a series of pointers. The pointers indicate relationships among data items. *Source:* Modified from R.G. Healey, "Database Management Systems," Chapter 18 in *Geographical Information Systems, Principles and Applications,* D.J. Maguire, M.F. Goodchild, and D.W. Rhind, Eds., Longman Scientific and Technical, Essex, England, © 1991, Figure 18.4, page 257. Used with permission.

As such, they allow much greater flexibility of search than do the hierarchical structures. Also, unlike the hierarchical structure, they reduce redundancy of data (e.g., coordinate pairs). Their major drawback is that, in very complex GIS databases, the number of pointers can get quite large, ultimately comprising a substantial portion of storage space. In addition, although linkages between data elements are more flexible, they must still be explicitly defined with the use of pointers. The numerous possible linkages may become an extremely tangled web, often resulting in confusion and missed and incorrect linkages. Novice database and GIS users often are overwhelmed by these conditions, although experienced users can become quite efficient with such systems and often prefer them over other types.

Relational Database Management Systems

The disadvantages of large numbers of pointers can be avoided by using another database structure: in relational database structures, the data are stored as ordered records or rows of attribute values called **tuples** (pronounced "tooples," to rhyme with quadruples) (Figure 4.8). In turn, tuples are grouped with corresponding data rows in a form collectively called **relations** because they retain their respective row positions in each column and are, of course, related to one another (Healey, 1991) (Figure 4.9). Each column then represents the data for a single attribute for the entire dataset. For example, you could have a column of quadrat numbers (a single attribute) organized numerically. In a separate column, you would have additional information pertaining to the collector, yet another showing collection date, and finally a column showing the site number that corresponds to each of the other columns. In this way each item in each column corresponds and can then be related to additional tables as well.

Relational systems are based on a set of mathematical principles called relational algebra (Ullman, 1982) that provides a specific set of rules for the design

Figure 4.8 **Elements of relational tables.** All relational database systems share these elements.

and function of these systems. Because relational algebra relies on set theory, each table of relations operates as a set, and the first rule is that a table cannot have any row (tuple) that completely duplicates any other row of data. Because each of these rows must be unique, a single column or even multiple columns can be used to define the search strategy. Thus, as an example of using a single column to decide your search strategy, you might search for a social security

Quadrat no.	Collector	Collection date	Site no.
1	Smith	6/10/96	2
2	Jones	7/01/96	3
3	.	.	.
	.	.	.

Quadrats table

Quadrat no.	Species no.	Latitude	Longitude
1	3	35.19°N	106.10°W
1	4	35.19°N	106.10°W
2	2	35.20°N	.
.	.	.	.

Species code no. table

Species no.	Species name	Type
1	Conomyrma insana	Generalist
2	Pogonomyrmex rugosus	Specialist
.	.	.
.	.	.

Species table

Figure 4.9 **Relational database structure.** Note the tuples, relations, and primary keys. *Source:* Modified from R.G. Healey, "Database Management Systems," Chapter 18 in *Geographical Information Systems, Principles and Applications,* D.J. Maguire, M.F. Goodchild, and D.W. Rhind, Eds., Longman Scientific and Technical, Essex, England, © 1991, Figure 18.5, page 258. Used with permission.

number, telephone number, home address, and so on available in other columns of the same table by selecting a particular name from the first column. This search criterion is called a **primary key** for searching the other columns in the database (Date, 1986). No primary key row can have missing values because missing row values could result in permitting duplicate rows to be stored, thus violating our first rule.

Relational systems are useful because they allow us to collect data in reasonably simple tables, keeping our organizational tasks equally simple. When we need to do so, we can match data from one table to corresponding (same-row) data in another table by using a linking mechanism called a **relational join**. Because relational systems are predominant in GIS, and because of the rather large databases produced for GIS, this process is a common one and you should pay close attention to it. Any number of tables can be "related." The process is one of matching the primary key (column) in one table to another column in a second table. The column in the second table to which the primary key is linked is called a **foreign key**. Again, the related row values are assumed to be in the same positions to ensure that they correspond. This link means that all the columns in the second table are now related to the columns in the first table. In this way, each table can be kept simple, making accounting easier. You can link a third table by finding a key column in the second table that acts as the primary key to a corresponding key column (now called the foreign key) in the third table. The process can continue, connecting each of the simple tables to allow for quite complex searches while maintaining a very simple, well-defined, and easily developed set of tables. This approach eliminates the confusion found in database development using network systems.

To allow us to perform relational joins, each table must have at least one column in common with each other table we are trying to relate to. This redundancy allows the relational joins in the first place. Whenever possible, however, the amount of redundancy should be reduced. A set of rules called **normal forms** has been established (Codd, 1970) to indicate the forms your tables should take. We will discuss three basic normal forms; there have been some additions, but they are really more refinements than normal forms (Fagin, 1979).

The **first normal form** states that the table must contain columns and rows, and because the columns are going to be used as search keys, there should be only a single value in each row location. Imagine how difficult it would be to search for information by name if the name column had multiple values in each row location.

The **second normal form** requires that every column that is not a primary key be totally dependent on the primary key. This simplifies the tables and reduces redundancy by imposing the restriction that each column of data be findable only through its own primary key. If you wish to find a given column using other relationships, you can use the relational join rather than placing the column over and over again in separate tables to make sure it can be found.

The **third normal form**, which is related to the second normal form, states that columns that are not primary keys must "depend" on the primary key, whereas the primary key does not depend on any nonprimary key. In other words, you must use the primary key to find the other columns, but you don't need the other columns to search for values in the primary key column. Again, the idea is to reduce redundancy—to ensure that the smallest number of columns are produced.

The rules of the normal forms were summed up by Kent (1983), who indicated that each value of a table represents something important "about the primary key, the whole primary key, and nothing but the primary key." For the most part, these rules are highly useful and should be rigorously enforced. There are always situations, however, when rigid enforcement will be impossible or will hamper system performance (Healey, 1991).

GRAPHIC REPRESENTATION OF ENTITIES AND ATTRIBUTES

Thus far we have concentrated on data structures that have little to do with the graphic representation of cartographic or geographic objects as we have envisioned and abstracted them. Although all three of the systems mentioned could be used to manage graphics, they tell us little about how the graphics themselves will be represented in the GIS. We know that the human mind is capable of producing a graphic abstraction of space and objects in space. This representation is actually quite sophisticated, as you will see when we attempt to make the jump to computer handling of graphic devices. A primary difficulty appears because our graphic devices contain an implied set of relationships about the elements contained on the paper. Lines are connected to other lines and together are linked to create areas or **polygons**. The lines are related to each other in space through angles and distances. Some are connected, whereas others are not. Some polygons have neighbors, whereas others are isolated. The list of possible relationships that can be contained on a graphic diagram is virtually endless. We need to be able to find a way to represent each object and each relationship between objects as an explicit set of rules and to permit the computer to "recognize" that all these points, lines, and areas represent something on the earth, that they exist in an explicit place in space, and that those explicit locations are also related to other objects within space in an absolute as well as a relative sense. We may even want to explain to the computer that a polygon has an immediate neighbor to its left and that this neighbor may share points and lines. In other words, we need to create a language of spatial relationships.

There are two fundamental methods of representing geographic space. The first method **quantizes** or divides space as a series of packets or units, each of which represents a limited but defined amount of the earth's surface. This **raster** method or field-based model can define these units in any reasonable geometric shape, as long as the shapes can be interconnected to create a planar surface representing all the space in a single study area. Although a wide variety of raster shapes are possible—for example, triangles or hexagons—it is generally simpler to use a series of rectangles or, more often, squares called **grid cells**. Grid cells or other raster forms generally are uniform in size, but this is not absolutely necessary, as we will see when we briefly review a less often used system called **quadtrees**. For the sake of simplicity, we will assume that all grid cells are the same size and that, therefore, each occupies the same amount of geographic space as any other. We will further restrict our discussion to square grid cells, again for simplicity.

Raster data structures do not provide precise locational information because geographic space is now divided into discrete grids, much as we divide a checkerboard into uniform squares. Instead of representing points with their absolute locations, they are represented as a single grid cell (Figure 4.10). The assumption is that somewhere inside that grid cell, we will find a point object. This is an additional form of change of dimensionality abstraction that requires us to illustrate a zero-dimensional object with a two-dimensional data structure. Similarly, lines, or one-dimensional objects, are represented as a series of connected grid cells. Again, we are changing our dimensionality from one-dimensional objects to two-dimensional data structures. Each point along the line is represented by a grid cell, meaning that any point along the line must occur somewhere within one of the displayed grid cells. It is easy to see that this form of data structure produces a rather stepped appearance when it is used to represent very irregular lines (Figure 4.10). This stepped appearance is also obvious when we represent areas with grid cells (Figure 4.10). All points inside the area that is bounded by a close set of lines must occur within one of the grid cells to be represented as part of the same area. The more irregular the area, the more stepped the appearance.

In grid-based or raster systems there are two general ways of including attribute data for each entity (object). The simplest is to assign a single number, representing an attribute (e.g., a class of land cover), for each grid cell location. By positioning these numbers, we ultimately are allowing the position of the attribute value to act as the default location for the entity. For example, if we assign a code number of 10 to represent water, then list this as the first number in the X or column direction and the first in the Y or row direction, by default the upper left grid cell is the location of a portion of the earth representing water. In this way, each grid cell can hold only a single attribute value for a given map. An alternative approach, actually an extension of the one just discussed, is to link the grid cells to a database management system, with the result that more than one attribute can be represented by a single grid cell. This approach is becoming more prevalent because it reduces the amount of data that must be

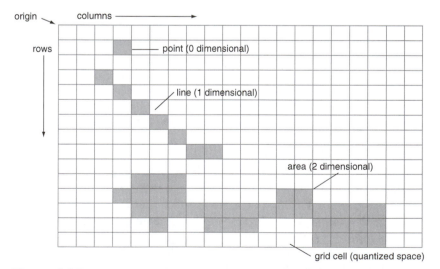

Figure 4.10 Basic raster graphic data representation. The structure illustrates points, lines, and areas as quantized units of geographic area. Grid structures do not allow precise locational information.

stored and because it can easily be linked to other data structures that also rely on **database management systems** to store, search, and manipulate data.

Although absolute location is not explicitly part of the raster data structure, it is implied by the relative locations of the grid cells. Thus a line is represented by grid cells at particular locations relative to one another, and areas are represented by the grid cells that form as a result of being adjacent to one another or connecting at points of one another. As you may have surmised, the larger the grid cell, the more land area is contained within it—a concept called **resolution**. The more land area contained—that is, the coarser the resolution of the grid cell—the less we know about the absolute position of points, lines, and areas represented by this structure.

Raster structures, especially square grid cells, are pieced together to represent an entire area. To do this we must assume that the area is more or less flat. In other words, the normal coordinate system we are using is a Cartesian coordinate system (see Chapter 3), and the grid cells themselves approximate the locations in space. As we have seen, Cartesian coordinate systems employ a map projection to permit an approximation of the three-dimensional shape of a portion of the earth. The grid system representation may have embedded in it a coordinate system that better approximates absolute location. For example, the pixels from satellite imagery (see Chapter 3) have an associated projection, and a more exact grid can be placed on them for reference. In general, however, exact measurements based on any raster structure are difficult. When exact measures are needed, therefore, raster structures are less often used than are other types.

Raster data structures may seem to be rather undesirable because of the lack of absolute locational information. However, quite the contrary is true. Raster data structures have numerous advantages over other structures. Notably, they are relatively easy to conceptualize as a method of representing space. Some of you may have seen the character printouts of cartoon characters that were so prevalent on computer-produced calendars in the 1960s and 1970s. If not, you are no doubt familiar with television, a technology that is quite analogous to the raster representation of earth features in a GIS. Few of us have difficulties identifying actors or other images on the TV screen, even though everything is represented by a series of dots or pixels. In fact, the relationship between the pixel used in much remote sensing and the grid cell used in GIS allows data from satellites to be readily incorporated into raster-based GIS without any changes. This is another advantage of the raster data structures over the alternatives. Another nice feature of grid-based systems, as you will see in more detail in Chapter 12, is that many functions, especially those involving surfaces and overlay operations, are simple to perform with this type of data structure.

Among the major disadvantages of the raster data structure are the already mentioned problem of reduced spatial accuracy, which decreases the reliability of area and distance measures, and the need for large storage capacity, associated with having to record every grid cell as a numerical value. The latter problem is less serious than it once was because of vast improvements in computer hardware storage capabilities. In addition, as you will see later in this chapter, the storage needed can be reduced by compacting groups of grid cells into more integrated forms. Although storage is no longer a major limitation to the use of raster structures, even the fastest computers can be slowed to a crawl if highly complex calculations are performed on very large raster databases.

The second method of indicating geographic space, called the **vector** or polygon-based model, allows us to give specific spatial locations explicitly. Here it is assumed that geographic space is continuous rather than being quantized as smaller discrete grids. This perspective is acquired by associating points as a single set of coordinates (*X* and *Y*) in coordinate space, lines as connected sequences of coordinate pairs, and areas as sequences of interconnected lines whose first and last coordinate points are the same (Figure 4.11). Anything that has a single *(X, Y)* coordinate pair not physically connected to any other coordinate pair is a point (zero-dimensional) entity.

The vector data structure is much more representative of dimensionality as it would appear on a map. To give it the utility of a map, however, we combine the entity data with associated attribute data kept in a separate file, perhaps in a database management system, and then link them together. This extends the structure beyond a simple graphic caricature of the objects, making it more maplike and more representative of the earth surface we are modeling. Whereas in the raster structure we explicitly stored the attribute and implied its location on the basis of its position, in vector representation we take a quite different approach, explicitly storing the entities without their attributes and relying on the link to the separate attribute database.

In vector data structures, a line consists of two or more coordinate pairs, again storing the attributes for that line in a separate file. For straight lines, two coordinate pairs are enough to show location and orientation in space. More complex lines will require a number of line segments, each beginning and ending with a coordinate pair. For complex lines, the number of line segments must be increased to accommodate the many changes in angles. The shorter the line segments, the more exactly they will represent the complex line. Thus we see that, although vector data structures are more representative of the locations of objects in space, they are not exact. They are still an abstraction of geographic space.

Although some lines act alone and contain specific attribute information that describes their character, other, more complex collections of lines, called **networks**, add a dimension of attribute characters. Thus, not only does a road network contain information about the type of road or similar variables, but it will also indicate, for example, that travel is possible only in a particular direction. This information must be extended to each connecting line segment to advise the user that movement can continue along each segment until the attributes change—perhaps when a one-way street becomes a two-way street. Other codes linking these line segments might include information about the **nodes** that connect them. For example, one node might indicate the existence of a stop sign, a traffic signal, or a sign prohibiting U-turns. All these attributes must be connected throughout the network so that the computer knows the inherent

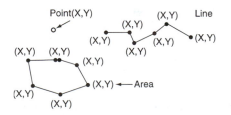

Figure 4.11   Basic vector graphic data representation. The structure shows points as individual coordinate pairs, lines as groups of two coordinate pairs, and areas as connected lines having identical beginning and ending coordinates.

real-world relationships that are being modeled within the network. Such explicit information about connectivity and relative spatial relationships is called **topology**, a topic we will return to when we look at the vector data models we can produce from the basic vector data structure.

Like line entities, area entities can be produced in the vector data structure. By connecting pairs of coordinates into lines and organizing the lines into a looping form, where the first coordinate pair on the first line segment is the same as the last coordinate pair on the last line segment, we create an area or polygon. As with point and line entities, the polygon will also have associated with it a separate file that contains data about the attributes or characteristics of the polygon. Again, this convention improves the simple graphic illustration of area entities, making it possible for them to better represent the abstraction of area patterns we observe on the earth's surface.

Generally speaking, cartographers tend to prefer vector data structures because of their similarity to the graphical structures most commonly associated with analog maps. With some limitations, output from vector data structures resembles quite closely the hand-drawn map. But simple map output is not the major focus of GIS; rather, it is the ability to measure and analyze cartographically organized data. As we have already seen, even when these **graphic data structures** are physically viewed, there is a need to find a way to combine entities with their descriptive attributes. We have alluded to the use of external files or database systems in this connection. The data structures must be developed to allow such links either directly or indirectly. In addition, there are many other facets of graphic structure that allow the GIS to operate as an analyzable maplike system. We need to move from simple data structures to what are often termed data models, which will act more like the map in terms of capabilities to be analyzed. Some generic types will be illustrated for both raster and vector data structures.

GIS DATA MODELS FOR MULTIPLE MAPS

Although simple raster and vector data structures provide us with a means of depicting spatial phenomena in a single map environment, we still need to develop more complex versions, called data models, to accommodate the necessary interactions of objects in the database, to link entities and attributes, and to allow multiple maps to be analyzed in combination. We will look at raster data models first and then proceed to vector data models to illustrate some basic methods that can be used to allow us to perform the multiple analyses we would have performed without automation. Then we will go one step further to illustrate some of the ways of combining these data models into systems—in this case, GIS.

Raster Models

As indicated in our initial discussion of raster data structures, the simplest approach is to use grid cells to represent quantized portions of the earth. Each grid cell is represented by one and only one attribute value, again in its simplest form. To create a thematic map of grid cells, we collect data about a particular topic (theme) in the form of a two-dimensional array of grid cells, where each grid cell represents an attribute of the individual theme. This two-dimensional array (essentially a 2×2 matrix, as used in matrix algebra) is called a **grid**. We can use grids to separate individual types of subject matter (land use, vegetative cover, soil type, surface geology, hydrology, etc.) so that their respective attributes are easily identified. In addition, this approach forces us to focus our attention on the objects, patterns, and interrelationships of each theme without unnecessary confusion. Because we will most often be concerned about the relationships of one theme, such as soil type, with another theme, say vegetation, we create a separate map for each additional theme. We can then stack these maps into a form of three-dimensional structure, where the combinations of all the themes might adequately model all the important attributes for an entire area, based on what we have selected as our geographic filter or study framework. As such, if we are interested only in physical phenomena, each of the important components of the physical geography will be represented separately, but together they will give us a more complete, three-dimensional view of the study area.

There are a number of ways of forcing a computer to store and reference the individual grid cell values, their attributes, map names, and legends. Most of these systems are older and may be of historical significance, but we will focus on more current approaches. One of the most copied grid cell data models for GIS is called the **map analysis package (MAP)** model, named after the system developed by C. Dana Tomlin (Burrough, 1983) for his doctoral dissertation. In this data model (Figure 4.12), each thematic map is recorded and accessed separately by map name or title. This is accomplished by recording each variable, or mapping unit, of the map's theme as a separate number code or label, which can be accessed individually when the map is retrieved. The label corresponds to a portion of the legend and has its own symbol assigned to it. In this way, it is easy to perform operations on individual grid cells and groups of similar grid cells. The resulting changes in value require rewriting only a single number per mapping unit, thus simplifying the computations. The overall major improve-

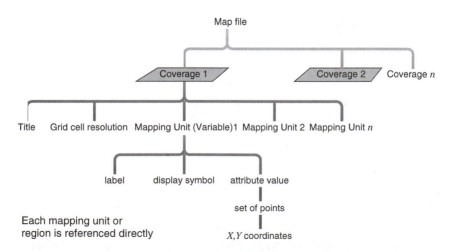

Figure 4.12   The map analysis package (MAP) GIS data model.

3	3	3	2	2	1
3	3	2	4	4	4
4	0	0	0	4	4
0	0	0	4	4	1
0	0	4	4	1	1
3	3	3	4	4	1

Value	Count	Type	Condition	Farmer
0	8	Fallow	Fair	McMillan
1	5	Barley	Fair	Denmeyer
2	3	Wheat	Poor	Bushee
3	8	Corn	Excellent	Denmeyer
4	11	Oats	Good	Denmeyer

Figure 4.13   Extended raster data model. Note how each grid cell value has additional information stored in the database management system tables.

ment is that the MAP method allows ready manipulation of the data in a many-to-one relationship of the attribute values and the sets of grid cells.

Although raster GIS have traditionally been developed to allow single attributes to be stored individually for each grid cell, most today have evolved to include direct links to existing database management systems (Figure 4.13). This approach extends the utility of the raster GIS by minimizing the number of maps and substituting multiple variables for each grid cell in each grid. Such extensions to the raster data model have also allowed direct linkage to existing GIS programs that use a vector graphic data model. Because such integrated raster/vector systems include modules that convert back and forth from raster to vector, the user can operate with all the advantages of both data models. The conversion process is often quite transparent, allowing the user to perform the analyses needed without concern for the original data structure. This feature is particularly important because it strengthens the relationship between traditional digital image processing software, used to manipulate grid cell–based, remotely sensed data, and GIS software. Many software systems already have both sets of capabilities, and still more such systems are likely in the future. Together with the linkage with existing statistical packages, we are rapidly approaching systems that operate with a superset of spatial analytical techniques, resulting in a maturing of automated geography.

Compact Storing of Raster Data

Before we leave our discussion of raster data models, we will look at some common methods of storing raster data that result in substantial savings in disk space. These require a data model similar to the MAP data model that allows groups of grid cells to be accessed directly, because the compactions generally operate by reducing the information content of these groups of cells to the absolute minimum needed to represent them as a unit. Compact methods for storing raster data certainly operate under the storage and editing subsystem of a GIS, but they can also be applied directly during the input phase of the GIS operation. We will revisit these methods in the next chapter when we discuss data input. The approaches discussed in this section are illustrated in Figure 4.14.

The first method of compacting raster data is a process called **run-length codes**. In the old-fashioned method, raster data were input into a GIS by preparing a clear acetate grid, referencing it, and overlaying it on the map to be encoded. Each grid cell has a numerical value corresponding to a category of data on the map that must be input (generally typed) into the computer. For a map of perhaps 200×200 grid cells, someone would have to type 40,000 numbers into the computer. If your instructor hears you giggling at this point, don't be surprised if you find yourself doing it, as an exercise in historical GIS, or perhaps a lesson in humility. If you have access to a raster-based GIS, you might try it anyway. As you begin typing, you will quickly see patterns emerging from the data that present opportunities for reducing your workload. Specifically, there are long strings of the same number in each row. Think how much time you could save if, for a given row, you could just tell the computer that starting at column 8 all the numbers are 1s, representing some map variable, until you get to column 56, then at column 57 the numbers are 2s until the end of the row.

Figure 4.14 Compact data models. Methods of compacting raster data to preserve storage: *(a)* run-length codes, *(b)* raster chain codes, *(c)* block codes, and *(d)* the unique structure called quadtrees.

Indeed, you could also save a great deal of space by simply giving starting and ending points for each string and the value that should be stored for that string. That's the run-length codes process.

This method limits you to operation on a row-by-row basis. If you could tell the computer to begin at a single grid cell with value 1, then go in a particular direction, say vertically, 27 grid cells, and then change to a different grid cell value, you could therefore code strings in any direction. But the principle can be extended even further. Suppose you see large groups of grid cells that represent an area. If you started at one corner, giving its starting position and grid cell value, then moved in cardinal directions along the area, storing a number representing the direction and another indicating the number of grid cells moved, you could then store whole areas using very few numbers. In this way you would save even more space and typing time as well. This method relies on so called **raster chain codes**, which literally run a chain of grid cells around the border of each area. In other words, you assign origins, based on X and Y position, a grid cell value for the entire area, and then directional vectors, showing where to move next, where to turn, and how far to go. Usually, the vectors include nothing more than the number of grid cells and the vector direction based on a simple coding scheme, where 0, 1, 2, and 3 could indicate north, south, east, and west, respectively.

The two remaining approaches to reducing the storage necessary for grid-based systems both rely on a square collection of grid cells as the primary unit of storage. The first, called **block codes**, is a modification of run-length codes. Instead of giving starting and ending points, plus a grid cell code, we select square groups of cells and assign a starting point, say the center or a corner, pick a grid cell value, and tell the computer how wide the square of grid cells is, based on the number of cells. As you can see, this is really a two-dimensional run-length code. Each square group of grid cells, including individual grid cells, can be stored in this way with a minimum amount of numbers. If your grid has very few large square groups of cells, this method is not a major improvement for storage. But then, run-length codes also become somewhat cumbersome if there are few long runs or strings of the same value. Most thematic maps, however, have fairly large numbers of these groups, and block code methods are very effective methods of reducing storage.

Quadtrees, our final method of compact storage, is a somewhat more difficult approach, and your instructor may elect not to cover it. Still, at least one commercial system called Spatial Analysis System (SPANS), from Tydac, and one experimental system called Quilt (Shaffer et al., 1987), are based on this scheme. Like block codes, quadtrees operate on square groups of cells; in this case, the entire map is successively divided into uniform square groups of grid cells with the same attribute value. Starting with the entire map as entry point, we then divide the map into four quadrants (NW, NE, SW, and SE). If any of these quadrants is homogeneous (i.e., contains grid cells with the same value), that quadrant is stored and no further subdivision is necessary. Each remaining quadrant is further divided into four quadrants, NW, NE, SW, and SE, and again, each quadrant is examined for homogeneity. All homogeneous quadrants are again stored, and each of the remaining quadrants is further divided and tested in the same way until the entire map is stored as square groups of cells, each with the same attribute value. In the quadtree structure, the smallest unit of representation is a single grid cell (Burrough, 1983).

Systems based on quadtrees are called variable resolution systems because they can operate at any level of quadtree subdivision. If you do not need high resolution for your computations—in other words, if detail is not essential— you can use a rather coarse level of resolution (or quadtree subdivision) in your analysis. Thus, users can decide how fine the resolution needs to be for various manipulations. In addition, because of the compactness of storage from this method, very large databases, perhaps continental or even global scales, can be stored in a single system. The major difficulty with the quadtree structure lies in the method by which it separates the grid cells into regions. In block codes, the decision is based entirely on the existence of square groups of homogeneous grid cells, regardless of where they were located on the map. With quadtrees, the subdivision is preset to the four quadrants (NW, NE, SW, SE), as a result of which some otherwise homogeneous regions lie in two or more different quadrants. This produces computational difficulties for analysis of shape and pattern that must be overcome through rather complex computational methods, which are beyond the scope of this book. GIS software using the quadtree data model operates under workstation and PC platforms and uses multiple operating systems. Such programs are in use worldwide and offer some interesting opportunities, especially for those needing to use very large databases.

Commercial Raster Compaction Products

With the advent of the Internet and the rapid increase in downloading of raster imagery, a vast array of new methods of data compression techniques and file types has arisen. The intent is not to review all of them here. Rather, we focus on one method that seems to be having a profound impact on the GIS and remote sensing community. Called **MrSID**, this image compressor is both very common in the industry and extremely powerful in its ability to compress files so that they can be passed easily from user to user over the Internet.

MrSID stands for Multiresolution Seamless Image Database. Originally developed at Los Alamos National Laboratory, it is now a commercial package developed and marketed exclusively by LizarTech Corporation. Although the specific techniques are proprietary, it is important to be aware of MrSID, especially if your GIS operations use very large datasets or if your GIS delivery is going to be on-line.

Among the benefits of MrSID is its ability to handle very larger images and to display them in seconds. It allows instantaneous browsing of these large images and reduces the storage space needed to archive them. Typical compression ratios are 15–20:1 for grayscale and 30–50:1 for color images without loss of resolution. In addition, the file format is compatible with any platform or operating system.

As GIS and remote sensing data become both more voluminous and more commonplace, the need for such technology becomes increasingly important. Most GIS software already has the ability to work with JPEG, TIFF, GIF, and many other common graphics formats. It will not be long before virtually all commercial and most noncommercial GIS software will operate fully with MrSID.

Vector Models

Vector data structures allow the representation of geographic space in a way that is somewhat more intuitive and certainly more reminiscent of the familiar analog map. You also remember that they represent the spatial location of items explicitly, most often storing the attributes in another file for later access. Since the relationships between individual entities are implicit rather than explicit, the space intervening between graphical entities does not have to be stored. Vector data structures can be put together into a vector data model in several ways, enabling us to examine the relationships between variables in a single map or among variables in different maps. We will look at a range of these models using three basic types as examples: **spaghetti models**, **topological models**, and vector chain codes. Although there are others, and many variations on each type, these should give you a sufficient overview of what is available for vector GIS systems.

The simplest vector data structure, called the spaghetti model (Dangermond, 1982) (Figure 4.15), is essentially a one-for-one translation of the graphical image we would normally encounter on a map. It is probably the one most of us consider the most natural or most logical, primarily because the map is maintained as the conceptual model. Although the name may seem odd, it is actually quite representational. If you imagine covering each graphic object on our analog map with a piece of spaghetti *(al dente),* you will have a pretty good idea of how the model works. Each piece of spaghetti acts as a single entity—very short ones for points, longer ones for straight-line segments, and collections of line segments that come together at the beginnings and endings of surrounding areas. Each entity is a single, logical record in the computer, coded as variable-length strings of *(X, Y)* coordinate pairs.

Data Structure

Feature	Number	Location
Point	5	x,y (single pair)
Line	16	(string of x,y coordinate pairs)
Polygon	25	(closed loop of x,y coordinate pairs where first and last pair are the same)
	26	(closed loop sharing coordinates with adjacent polygons to form a data structure)

Figure 4.15 Spaghetti vector data model. There is no explicit topological information, but the model is a direct translation of the graphic image. *Source:* Figure derived from Environmental Systems Research Institute, Inc. (ESRI) drawings and tabular data.

If we imagine a large collection of pieces of spaghetti, each straight piece has a beginning and an ending set of coordinates. Any polygons that lie adjacent to each other must have separate pieces of spaghetti for adjacent sides. That is, no two adjacent polygons share the same string of spaghetti. Instead, each side of each polygon is uniquely defined by its own set of lines and coordinate pairs. Of course, adjacent sides of polygons, even though they are recorded separately in the computer, should have the same coordinates.

Because the data model is conceptualized as a one-for-one translation of the analog map, the relationships (topology) between and among objects—for example, the locations of adjacent polygons—are implied rather than being explicitly coded in the computer. In addition, all relationships among all objects must be calculated independently. A result of this lack of explicit topology is enormous computational overhead, making measurements and analysis difficult. Because it so closely resembles the analog map, however, the spaghetti model is relatively efficient as a method of cartographic display and is still used quite often in computer-assisted cartography when analysis is not the primary objective. The representation is quite similar as well to that found in many plotting devices, making the translation of the spaghetti model to the plotter language easy and efficient. Plotting of spaghetti data models is usually quite fast compared with some other models.

Whereas absence of topology prevents the spaghetti structure from offering ease of analysis, the next vector data model includes it. As you might have guessed, we call models of this type topological models (Dangermond, 1982) (Figure 4.16). To allow advanced analytical techniques to be performed easily, we want to provide the computer with as much explicit spatial information as possible. Just as a numerical coprocessor in a PC incorporates a great many precalculated mathematical operations, the topological data model incorporates solutions to some of the more often used operations used in advanced GIS analytical techniques. This is done by explicitly recording adjacency information into the data structure in order to eliminate the need to determine it for multiple operations. Each line segment, the basic logical entity in topological data structures, begins and ends when it either contacts or intersects another line or when a change in direction of the line occurs. Each line then has two sets of numbers: a pair of coordinates and an associated node number. The node is more than just a point; it is the intersection of two or more lines, and its number is used to refer to any line to which it is connected. In addition, each line segment, called a link, has its own identification number that is used as a pointer to indicate which set of nodes represents its beginning and ending. Polygons, composed of these links, also have identification codes that relate back to the link numbers. Each link in the polygon now is capable of looking left and right at the polygon numbers to see which two polygons are adjacent to each other along its length. The "left and right polygons" are also stored explicitly, so that even this tedious step is eliminated. This design feature allows the computer to know the actual relationships among all its graphical parts. In other words, we have a vector data model that more closely approximates how we as map readers identify the spatial relationships contained in an analog map document.

A number of topological data models have been developed and are in common use. All are slightly different, and we shall look at some of the more common ones to see how the implementations of the data model might be performed. Perhaps

Topologically Coded Network and Polygon File

Link #	Right polygon	Left polygon	Node 1	Node 2
1	1	0	3	1
2	2	0	4	3
3	2	1	3	2
4	1	0	1	2
5	3	2	4	2
6	3	0	2	5
7	5	3	5	6
8	4	3	6	4
9	5	4	7	6
10	4	0	7	4
11	0	5	5	7

X,Y Coordinate Node File

Node #	X Coordinate	Y Coordinate
1	19	6
2	15	15
3	27	13
4	24	19
5	6	24
6	20	28
7	22	36

Figure 4.16 **Topological vector data model.** Note the inclusion of explicit information concerning connected points, lines, and polygons. *Source:* Figure derived from Environmental Systems Research Institute, Inc. (ESRI) drawings and tabular data.

the best-known topological data model is the **GBF/DIME** (geographic base file/dual independent map encoding) model created by the U.S. Bureau of the Census to automate the storage of street map data for the decennial census (U.S. Department of Commerce, Bureau of the Census, 1969) (Figure 4.17*a*). In this case, the straight-line segments represent streets, rivers, railroad lines, and so on (Peuquet, 1984). In this topological data structure, each segment ends when it either changes direction or intersects another line, and the nodes are identified with codes. In addition to the basic topological model, the GBF/DIME model assigns a directional code in the form of a From node and a To node (i.e., low-value node to higher-value node in a sequence). This approach makes it easy to check for missing nodes (see Chapter 6) during the editing process. If, for example, you want to see whether a polygon is missing any links, simply match the To node of one line to the From node of the preceding link. If the nodes do not completely surround an area, a node is missing (Peuquet, 1984).

As an additional useful feature of the GBF/DIME system, both the street addresses and UTM coordinates for each link are explicitly defined, permitting street addresses to be accessed by geographic coordinates. However, this data model suffers the same basic problem that dogs the generic topological model and the spaghetti model as well. Since there is no particular order in which the line segments occur in the system, to search for a particular line segment, the program must perform a tedious sequential search of the entire database. As you remember, this is the slowest possible way to search for records in a computer. Moreover, the GBF/DIME system is based on the idea of graph theory: it

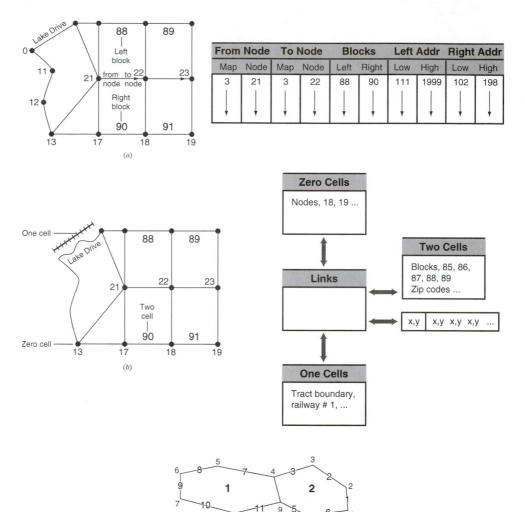

Figure 4.17 DIME files and TIGER files. Example of topological vector data models: (a) the GBF/DIME model, (b) the TIGER model, and (c) the POLYVRT model. *Source:* Parts (a) and (b) modified from K.C. Clark, *Analytical and Computer Cartography,* © 1990, Prentice-Hall, Inc., Englewood Cliffs, NJ. Used with permission.

doesn't matter whether the line connecting any two points is curved or straight. Thus, a side of a polygon serving to indicate a curved river boundary would be stored not as a curved line but rather as a straight line between two points, and the resulting model would lack the geographic specificity we expect of an analog map.

Some of the problems associated with the GBF/DIME system have been eliminated with the development of another system, the topologically integrated geographic encoding and referencing system (or **TIGER**) (Marx, 1986), designed for use with the 1990 U.S. census (Figure 4.17b). In this system, points, lines, and areas can be explicitly addressed. Therefore census blocks can be retrieved directly by block number rather than by relying on the adjacency information contained in the links. In addition, because the model does not rely on graph theory, real-world features such as meandering streams and irregular coastlines are given a graphic portrayal more representative of their true geographic shape (Clarke, 1990). Thus TIGER files are more generally useful for noncensus-related research.

Another data model, developed by Peucker and Chrisman (1975) and later implemented at the Harvard Laboratory for Computer Graphics (Peuquet, 1984), was called the **POLYVRT** (POLYgon conVeRTer) model (Figure 4.17c). As with the TIGER system, it eliminates the storage and search inefficiencies of the basic topological model by separately storing each type of entity (points, lines, and polygons). These separate objects are then linked in a hierarchical data structure with points relating to lines, which, in turn, are related to polygons, all through the use of pointers. Each collection of line segments, collectively called **chains** in this model, begins and ends with specific nodes (intersections between two chains). And as in the GBF/DIME system, each chain contains explicit directional information in the form of To–From nodes as well as left–right polygons (Figure 4.17c).

Like TIGER, POLYVRT has the advantage of allowing selective retrieval of specific entity types: you can select points or lines or polygons at will by identifying them based on their codes (which, of course, are connected to records of their attributes). An additional advantage of POLYVRT is that because lines bounding polygons are explicitly recorded as chains of individual line segments, the individual line segments do not have to be accessed to find the beginning and ending of a particular polygon. Instead, the chains can be accessed directly, saving time for searches (Peuquet, 1984). Because in POLYVRT chain lists bounding polygons are explicitly stored and linked through pointers to each polygon code, the size of the database is largely controlled by the number of polygons rather than by the complexity of the polygon shapes. This makes storage and retrieval operations more efficient, especially when highly complex polygonal shapes found in many natural features are encountered (Peuquet, 1984). The major drawback of POLYVRT is that it is difficult to detect an incorrect pointer for a given polygon until the polygon has actually been retrieved, and even then you must know what the polygon is meant to represent.

Despite the efficiencies of topological data structures, advances in computer processor speeds and massive increases in the capacity of storage devices may have reduced the need for explicit topology. The shapefile is a newer, nontopological data structure that stores the geometry and attribute information for geographic features in a dataset. The geometry for each feature is stored as a shape comprising a set of vector coordinates and is linked to their attributes.

Value	Shape type
0	Null shape
1	Point
3	PolyLine
5	Polygon
8	MultiPoint
11	PointZ
13	PolyLineZ
15	PolygonZ
18	MultiPointZ
21	PointM
23	PolyLineM
25	PolygonM
28	MultiPointM
31	MultiPatch

Figure 4.18   Examples of the shape types of geographic features in a dataset for a shapefile.

There are 14 shape types, each describing particular entities or entity combinations (Figure 4.18).

The shapefile generally has lower processing overhead than its topological counterpart. This results in faster drawing speed and editing, allows for the handling of overlap and noncontiguous features, reduces disk space requirements, and makes the files themselves easier to read and write. Shapefiles are usually three separate and distinct types of files: main files, index files, and database tables. The main file (e.g., counties.shp) is a direct-access, variable-record-length file that contains the shape as a list of vertices. The index file (e.g., counties.shx) contains character length and offset (spaces) information for locating the values, as well as a dBASE table (e.g., counties.dbf) that contains the attributes that describe the shapes.

Compacting Vector Data Models

When we looked at raster data models, we discovered that data can be compacted to reduce storage space in a number of ways. Although vector data models are generally more efficient at storing large amounts of geographic space, it is still necessary to consider reductions. A simple codification process developed more than a century ago by Sir Francis Galton (1884) is relatively similar to the compaction technique we are about to see. It might be useful to travel back in time to accompany the English scientist as he tried to develop a shorthand scheme for recording directions during geographic excursions. The form Galton devised is simplicity itself. He simply applied eight numerical values: one for each of the cardinal compass directions and one for each of the intermediate directions, northeast, southeast, southwest, and northwest (Figure 4.19).

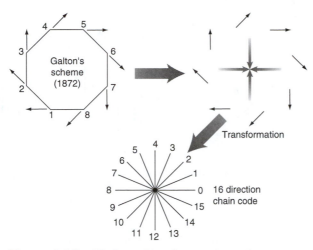

Figure 4.19 Chain codes. Comparison of compact models for direction and path finding developed by Sir Francis Galton and refined in the Freeman–Hoffman chain codes. Note the strong similarity between the older and more recent models.

A surprisingly similar coding scheme, developed in our time, is known as **Freeman–Hoffman chain codes** (Freeman, 1974) (Figure 4.19). Eight unique directional vectors are assigned the numbers 0 through 7. As Galton had done for ground navigation on his journeys, the Freeman–Hoffman method assigned these vectors in the same four cardinal directions and their diagonals. By assigning a length value to each vector, individual line entities can be given a shorthand to show where they begin, how long they are, in which direction they are drawn, and where the vector changes direction. There are many variations on this theme, including increasing the codes to 16 (Figure 4.19) or even 32 values rather than 8, to enhance accuracy. But the result is the same—reduced storage for vector databases.

Although the chain-code models produce significant improvements in storage, they are essentially compact spaghetti models and contain no explicit topological information. This limits their usefulness to storage, retrieval, and output functions because of the analytical limitations of nontopological data structures. In addition, the way the lines and polygons are encoded as vectors, performing coordinate transformations, especially rotation, leads to heavy cost computing overhead. Chain-code models are good for distance and shape calculations because much of this information is already part of the directional vectors themselves. In addition, because the approach is so similar to the way vector plotters operate (see Chapter 14), the models are efficient for producing rapid plotter output.

A Vector Model to Represent Surfaces

We have largely ignored surfaces thus far, even though they are a fundamental feature we will want to model with a GIS. They differ greatly, however, in their manner of representation, especially in vector. In raster, the geographic space

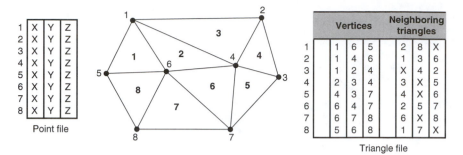

Figure 4.20 **TIN model.** In a vector system, surfaces are represented by connecting points of known elevation into triangulated flat surfaces. The model is called a triangulated irregular network (TIN) model, a specific form of tessellation.

is assumed to be discrete in that each grid cell occupies a specific area. Within that discretized or quantized space, a grid cell can have encoded as an attribute the absolute elevational value that is most representative of the grid cell. This might be the highest or lowest value, or even an average elevational value for the grid cell. As such, the existing raster data structures are quite capable of handling surface data.

In vector, however, the picture is quite different. Much of the space between the graphical entities is implied rather than explicitly defined. To define this space explicitly as a surface, we must quantize the surface in a way that retains major changes in surface information and implies areas of identical elevational data. For a simple way to envision this, consider how mineralogists or crystallographers describe minerals. Each mineral is said to have a series of smooth faces connected by points and lines that show major changes in its structure. In a similar way, we can envision a topographic surface to be a crude mineral, with flat surfaces, edges, and vertices (Figure 4.20). Thus we can model a surface by creating a series of either regularly or irregularly placed points that act as vertices. Each point has an explicit topographic value. We connect any three points to represent an area of uniform topography, thereby creating what essentially amounts to a crystallographic model of our surface.

This **tessellation** model, called a **triangulated irregular network (TIN)**, allows us to record topographic data as points in a regular or irregular grid. Then when a three-dimensional view is necessary, the grid can be converted to the TIN or crystalline shape. In addition, the point data can be used for performing the typical processes of representing the surface as a series of contour lines, with all the interpolation procedures needed for surface analysis. This particularly elegant means of representing surfaces was, in fact, used as the primary data structure for earlier systems that relied heavily on surface data (DeMers and Fisher, 1991).

Hybrid and Integrated Systems

We have discussed systems in order of complexity, moving from file structures to database management systems to spatial data models. We now need to move one step further to complete systems. Most raster systems are so simple that the data model itself provides a relatively complete description. In vector systems, however, there are two primary approaches to integrating the graphics elements of the data model to a database management system. It is useful to examine these two models not only because of the basic differences in approach but also because of the overall prevalence of vector GIS systems in the marketplace. The two major forms of vector GIS are **hybrid system** and **integrated systems**; both accommodate the linkage between spatial entities and their attributes.

The hybrid GIS data model is an acknowledgment that although graphic data structures and models are efficient at handling entity data, they lack the computational sophistication to manage the attribute data with the same efficiency (Aronson, 1985; Morehouse, 1985). Alternatively, database management systems are well recognized for their ability to manage attribute types of data but are not well adapted to graphics entities. It seems only logical that these two technologies, if linked through software, would provide the best of both modes. To implement this approach, the coordinate and topological data required for graphics are stored as a separate set of files (Figure 4.21). The attribute tables, carrying all the necessary attribute data for each graphic entity, are also stored separately within existing commercial database management system software. Linkage is performed by storing identification codes (i.e., polygon identification codes) as a column of data in the attributes database. In this way, the column is directly associated with the attribute codes contained in the table. Because the hybrid model allows raster and vector data types to be operated on in the same system, this approach is sometimes used to link grid-based modules to the database management system within the overall GIS structure. Moreover, since multiple attributes can be stored in the database management system, analytical capabilities are increased and storage space is saved. Hybrid systems include the CAD-based INTERGRAPH IGDS/DMRS and MICROSTATION-32, the vector/topology-based ARC/INFO, GEOVISION, and INTERGRAPH MICROSTATION GIS, and at least one quadtree-based system, SPANS. An examination of these systems is beyond the scope of this text. Instead the reader is directed to GIS reference works—for example, the chapter on database management systems by Healey (1991), which appears in the two-volume work edited by Maguire, Goodchild, and Rhind (1991).

The second major type of GIS model is the integrated data model, which uses another spatial database management system approach. In this case, the GIS serves as the query processor, but it is more closely integrated with the database management system than with the hybrid system (Guptill, 1987; Morehouse, 1989). Usually based on vector/topological data models, the integrated system stores map coordinate data (entities) as relational tables, together with

Figure 4.21   Hybrid vector geographic information system (GIS). Pointers connect the entity files to the database management system that contains the attribute information.

separate tables containing topological data as separate tables within the same database (Figure 4.22). Attributes are also stored in the same database and can be placed in the same tables as the graphic entities. Or, as noted earlier, they can be stored as separate tables and accessed through the use of relational joins (Healey, 1991).

Coordinate information can be stored as relational tables in two ways. The first records individual *(X, Y)* coordinate pairs, representing points as well as line and polygon terminators and vertices as individual atomic elements or rows in the database. This approach conforms to Codd's normal forms, but it makes search quite difficult because each element must be recomposed from its atomic format to create whole polygons or groups of polygons. Most GIS applications access large groups of entity elements for display purposes, a function used more often than users might think as they review the results of intermediate analytical steps. To avoid this approach, the integrated model could code whole strings or collections of coordinate information in the tables. As such, a single polygon could be described with its ID code in one column and a list of lines, addressed by code, in another. Then the lines, which would be identified by code in a separate column of a line table, would describe the polygon's locations with a number of coordinate pairs. This approach reduces overhead for retrieval and display purposes, but it violates the first normal form. From the user's perspective, this is generally not a serious problem, and grouping these nonatomic strings of data as one-dimensional arrays provides the advantage of enhanced system performance (Dimmick, 1985) while more closely following the rules of first normal form (Sinha and Waugh, 1988).

The choice of hybrid versus integrated system for the vast majority of users is less technical than pragmatic. Each has some advantages over the other, but especially as we move toward higher-powered workstations, networking, and distributed computing, both can supply a wide spectrum of analytical power. For most of us, at least as novices, the choice of system will be made for us. Those in the enviable position of choosing their system want to decide on which one best fits present equipment and future networking needs. Both system types will continue to improve, and it will be necessary to ask the vendors for detailed specifications and even performance evaluation tests on given hardware configurations.

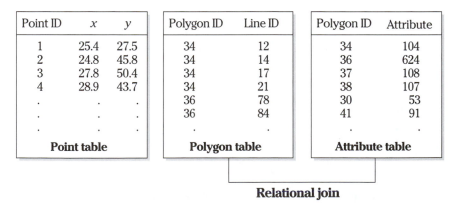

Point ID	x	y
1	25.4	27.5
2	24.8	45.8
3	27.8	50.4
4	28.9	43.7
.	.	.
.	.	.
.	.	.
Point table		

Polygon ID	Line ID
34	12
34	14
34	17
34	21
36	78
36	84
.	.
Polygon table	

Polygon ID	Attribute
34	104
36	624
37	108
38	107
30	53
41	91
.	.
Attribute table	

Relational join

Figure 4.22   Integrated geographic information system (GIS). Note how a single database can be configured to contain separate files for entities and attributes.

In addition to the typical high-level models already discussed, a third, called **object-oriented database management systems**, is emerging as an alternative approach. This model is an extension of the integrated GIS model that incorporates a spatial query language (Healy, 1991) and reflects recognition of the importance of being able to access both the cartographic database and the operations to be performed. Conceptually, such a GIS is nearly identical to the object-oriented approach to computer programming (Aronson, 1987).

There is little agreement about the exact meaning of object oriented, but we know that an object is an entity that has a condition or state that is represented by both local variables and a set of operations that will act on the object. Because it belongs to a set of objects and operations, each individual object can be thought of as a member of that set (i.e., a set defined by local variables and operations at the same time). Each of these classes of objects belongs to a larger group called a superclass and inherits properties from that superclass—in the same manner that humans (as a class of objects) inherit characteristics of a larger set called mammals. For GIS the concept might be exemplified by an object class called polygon that gives to each polygon in the database its many properties (e.g., local variables like number of polygons; lists of nodes, arcs, and areas; operational procedures for calculating centroids, drawing polygons, overlaying polygons, etc.) (Figure 4.23).

In addition, in the GIS context the object class called polygon acts as a superclass for another set of objects called land parcel. Thus, land parcel objects also inherit the variables and operational instructions for the superclass polygon, as well as having many of their own characteristics (e.g., categories of land parcels, values, owners, and operational procedures allowing transfer of ownership, rezoning operations). This explicit linking of variables and operations, together with the property of inheritance, is meant to more closely resemble the way actual geographic queries are performed. It also provides a method by which changes in one set of objects can be immediately reflected by changes in related objects.

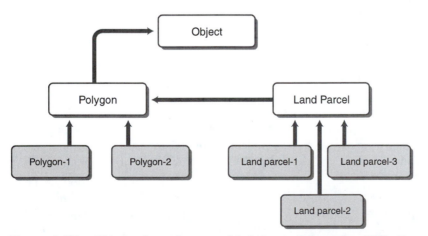

Figure 4.23   Object-oriented geographic information system (GIS). Example of a hierarchy of object classes as they might be configured by an object-oriented GIS.
Source: Modified from R.G. Healey, "Database Management Systems," Chapter 18 in *Geographic Information Systems, Principles and Applications,* D.J. Maguire, M.F. Goodchild, and D.W. Rhind, Eds., Longman Scientific and Technical, Essex, England, © 1991. Used with permission.

An older object-oriented GIS is Intergraph's TIGRIS system (Herring, 1987), which is based on object-oriented programming rather than the newly developing object-oriented database management systems. Current systems that rely to one degree or another on the object model include ESRI's Map Objects and GE Smallworld's GIS. ESRI's new **geodatabase** (short for geographic database) represents a radical change in data models for that company. This system is a shift from the hybrid ArcInfo model, also known as the coverage model, to an integrated software system which it now calls ArcGIS to indicate a clear shift to an alternative data model. One major change in this software system is that it has added object orientation, with all the flexibility that entails. This new model allows raster, vector, address, measures, CAD, and a wide variety of file types to be stored together (integrated system) in a commercial off-the-shelf DBMS. It also supports intelligent features, rules, and relationships. The geodatabase model supports, as standard, a wide selection of objects (rows in the database tables) and features (objects with geometry). All of these suggest that object-oriented GIS models are likely to increase in number and that as the ability to model with the GIS software becomes easier, the model complexity will increase. It also suggests that the idea of GIS as out-of-the-box software is being replaced with the concept of GIS as toolkit. Under this concept, the users will select the objects they need for particular tasks and will design the software and its user interface for particular applications. It also suggests that students of GIS will, of necessity, need to learn to program, particularly to program in object-oriented languages so that their effectiveness with the new generation of GIS software will be the best possible.

Terms

block codes
chains
computer file structures
database
database management system
database structures
direct files
first normal form
foreign key
Freeman–Hoffman chain codes
GBF/DIME
geodatabase
graphic data structure
grid
grid cells
hierarchical data structures
hybrid systems

indexed files
integrated systems
inverted files
map analysis package (MAP)
MrSID
networks
network systems
node
normal forms
object-oriented database management systems
ordered sequential files
polygon
POLYVRT
primary key
quadtrees
quantize
raster
raster chain codes

relational database structures
relational join
relations
resolution
run-length codes
second normal form
simple list
spaghetti model
tessellation
third normal form
TIGER
topological models
topology
triangulated irregular network (TIN)
tuples
vector

Review Questions

1. Explain the fundamental difference between a simple set of graphics and a map in terms of how each represents our environment. What is so difficult about transferring a map to a computer?

2. Since we're mostly going to use software, not program it, why do we need to know about basic computer file structures, database structures, and graphic data structures?

3. What is the difference between simple list file structures and ordered sequential file structures? Which is more efficient at adding records? Which is more useful for sorting and retrieving records? Give an example of how an ordered sequential file structure works.

4. What are indexed files? How do they differ from ordered sequential files? What are the advantages of indexed files? What is the difference between direct and inverted indexed file structure? Which is more efficient at handling data retrieval?

5. What is a hierarchical database structure? How does it work? Give an example. What are its limitations, especially for GIS?

6. What are network database structures? How do they keep track of records without a hierarchical structure? What are their advantages or disadvantages over hierarchical systems?

7. What is a relational database management system? How does it work? What advantages or disadvantages might it have over database management systems of other types?

8. What is a primary key? a tuple? a relation? a foreign key? a relational join?

9. What are normal forms? List the first three normal forms and describe the restrictions they place on the database management system.

10. Describe the process of quantizing space into equal-sized rasters called grid cells. What impact does grid cell size have on the locational accuracy? How would you store points, lines, and polygons using a raster system?

11. What are some possible advantages and disadvantages of using a raster GIS as opposed to vector?

12. Describe the vector data structure. How does it differ from raster in its ability to locate objects in space? How does it deal with the space between objects and other spatial relationships, as opposed to raster?

13. Describe the three raster GIS models. Explain the advantages and disadvantages of each.

14. Describe the methods of compact storage for raster models. Why are they needed anyway? What is the major problem with quadtree representation of earth features as squares, compared to the block codes method?

15. Describe the spaghetti vector data model. What are its advantages and disadvantages?

16. Describe the general topological vector data model. How does it differ from the spaghetti model? How is this difference achieved? What are some examples of topological models? Describe their differences. What advantages or disadvantages does each have?

17. What is a common method of compacting vector data models? How does it work? What nineteenth-century geographer developed a similar scheme?

18. Describe the TIN model. How does it quantize space differently from raster models? Why is it necessary to develop such a model for vector GIS?

19. What is the major difference between hybrid and integrated GIS systems? Diagram data storage and access in each system.

20. Explain, in general terms, what an object-oriented GIS is and indicate its potential advantages over other systems.

References

Aronson, P., 1985. "Applying Software Engineering to a General Purpose Geographic Information System." In *Proceedings of AUTOCARTO 7.* Falls Church, VA: ASPRS, pp. 23–31.

Aronson, P., 1987. "Attribute Handling for Geographic Information Systems." In *Proceedings of AUTOCARTO 8.* Falls Church, VA: ASPRS, pp. 346–355.

Beyer, R.I., 1984. "A Database for a Botanical Plant Collection," In *Computer-aided Landscape Design: Principles and Practice,* B.M. Evans, Ed., Scotland: Landscape Institute, pp. 134–141.

Burrough, P.A., 1983. *Geographical Information Systems for Natural Resources Assessment.* New York: Oxford University Press.

Clarke, K.C., 1990. *Analytical and Computer Cartography.* Englewood Cliffs, NJ: Prentice Hall.

Codd, E.F., 1970. "A Relational Model of Data for Large Shared Data Banks." *Communications of the Association for Computing Machinery,* 13(6):377–387.

Dangermond, J., 1982. "A Classification of Software Components Commonly Used in Geographic Information Systems." In *Proceedings of the U.S.–Australia Workshop on the Design and Implementation of Computer-Based Geographic Information Systems,* Honolulu, HI, pp. 70–91.

Date, C.J., 1986. *An Introduction to Database Systems,* Vol. II. Reading, MA: Addison-Wesley.

DeMers, M.N., and P.F. Fisher, 1991. "Comparative Evolution of Statewide Geographic Information Systems in Ohio." *International Journal of Geographical Information Systems,* 5(4):469–485.

Dimmick, S., 1985. *Pro-FORTRAN User Guide.* Menlo Park, CA: Oracle Corporation.

Fagin, R., 1979. "Normal Forms and Relational Database Systems." In *Proceedings of the ACM SIG-MOD International Conference on Management of Data,* pp. 153–160.

Freeman, H., 1974. "Computer Processing of Line-Drawing Images." *Computing Surveys,* 6:57–97.

Galton, F., 1884. *Art of Travel; or, Shifts and Contrivances Available in Wild Countries.* New York: David & Charles Reprints.

Guptill, C., 1987. "Desirable Characteristics of a Spatial Database Management System." In *Proceedings of AUTOCARTO 8.* Falls Church, VA: ASPRS, pp. 278–281.

Healey, R.G., 1991. "Database Management Systems." In *Geographical Information Systems: Principles and Applications,* D.J. Maguire, M.F. Goodchild, and D.W. Rhind, Eds. Essex: Longman Scientific & Technical.

Herring, J.R., 1987. "TIGRIS: Topologically Integrated Geographic Information System." In *Proceedings of AUTOCARTO 8.* Falls Church, VA: ASPRS, pp. 282–291.

Kent, W. (1983), "A Simple Guide to Five Normal Forms in Relational Database Theory." *Communications of the Association for Computing Machinery,* 26(2):120–125.

Maguire, D.J., M.F. Goodchild, and D.W. Rhind, Eds., 1991. *Geographical Information Systems: Principles and Applications.* Essex: Longman Scientific & Technical.

Marx, R.W., 1986. "The TIGER System: Automating the Geographic Structure of the United States Census." *Government Publications Review,* 13:181–201.

Morehouse, S., 1985. "ARC/INFO: A Geo-relational Model for Spatial Information." *Proceedings of AUTOCARTO 8.* Falls Church, VA: ASPRS, pp. 388–397.

Morehouse, S., 1989. "The Architecture of ARC/INFO." In *Proceedings of AUTOCARTO 9.* Falls Church, VA: ASPRS, pp. 266–277.

Peucker, T., and N. Chrisman, 1975. "Cartographic Data Structures." *American Cartographer.* 2:55–69.

Peuquet, D.J., 1984. "A Conceptual Framework and Comparison of Spatial Data Models." *Cartographica,* 21:66–113.

Shaffer, C.A., H. Samet, and R.C. Nelson, 1987. "QUILT: A Geographic Information System Based on Quadtrees." College Park: University of Maryland, Center for Automation Research, CAR-TR–307, CS-TR–1885, DAAL02–87-K–0019.

Sinha, A.K., and T.C. Waugh, 1988. "Aspects of the Implementation of the GEOVIEW Design." *International Journal of Geographical Information Systems,* 2:91–100.

Ullman, J.D., 1982. *Principles of Database Systems.* Rockville, MD: Computer Science Press.

U.S. Department of Commerce, Bureau of the Census, 1969. *The DIME Geocoding System.* In Report No. 4, Census Use Study. Washington, DC: U.S. Government Printing Office.

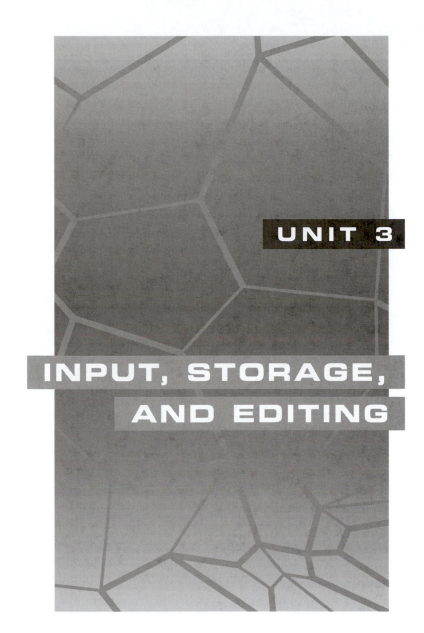

UNIT 3

INPUT, STORAGE, AND EDITING

CHAPTER 5

The Input Subsystem

The most important tasks associated with GIS are selecting, acquiring, and converting data into a compatible digital format the software can use. The many decisions that are made prior to GIS analysis will have a profound long-term impact on the quality and utility of the GIS. These decisions include a determination of what data need to be included in the database, where they can be obtained, their quality, costs, methods of input, documentation for future users, and many more. While most of us view GIS as a tool for analysis, and are often impatient to get started analyzing them, the most time intensive and costly operations deal with the many aspects of data entry.

GIS data come in two general forms—primary and secondary. Often referred to as **primary data acquisition** and **secondary data acquisition**, this terminology is somewhat misleading because we actually do more than just acquire the data. Primary data refer to data that are created in-house, while secondary data are obtained from outside data providers. Most operational GIS will use some combination of primary and secondary sources and will acquire both analog and digital data, at different scales and projections, in many different formats. We must, therefore, convert them to data models and data structures that are compatible with our own GIS software package. This allows us to store, edit, retrieve, analyze, and output all of them regardless of their original form or source.

In the following pages we will examine primary data input first and then secondary data input. Next we will look closely at the role of data documentation to describe the many aspects of our data. This particular topic is becoming increasingly important, especially because much GIS data are now being used for long-term projects and are often shared among different users both within single organizations and among different organizations.

LEARNING OBJECTIVES

When you are finished with this chapter you should be able to:

1. Define the difference between primary and secondary data acquisition and describe the unique problems of each.

2. Know the four primary functions of GIS data input.

3. Know how to decide on the input devices you will need.

4. Be aware of the potential problems of converting to and from different data structures.

5. Understand the transformations that take place during data input.

6. Know the procedures for translation, rotation, and scale change necessary for modifications in GIS data layers.

7. Understand the procedures of map preparation and the importance of those map preparation procedures in map input.

8. Have a feel for what data and how much data need to be input and why they need to be input to a GIS.

9. Understand the relationship between input scale and projection on GIS error.

10. Be familiar with the basic procedures for digitizing and the importance of attribute data to the overall quality of the GIS database.

11. Be familiar with the four basic methods of raster data input, with their advantages and disadvantages, and with the data types that should be input with the respective methods.

12. Describe what metadata are in detail and discuss their role integrity and data sharing.

13. Understand some of the technical problems of using aerial photography for GIS input.

14. Be aware of the potential for and the problems of external databases for GIS input.

PRIMARY DATA

Primary data are those that are developed by the individual or organization that intends to use them or, in some cases, are contracted out (outsourced) with specific guidelines about how they will be prepared. Because the individuals or organizations are going to be the primary users, they are the most familiar with the necessary data quality, scales and resolutions, classifications and aggregation levels, and all other aspects of the data. Primary data are generally going to experience a higher level of quality control than secondary data. They may also require both intensive collection (e.g., field collection) and conversion (e.g., digitizing) than secondary data. As a result, they are often better quality for the specific applications and more costly to produce.

There are many forms of primary data ranging from field observations to biological specimen sampling, human interviews and surveys, aerial photographs and satellite remote sensing, field mapping, GPS surveys, police calls and accident reports, and literally thousands more. The specific methods and techniques that are applied are as varied as the organizations and individuals that use them. It would require volumes to summarize all the guidelines and criteria for all possible primary data collection methods. You need to consult your own subject-specific literature to become familiar with them. Some of these primary data sources will create digital files that are readily accessible by the GIS software you use, and the methods are frequently well explained by the acquisition hardware documentation. We will limit ourselves here to the conversion of analog mapped data to the digital form through the processes of digitizing or scanning.

INPUT DEVICES

Many different types of devices have been and are being used for general input of any data into a computer. For inputting spatial data manually, the use of a digitizer is standard. The digitizer is a somewhat more advanced and much more accurate version of an input device used in nearly all modern PCs—the mouse, which the user moves freely along pretty much any surface. Inside the mouse are sensors that respond to the motion of a rubber ball encased in the frame of the device. All the locational information is contained within the mouse itself. To enhance the accuracy of such a device, the degree of movement for a digitizer is recorded through the use of an electronically active grid within the digitizing tablet. A mouselike device, often called a puck, is connected to the tablet and is moved along the tablet to different locations on a map that is attached to the tablet. Digitizing pucks contain a crosshair device, encased in glass or clear plastic, that allows the operator to place the puck exactly over individual map elements. In addition, the puck has buttons (the number depends on the sophistication of the digitizer system) that indicate the beginnings and endings of lines or polygons, or explicitly define left and right polygons, and so on. How the digitizer buttons are used is largely determined by the GIS software requirements.

Digitizing tablets come with rigid or flexible tablets and can range from small page-sized formats to very large formats capable of accepting good-sized maps with room to spare (Figure 5.1). Some of the larger format digitizers also have adjustable stands that can vary the elevation of the tablet from the ground and

Figure 5.1   Large-format digitizing table.
Most large tablets allow the user to raise and lower the table and adjust the angle.

change the angle of the tablet to make digitizing easier for the operator. Sizes and formats are determined in part by the general size of documents to be input to the GIS and in part by budgetary constraints. Generally, the smaller the digitizer the lower the cost. In addition, there is a direct relationship between cost and digitizer accuracy. The lower-priced digitizers frequently give adequate results, and as technology improves, the prices of these devices will continue to drop. Modern digitizers can provide **resolutions** of 0.001 inch, with an accuracy that approaches 0.003 inches for an area of 42 inches by 60 inches (Cameron, 1982).

Factors that will prove useful in selecting a digitizer include **stability**, **repeatability**, **linearity**, resolution, and **skew**. *Stability* deals with the tendency of the exact reading of the digitizer to change as the machine warms up. For the first-time digitizer operator, it can be most disconcerting to watch the values change while the puck sits in one place. The simplest solution is to allow the digitizer table to come up to operating temperature before using it. If the drift continues when the tablet is warm, the tablet may need to be repaired or replaced.

Repeatability is a synonym for precision. If you place the puck in the same exact location twice, how close will the first and second readouts be? Good digitizers should be repeatable to about 0.001 inch (Cameron, 1982). *Linearity* is a measure of the ability of the digitizer to be within a specified distance (tolerance) of the correct value as the puck is moved over large distances. A linearity of 0.003 inch measured over 60 inches is common with today's equipment. *Resolution* is the ability of the digitizer to record increments of space. In other words, the smaller the units of measure it can handle the better its resolution. This is not unlike the resolution of a camera or a remote sensing system. Resolutions of 0.001 inch are quite good but may prove to be unnecessary for much GIS work. Finally, *skew* is a measure of the squareness of the results on a tablet: Do coordinates located at the four corners of your digitizer produce a true rectangle, as intended? Some portions of the digitizer may begin to wear out, especially toward the edges, reducing the ability to digitize on the entire tablet, perhaps even compromising the quality of the input.

Scanners are more advanced input devices that perform much of the work of the entity input process with little help or intervention from the user. Scanners come in three general types: line-following scanners, flatbed scanners, and drum scanners. Line-following scanners, as the name implies, are placed on a line and move on small wheels, tracking by a laser or other guiding mechanism. At different intervals (either time intervals or distance intervals), these devices send a signal to the computer that records the digitizer coordinates at each of the sample locations. Line-following scanners require more technician intervention because they must be manually placed on each new line to keep the scanning process going. They also suffer from two shortcomings. First, complete automation requires the device to use a simple time or distance measure within which to record or sample the lines followed. As you remember from Chapter 2, however, this is not always the best way to abstract a line. For very complex lines, more points should be sampled than is necessary for relatively straight lines. Additional operator intervention is required to ensure that sufficient points are sampled. The second problem is one that is more difficult for the line scanner to handle. Say you are using a line scanner on a topographic map to digitize contour lines that converge along a cliff (a conventional method of depicting cliff faces with contour lines). When the lines diverge on the opposite side of this feature, however, it is quite common for the scanner to get confused about where it should go next. This problem can get even worse if the topography under an overhang is represented by dashed lines, which the scanner may not be able to find because of the gaps in the lines or because the color is lighter and has less contrast than the original contour line.

Flatbed scanners are more commonly available today because of major reductions in cost. These scanners range from small units, found in office supply stores, that are typically designed for scanning business cards, photographs, or documents, to large commercial scanners (Figure 5.2). Using a charged coupled device (CCD) scanning technology not unlike that commonly available in digital cameras and remote sensing devices, the commercial flatbed scanner most often scans the map or image one line at a time and then compiles all the lines as a composite raster image. Depending on the amount of memory and the quality of the original image or map, these devices can produce very good resolutions, certainly far better than the smaller office variety. Of course, the increased resolution results in very large images that can range up to many gigabytes.

The third basic form of scanner, the drum scanner, uses a rasterlike approach much like the flatbed scanner, but the mechanics are somewhat different. The map document is placed on a drum that rotates at the same time that a sensor device moves along the frame at right angles to the direction of rotation (Figure 5.3). In this way, the entire document is scanned one line at a time. As with flatbed scanners, each location on the map or image is recorded, even if no graphical objects are present. The result is a detailed raster image of the entire document.

Both drum scanners and flatbed scanners can give monochromatic or color output. For color, each of the three primary colors is scanned, either individually and then recombined, or all at once depending on the technology used.

Figure 5.2　**Large-format flatbed scanner.**

Figure 5.3 Principles of map scanning. Diagram of a drum scanner showing the rotation of the drum and the movement of the scanning device along the drum.

Whether the output is monochromatic or color, they can be converted to vector format (usually for cartographic scanning) or can be retained as raster (usually for imagery).

Scanning has taken on a new importance primarily for the use of aerial photographic imagery as both a GIS thematic map or coverage, and as a background from which onscreen or heads-up digitizing can take place. In many situations, scanned aerial photography can provide a much improved method of producing additional themes by simply tracing the data directly as observed on the aerial photography. In addition, for the GIS analyst, the digital photography also provides context for the digital cartographic data, thus enhancing the visualization and improving understanding (Figure 5.4).

The costs of scanning came down considerably in the 1990s. With renewed interest in scanned imagery products, such as the the digital orthophotoquads (DOQs) and digital raster quads (DRQs) now commonly available from the U.S. Geological Survey and commercial sources, the importance of scanning is certain to increase. The technology is improving rapidly, the amount of editing necessary is decreasing, and the uses of the scanned products are increasing.

Figure 5.4 Scanned aerial photograph. The relationship between the point and line features and the natural environment is much clearer when they appear in context.

Still, remember that scanning, especially scanning of complex maps, is not a panacea for digitizer woes. Even with large-format commercial scanners, the editing process will require human intervention. And these scanners are still more expensive than most small GIS operations can afford. Thus scanning the occasional piece of large cartographic and imagery data will typically be performed by professional scanning firms for the near future. Trade shows are showing more and more large-format scanning devices at prices that are far below those of only a few years ago. Beyond the technical specifications of these devices, you need to examine the volume of scanning that you or your company is likely to perform before deciding on the purchase of one of these devices. Still, you are likely to be doing manual digitizing for many maps for which the large-format scanning devices are either unsuitable or for which the cost is prohibitive.

Raster, Vector, or Both

No matter which approach is chosen for GIS data input, it is necessary to determine at the outset whether you will be using a raster or vector GIS, as well as whether your GIS is capable of converting back and forth within these data types. Some programs, especially those that evolved primarily to handle remotely sensed data, operate predominantly on the grid data structure, whereas others operate primarily on vector data. Most of the commercial programs allow interconversion. In addition, data types among GIS software are becoming more alike in general format as the clear separation between the two is becoming less profound.

The type of input device will also have an impact on how you will operate on the data and on whether format changes will be needed. As we have seen, line-following scanners tend to produce vector output, whereas drum scanners will produce raster output. Hand digitizers are largely controlled by the software system that drives them, providing output in the appropriate format. Although conversion between raster and vector is fairly common, there are some things you should remember. Most often, when you convert from vector to raster, the results are visually satisfactory, but the techniques for the conversion can produce results that are not satisfactory for the attributes each grid cell represents. This is particularly true along the edges of areas, where the user seldom knows the decision rules concerning how the space is to be quantized. (See the later discussion on raster input.) Alternatively, by converting from raster to vector, you may preserve the vast majority of the attribute data, but the visual results will often reflect the blocky, steplike format of the grid cells from which conversion proceeded. Some algorithms are available for smoothing this blocky appearance with the use of a mathematically based graphics technique called **splines**. Without going into the gory details, this is simply a graphical approach that smooths out jagged lines and sharp edges. If you have access to a CAD system, you can sample this effect by using the spline feature to enter a stick person on the screen. The result will look much like the Pillsbury Doughboy, with fat, rounded surfaces.

Before entering data with your system, especially if you need to convert back and forth, try to find out how the conversion takes place by talking to the vendor or consulting the documentation. You may want to develop a small database against which to compare test results. This check may not allow you to determine the actual algorithms used but will indicate the nature of the results. You can then determine which system is the most appropriate for your needs.

Reference Frameworks and Transformations

Digitizers are designed to input existing maps, although they can also be used to input data from aerial photography and other analog products of remote sensing equipment. Later in this chapter we will discuss the role of remote sensing in data input; here we will concentrate instead on the typical cartographic document from which we will produce a cartographic database. As we have seen, maps are representations of a three-dimensional reference globe projected onto a flat surface. That is, our geographic data have already been transformed from their original form as part of a spherical planet to a two-dimensional surface, with all the accompanying shape, area, distance, and angular deformations. When we digitize this map, we reduce this sophisticated projection to a set of Cartesian (in this case digitizer) coordinates. Before we do so, we will normally provide the GIS software with information about the type of projection used and even specific information about the grid system and the zone or zones of origin within which it was produced. This preparation will allow us to transform the map to its original projection after it is input. In fact, the GIS will produce a number of transformations of this kind as we project from the Cartesian coordinates on the digitizer to the two-dimensional map projection coordinates, and from there through a process called the **inverse map projection** to three-dimensional latitude and longitude coordinates. We will eventually need to reverse the process to produce Cartesian coordinates for the output device (Figure 5.5).

To perform these various projections and manipulations, the GIS software, in large part within the input subsystem, will need to perform a number of graphical manipulations. The three primary processes that must occur, often simultaneously, are **translation**, **rotation**, and **scale change**. Translation is simply the movement of parts or all of a graphic object to a different location on the Cartesian surface. This is done by adding or subtracting the coordinate values necessary for the X and Y coordinates of the object (Figure 5.6a). In other words, the new X coordinate X' for each graphic object will be equal to the original X coordinate plus some value T_X, and the new Y coordinate Y' for each graphic object will be equal to the original Y coordinate plus some value T_Y

$$X' = X + T_X \quad \text{and} \quad Y' = Y + T_Y$$

where

T_X and T_Y can be either positive or negative

Scale change is also relatively useful because of the need to compare differently scaled maps and to output in different scales as well (Figure 5.6b). This is done by multiplying the overall X coordinate extent (e.g., the two sets of X coordinates for a line segment) by a scale factor s_X, and each set of Y coordinates by a Y scale factor s_Y:

Figure 5.5   Geographic information system (GIS) coordinate transformations. Transformation steps from Cartesian (digitizer) coordinates to two-dimensional projected map coordinates, through an inverse map projection to longitude and latitude coordinates. To produce map output, the process is reversed through the projection process and finally to the necessary Cartesian (display device) coordinates.   *Source:* Duane F. Marble, Department of Geography, Ohio State University, Columbus, OH.

$$X' = X\,s_X \quad \text{and} \quad Y' = Y\,s_Y$$

where

S_X and S_Y represent the amount or percentage of scale change

Rotation is used frequently during the process of projection and inverse projection. It is accomplished by using basic trigonometry (Figure 5.6c). For X coordinate locations, the new X location X' is found by multiplying it by the cosine of the new angle (θ) and then adding that value to the original Y coordinates multiplied by the sine of theta (sin θ). The new Y coordinate locations Y' are found by multiplying the negative of the original X value by the sine of the angle and again adding that to the product of Y coordinate and sin θ:

$$X' = X \cos \theta + Y \sin \theta \quad \text{and} \quad Y' = -X \sin \theta + Y \sin \theta$$

where

θ is the angular displacement wanted

All necessary transformations can be performed with these three basic types of graphical manipulation.

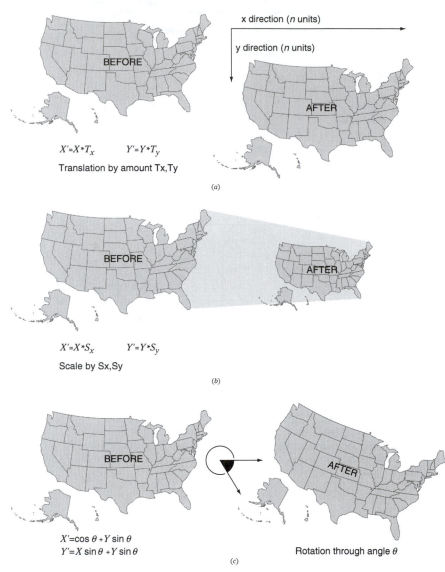

x direction (*n* units)

y direction (*n* units)

$X'=X*T_x$ $Y'=Y*T_y$

Translation by amount Tx,Ty

(a)

$X'=X*S_x$ $Y'=Y*S_y$

Scale by Sx,Sy

(b)

$X'=\cos\theta+Y\sin\theta$
$Y'=X\sin\theta+Y\sin\theta$

Rotation through angle θ

(c)

Figure 5.6   Translation, scale change, and rotation. The three basic graphic transformations necessary to make the projection transformations: *(a)* translation, or object movement within the coordinate space; *(b)* scale change, to adjust the object size; and *(c)* rotation, to reorient the object within the coordinate space.

Map Preparation and the Digitizing Process

As we just saw, we cannot begin digitizing until we have provided the GIS software with pertinent information regarding the projection, the grid, and so on. This is a part of the map preparation process that is so often neglected, yet is so important to developing a useful database. Many software packages will require you to provide this information before you can begin, although some will allow you to input the information later. In either case, you should prepare the information beforehand and keep it handy so that you will always know what it is and where to find it. It is also a good idea, before you place your map on the digitizing tablet to begin the input process, to prepare the map by making appropriate marks directly on the map document or on a firmly attached clear plastic covering, to identify the exact locations you will be digitizing (Figure 5.7). Remember, there will be many curved lines, and you will have to reduce the curves to short, straight-line segments. Although many prefer to digitize freehand, eliminating this useful step, if you know all the points that will be digitized (which are the beginnings and which are the endings of polygons, which are nodes, etc.), you will not need to perform this tedious procedure while you are digitizing.

Because digitizing is monotonous work, you probably will digitize the map in portions rather than all at once. This is all the more reason to prepare the map ahead of time by outlining directly on the map portions you expect to be digitizing at each session. However, because you will probably be digitizing in multiple sessions, and sometimes you will have to remove the map to allow others to use the equipment, you will need to tell the computer software where your map area is and what its coordinates are. These first points, called **registration**

Figure 5.7   Map preparation. Notice the markings used to illustrate where the tic marks are, locations of digitizing points, the order of digitizing, and the start and ending for each polygon.

points or **tic marks**, will be entered in digitizer inches as well as in map coordinates. They also should be marked on the map as part of your map preparation, so that you, as well as the computer, know your starting points. The registration points provide an outside frame for the document and should be outside any graphic object you will digitize, including a neat line if you will include that in your database. Usually three points, located on the corners of a rectangle, are needed to define the map area for the software. Some software can get by with only two points if they are located on a diagonal. In this case, the software assumes that the outside boundary is a rectangle and infers the other two corners. No matter which technique you are required to use for your software, the accurate location of these registration points or tick marks is absolutely essential to ensure good quality. Special care should be taken to locate your tick marks precisely. It is a good idea to double-check these because if registration marks are misplaced, virtually all the remaining digitizing will be erroneous.

Most GIS digitizing software will assist you in ensuring the accuracy of your tic marks during map registration prior to digitizing, and will subsequently result in more accurate maps. As you select the tic marks, the software will provide you with an error measurement using a method called **root mean square (RMS) error**. Mathematically, this is the spatial equivalent of standard deviation in that it provides a measure of deviation or discrepancy between the known point locations and their digitized locations. The lower the RMS error, the more accurate the digitizing or transformation (Walker, 1993). Because RMS error is expressed as a single number, it does not provide information about each tic and its real coordinates. Rather, it indicates that the actual error might vary across the map depending on the number, placement, and accuracy of the tics used to register the map.

Other map preparations include a clear definition of the order in which you intend to digitize; a systematic approach to identify the portions of the map to be digitized at each session has been mentioned. It is also a good idea to develop a method of identifying which areas (sections, lines, points, etc.) have already been digitized. Taking periodic breaks to mark the map with a felt-tip pen helps you to keep track of your progress as well as giving you a rest. Your software may require you to identify nodes, left and right polygons, and so on, depending on its sophistication and the graphic data model it uses. These should be placed directly on the map document as well, to spare you from having to stop frequently to retrieve the information.

Most digitizing software provides editing capabilities to help you identify your mistakes. Some software even allows you to use the editing system to do the digitizing, thus allowing you to edit as you digitize. In the next chapter, we will discuss the types of error that can occur and how to correct them. In the meanwhile, most digitizing software includes a feature that allows for the occasional shaky hand, namely, a sort of fudge factor, sometimes called a **fuzzy tolerance** (it can have other names). The inclusion of this feature reflects the assumption that you will not be able to place the crosshairs of the digitizing puck over exactly the same location twice, as required to digitize perfectly the beginning and ending of a polygon. This is because humans generally do not have the manual dexterity to execute such precise small-scale motions, and, of course, the digitizer itself has limits. The fuzzy tolerance may have to be set before digitizing, or it may be set afterward during the editing process. In any event, you should be sure to set the tolerance to a level that is reasonable for the detail of the map as well as for the shakiness of your hand. Small fuzzy tolerances are less forgiving of digitizing errors and may result in gaps between points that were meant to be connected. Alternatively, a fuzzy tolerance that is too large will result in the merger of points and lines that are real because the software assumes the gaps were simply mistakes. Marble et al. (1990) discusses the entire digitizing system, especially with regard to organizations and commercial operations.

Final map preparations deal primarily with the tendency of the source material to shrink and swell with changes in temperature and humidity. A stable material, such as plastic, is preferred to paper for digitizing. Although plastic products also will shrink and swell with changes in temperature, they react far less than paper. In addition, plastic products are far less affected by changes in humidity. Although this matter of digitizing media may seem trivial, look around your room for especially large paper posters attached with thumbtacks or staples along the borders. If you place your hand in the center of these posters, you will find as much as an inch or two of play in the paper. The whole poster may be sagging pitifully from the thumbtacks. This is most likely not because the poster was stretched tightly when it was hung, but because it has expanded as a result of changes in temperature and humidity and is being pulled downward by the force of gravity.

There are several ways to limit the amount of digitizing error based on media fragility. First, the room should be air-conditioned to maintain a standard cool temperature and low humidity level. The material you are about to digitize should be allowed to stay in the room for several hours, unrolled. (It is a good idea to avoid using folded maps because creases reduce the accuracy of such documents severely.) Once acclimated to the room conditions, the map can be taped to the digitizer tablet using a removable tape material, such as drafting tape. Do not use masking, clear plastic, or other very sticky tapes that will tear the document or even remove information when the document is moved. In addition, very sticky tape may make it difficult to remove the document, stressing it unnecessarily and possibly resulting in stretched media. When you tape the map to the digitizer, place it several inches from the edges to avoid any possible skew (horizontal distortion), which is most likely to occur there. The map should also be placed so that it does not cause you to stretch to locate the puck on the map objects, since this can put stress on the document and can also limit your freedom of movement, adding more error to your database. When you digitize a map in multiple sessions, be sure to keep the document in the climate-controlled room to avoid expansion and contraction.

WHAT TO INPUT

Now that we have some basic guidelines on how to digitize, especially how to avoid errors during digitizing, we can begin to select the appropriate data for input. Most texts, and even most software manuals, give little guidance in this regard. It is much like beginning a journey by hearing a guide warn you to be sure all your equipment is packed carefully, without giving you a clue as to what the equipment is. Each journey into the digital world is unique, however; each environment requires different coverages, and each purpose for which we journey into the GIS demands a separate set of criteria. We will attempt to sort this out to provide a simple set of guidelines, applicable under any circumstance.

A major factor guiding what cartographers put on the map and how it is produced is the target audience—the user. The same can be said for producing cartographic and geographic databases for map analysis within a GIS. Historically, a common practice in many GIS systems, including those designed for whole states, has been to input everything (DeMers and Fisher, 1991; Fisher and DeMers, 1990). As we will examine in Chapter 15, this very often destines a system to failure. Thus *rule one* is to determine why you are building the GIS database in the first place. This will, at least, limit the input to coverages that are likely to be used. Although a really nice map of quaternary geology may seem natural as an input coverage, first because it exists, and second because of its quality, it is unlikely to be at all useful for a study of atmospheric pollution caused by factory smokestacks. What you should observe from this is that the themes should be directly related to the modeling and analysis you plan to perform and to the results you need. Anticipation of future needs may seem to be a good reason to input your quaternary geology map, but you would be better advised to keep such materials in a separate file, or store them for later input to the GIS, should the need arise.

The need to anticipate what themes will be required in the future does present a problem, particularly if either you or your client has only a larval view of what is to be done. A serendipitous approach to GIS may prove to be fun, but a GIS begun under uncertain guidelines is unlikely to produce reasonable results without substantial reworking, corrections, enhancements, and workarounds. Such an approach becomes costly. Perhaps databases should be developed without a very clear understanding of the intended product (sometimes called **spatial information product**) only when the primary objective is to determine possible relationship among coverage variables as part of initial scientific hypothesis formulation. This is not the way to proceed in the commercial environmental. So, *rule two,* which relates to the first, is that you must define your goals as specifically as possible before selecting the layers.

Even under very specific goals, with known spatial information products, there will be multiple ways of obtaining the available data in some cases. For example, coordinate locations and elevation variables can now be obtained with the use of GPS units. But they may also already be available on maps in a reasonably accurate format. Or data on current land use may be obtainable from ground surveys, aerial photographs, satellite output, airborne scanners, or any number of other sources. There is no easy answer to which should be used. But although there is no recipe for success, there is a recipe for failure, which leads us to *rule three:* avoid the use of exotic sources of data when conventional sources are available, especially if the latter provide a similar level of accuracy. You will need to define exotic for your own particular project. In general, I like to use a practical definition, applying the term to any unfamiliar data source. If you or another team member is familiar with a given set of data and can comfortably use it correctly, and if it increases the utility or accuracy of your database, it should be applied. If all your multiple data sources for a particular theme or coverage are in traditional form, then invoke *rule four:* use the best, most accurate data necessary for your task.

You should keep in mind that accuracy in this situation refers to the necessary accuracy, not an absolute accuracy. If you don't need 1-centimeter contours for your topography, for example, use topographic data that most closely matches your level of observation. Although having an extremely detailed map of any coverage may seem advantageous, it is costly to input; it also slows analysis and may even make analysis more difficult. An example of the use of 30-meter TM data from the LANDSAT satellite as compared to 80-meter MSS data from the same source might prove instructive. Suppose your purpose is to identify large fields of grain. Since the enhanced spatial resolution of 30-meter data has been known to produce many difficult-to-separate categories over an area that is essentially all grain fields, the better resolution would confuse your situation rather than simplifying it. And, of course, the computing and human resources needed to provide clarification would increase the overall cost of the system. Thus we have *rule five:* remember the law of diminishing returns when deciding on data accuracy levels.

Most thematic maps (e.g., USGS topographic maps) contain ancillary data on roads and other anthropogenic features that may very well be useful as input to the GIS. Whenever possible, and when the quality of the data dictates it, you should input these data as separate coverages from the same map sheet. That's *rule six.* This rule does not restrict the use of other sources when they are of superior quality or accuracy, but it has two advantages. First, because the data are on a single map, you don't need to go to multiple map sheets and then repeat all the preliminary steps in map preparation. Second, because the data are on the same map sheet they are already georeferenced, reducing the need to perform this sometimes difficult task later.

The final general rule, *rule seven,* is that each theme should be as specific as possible. A coverage should be as thematically specific as possible, without relying on an IMGRID-like binary system. The more specific a theme, the easier it is to search if you need to know something pertaining to data contained in a single coverage. In addition, when you perform operations like overlay (see Chapter 13), it is easier to keep track of the process if you are completely familiar with the data. Overlay operations are simplified for the computer program itself if there are no extraneous data in a given coverage.

We can summarize the seven rules in a very few simple statements. First, define your purpose. Then be sure your maps address the purpose. Use the most accurate maps needed for the purpose—not too accurate for your needs and not too inaccurate to do the job. Keep your coverages simple, and use the same map to obtain these simple coverages whenever reasonable and prudent, to avoid the need to georeference them. Above all, think about your project before you begin inputting data. Data input takes time and costs money.

HOW MUCH TO INPUT

Related to the types of data you will input is the question of how much. Again, to use our field trip analogy, as you prepare for your journey you will need to know how much food to pack, not just what kind. Too much food and you will find yourself carrying unnecessary weight for the whole trip; too little and your journey will end prematurely because you will have to return to a source of food. In like manner, too much data input and your GIS must bear the added weight of the data throughout the project life cycle; too little data and you may find yourself unable to answer questions about matters you had planned to cover.

As with preparing packs for a trip, the input of data into the computer is a sampling process. In vector GIS, each line you input will likely have some

Figure 5.8 Sampling a complex line. An example of sampling for digitizing a line. Points to be digitized are selected based on change of direction of the line. Each direction change indicates an additional piece of information the map contains.

curves. To produce a reasonable facsimile of the line using straight-line segments, you will decide thousands of times where to place the digitizer puck and where to record data. This process is very similar to that of line generalization, encountered in our earlier discussion of cartography. A simple rule of thumb is to take more samples (i.e., record more points) for very complex objects than for simple lines (Figure 5.8). The locations of a straight line can be recorded accurately with only two points: one at the beginning and one at the end. I have seen straight-line segments outlining the external boundaries of a standard quadrangle map with as many as 2000 digitized points. Not only does this clutter the computer with tons of unnecessary data points and slow computation later on, but it is highly unlikely that the straight line will appear straight when output.

Line and polygon complexity can be compared to information. The more the line changes direction, the more information it conveys. (This can also be true for surfaces, but we will discuss that later.) And the more densely packed the points, lines, and areas, the more information content the map has. The higher the information content, the higher the sampling rate needs to be. Shannon (1948) used this science, known as **information theory**, to examine the sampling of voice information for telephone communication to ensure that a complete voice signal was received. He discovered a useful rate of sampling that can easily be applied to the digitizing process. In general, in voice transmission technology, the smallest object that is to be included in a system should be recorded at least twice. So, for each change of line direction (i.e., for each piece of information), you should record at least two data points. This is all the more reason for careful map preparation. You should also keep in mind that because, for each entity input to the GIS, there is going to be a need for attribute data as well, a direct relationship exists between the complexity of the map, or the volume of data in the map, and the spatial data handling problems (Calkins, 1975).

The idea of information content can also be applied to raster data. Once again, the general rule is this: the smaller the object to be identified in your database, based on your modeling needs, the smaller the grid cells need to be (DeMers, 1992). This principle often determines the selection of grid cell size (resolution) for the entire database. Of course, information theory can also be applied to the input of raster data, but keep in mind that grid cells are two-dimensional rather than one-dimensional. The information content is a change in attribute type as we move across the map. Let us say that you want to use grid cells to represent farms displayed on a map. If the smallest farm is, say, 40 hectares, you would need to sample this area at least four times to ensure that it has been captured in your GIS. Stated differently, it means that the grid cells will need to be 10 hectares or smaller to allow the capture of the 40-hectare object in your GIS database. This also assumes, by the way, that four grid cells will surround the field itself. Suppose the field lies along a long, skinny riparian area. Although it is 40 hectares in total area, it is also spread out as a linear object, reducing the chances that the entire field will be input into your GIS. This aspect of the process is somewhat dictated by the method you used to input the grid cells. (We will cover that in a later section.) The general rule of thumb, however, remains the same: sample more for more information content.

Whether in raster or vector, sampling is dependent on the amount of area covered by the map and the use for which the data are input. Small-scale maps, those covering large amounts of space, contain a much more abstract view of the land surface being mapped. In addition, the lines and symbols on the map take up physical space, as we discussed in Chapter 3. The amount of error contained in a symbol is dependent on the scale of the map on which it is placed. Lines on small-scale maps take up more land space than same-sized lines on large-scale maps. This physical condition, called **scale-dependent error**, is an indication that the amount of error is directly related to the scale of the map and needs to be considered during the map preparation phase prior to digitizing.

METHODS OF VECTOR INPUT

As indicated earlier, numerous instruments are available to input vector data into a GIS. We will restrict our discussion to manual digitizing because of the frequency with which it occurs. After preparing your map and placing it on the digitizing tablet, you will need to use the digitizer puck to locate and record the registration marks. Some software requires these marks to be recorded in a specific sequence, whereas other packages do not. The software documentation and/or the software itself will indicate the requirements. In addition, the software package will indicate which numbered keys you need to enter for specific object types. Some numbered keys will be used to indicate the location of a point entity, others for beginning and ending of line segments, and still others for the closure of a polygon. Many digitizing errors, especially those made by novices, are due to pushing the wrong-numbered button.

The exact method of digitizing is also related to the data structure on which the software operates (Chrisman, 1987). Some (e.g., the POLYVRT structure) require you to indicate the locations of nodes, whereas others do not; others require you to encode explicit topological data while you digitize; still others will use software to build in the topology after the database has been produced. The rules are different for each package, and you will have to look over the appropriate documentation to determine these strategies ahead of time. This work can be considered to be part of the preparation for digitizing rather than the digitizing. Remember to keep the rules uppermost in your mind as you digitize to ensure that your lines don't become nodes and your nodes don't become just points. Some will find it useful to tape a crib sheet to the corner of the digitizing tablet until they become more familiar with the process.

In vector GIS, the attribute data are most often typed in using the computer keyboard. Although simply typing data in is extremely simple, the task requires the same care used for inputting the entities. This is true for two reasons. First, it is very easy to make typing errors (why else do word processors have spell-checkers?). And second, and perhaps the most potentially problematic, the attributes must be attached to the entities that are to be associated with each other. As we will see in the next chapter, these are some of the more difficult errors to detect because they cannot always be picked up visually and often do not appear until an analysis is under way. A good practice is to check attributes while they are being input, perhaps by taking frequent short breaks to look them over. Time spent doing this on input will save a great deal of time during editing.

METHODS OF RASTER INPUT

Raster data input and its vector counterpart involve different strategies. In the first place, as we have seen in our discussion of how much to input, we must decide how much area should be occupied by each grid cell. This decision must be made prior to digitizing or Mylar overlay input, to tell the digitizer how big the grid cells will be or to let the person know how big to make the Mylar squares. In addition, we must decide whether it is appropriate to use a method of encoding that shortens the process, such as run-length encoding or block encoding. Although compacting methods are good at reducing data volume, their use in encoding may be even more important because of the reduction of input time. Some raster GIS systems that either do not support digitizer input or support both keyboard and digitizer input methods have commands that allow for data to be input by typing attributes set up as strings or blocks. You can refer to your software documentation to determine what these commands are and how they are used. Once you have selected a method, you will need to decide how each grid cell will represent the different themes that will occur. Beyond the grid cell resolution, this may be the most important decision you will have to make.

Normally in raster data input, the digitizer records vectors that are then converted to raster format. The entities and attributes are generally entered simultaneously. This approach, requiring vector-to-raster conversion, can sometimes be iffy depending on the sophistication of the digitizing software used. The problem arises primarily in the raster-to-vector conversion process, which is often not well documented by the software vendor. Most often, difficulties extend from digitizing adjacent areas using vector lines, which are then converted to two separable polygons. In such cases, the software must decide which polygon will contain the grid cell through which the line runs. (Remember that in many grid-based systems a grid cell can contain only a single value.) The decision is sometimes based on the "last come, last coded" rule. That is, when the same line is digitized first for one polygon, and then again for the second polygon, the second polygon will be assigned to that grid cell. In a more computer-oriented approach, the assignment will depend on which value occurs first inside the computer. Although there are convenient methods of making such determinations computationally, they often produce unacceptable errors along edges, especially if the grid cell resolution is coarse. For very large raster databases, this level of inaccuracy may not present a substantial percentage of error compared to the entire database, but measurements of shape and area will be compromised. A better method, at least from the standpoint of the utility of the database, would be to allow the system to select from one of four systematic input methodologies outlined by Berry and Tomlin (1984).

In the first of these methods, the **presence/absence method**, for each grid cell on each coverage a decision is made on the basis of whether the selected entity exists within the given grid cell—hence the name "presence/absence" (Figure 5.9*a*). A major advantage of this method is that decisions are easy. No measurements are necessary. A simple Boolean operator either is or isn't there. Presence/absence is the best method—in fact the only useful method—for coding points and lines for grid systems because these entities do not normally take up a large portion of a cell's area. Thus, if a road crosses through a grid cell, its presence is recorded with an attribute code (a number); if it does not, it is ignored.

The second method of raster data input is called the centroid-of-cell method (Figure 5.9*b*). Here the presence of an entity is recorded only if a portion of it occurs directly at the central point of each grid cell. Clearly, this requires substantial calculations, since each central point will need to be calculated for

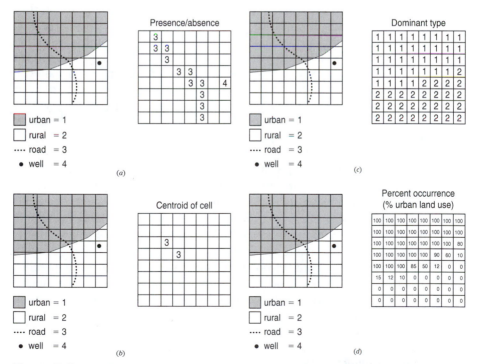

Figure 5.9 Methods of raster data input. The four basic methods of raster data input as defined by Berry and Tomlin (1984): *(a)* presence/absence, *(b)* centroid-of-cell method, *(c)* dominant type method, and *(d)* percent occurrence method.

each grid cell, and then the object will have to be compared with the location of that point. If you are using a clear Mylar grid overlay, this becomes a visual rather than a computational approach, but still you must evaluate each grid cell centroid against each entity. As you might imagine, the chances of point entities and line entities passing directly through the center of a given grid cell are minute compared to the chances they will occur anywhere else. Therefore, use of the centroid-of-cell method should be restricted to polygonal entities.

A more common method of coding area data is also generally considered to be the best method. This so-called **dominant type method** (Figure 5.9c) encodes the presence of an entity if it occupies more than 50 percent of the grid cell. Under most circumstances, the decision is straightforward, and the coding is reasonably representative of what is there. It seems logical that if you are restricted to a single category for each grid cell, the one that occupies the most space should be coded. Computationally, this requires the computer to determine the maximum amount of each polygon for each grid cell—again a computationally expensive strategy. Performed manually with a Mylar grid, the method is reasonably easy, since each grid cell can be given a quick visual inspection. In two cases, however, this approach may not be satisfactory. First, there may be highly irregular or elongate polygon shapes, and features that start out long and sinuous, like a stream, covering little of the grid cell, then become quite bulbous within the same cell. It is difficult to decide such cases by visual inspection, and computational approaches making use of a digitizer are likely to be preferable. In the second problem case, three or more polygon types converge in an irregular pattern within a single grid cell. With three or more polygons in a grid cell, the chances are slim that any one will occupy more than half of the area under the cell, and so you cannot use the 50-percent value as your cutoff point for input. Instead, the amount of area for each must be visually or computationally determined and the largest one encoded. This situation is likely to arise when you are encoding a map that is at a small scale but retains a high amount of areal detail—for example, a detailed land use or land cover map produced as a hard-copy map from satellite data at a scale of perhaps 1:1,000,000. Difficulties may also arise when you are trying to encode a map of physical features such as soils or vegetation. Natural features tend to be quite irregular and their polygons much more sinuous and intermingling. Although these examples do not mean that all detailed or complex maps are to be avoided as raster data input, they may suggest the use of centroid of cell as an alternative to simplifying the input process. This is especially true if the complexity occurs uniformly throughout the map.

A final method of raster data input, called **percent occurrence**, is also used exclusively for polygonal data (Figure 5.9d). The idea is to give more detail, not by coding just the existence of each attribute but rather by separating each attribute out as a separate coverage and then recording the percentage of the area of each grid cell it occupies. For example, a map of land use divided into urban and rural categories would be separated into two more specific themes, one urban and one rural. The percentages urban and rural would be recorded for each grid cell, with 0s entered to indicate nonappearance of a category. If the percentages are calculated accurately, the urban and rural maps should be perfect complements.

The percent occurrence method offers the advantage of more detailed data about each attribute. However, the disadvantages are numerous. First, for both automated and visual approaches, the amount of decision making is greatly increased. Second, we need to cross check both complementary themes to ensure that the percentages for each grid cell add up to 100 percent. Finally, as in the case of the dominant type method, we have the problem of multiple categories in a single grid cell. In this case, however, this problem may be even more severe because the more categories we have, the more themes we produce: a map of land use with 15 attributes explodes into 15 themes, each one representing a single attribute. This problem of data explosion is the same one we encountered in the GRID/LUNR/MAGI raster data model, which records each attribute separately, but only as 0s and 1s. Before the percent occurrence method is selected, the advantages of the additional data detail should be weighed carefully against the input problems.

Some raster GIS software allows a fifth method of raster data input beyond the four basic types. Known as **most important type method**, this approach employs the user's knowledge of the data being input into the system. In some cases, based on the purpose for which the data are to be used, retrieved, or analyzed, the user may make conscious deliberate decisions as to which category or class of data should be included for a given cell. If, for example, a set of land cover categories are to be included, and water is a category that is not going to be included in an analysis, any other type of category may take precedence over water when coding. In this way, all possible dry land uses will be included in the database and may even be overcounted, while the less important "water" category will be underrepresented because it is of limited, or in this scenario, no use.

No matter which of the five methods of raster data input is used, the choice usually falls to the software to determine. The GIS will either have its own predetermined (and often undocumented) rules for input, or, more commonly, it will ask the user to decide which method or methods are preferred during digitizing. In addition, when data are converted from vector forms to raster, the same decisions will apply. Some software will allow the user to employ a set of decision rules, while other software will make choices based on the algorithms contained within it. For quality control, it is best if the software allows the user to define the rules rather than relying on random or unspecified methods. The larger the grid cell size used, the more impact the method of raster data input or raster data conversion has on the final product and on any modeling that takes place subsequent to input or conversion.

REMOTE SENSING AS A SPECIAL CASE OF RASTER DATA INPUT

One could make a case for remotely sensed data being either primary or secondary data sources. Most commercial remote sensing systems are designed for a number of possible users. This might appropriately be considered secondary data sources given that the users are requesting data that have already been obtained and archived. In some cases, however, the user may request specific satellite or aircraft overflights, for predefined areas, at predefined times, with limited amounts of cloud cover, and so on. Or the users might have their own airborne or ground-based sensing systems for acquiring their data. Under such circumstances, the remotely sensed data would definitely be considered primary. Some users are currently employing some rather unique primary remote sensing data approaches for small study areas, including the use of tethered balloons with digital cameras attached to examine the movement of small mammals in small study areas, or sensors attached to booms for sensing crops frequently for experimental studies of growth patterns or the impacts of chemical applications. Whether primary or secondary, remotely sensed data are becoming increasingly available.

Currently, these data are not dominant compared to the many other sources of input, such as traditional cartographic products, digital elevation data, digital line graphs from topographic maps, digital land use data, and digital soils data. In addition, the raster appearance of remotely sensed data may give the impression that software designed to manipulate these raster data is by definition a GIS. Although many of the algorithms are shared by image processing and GIS, and both contribute to the field of automated geography, a GIS should not be viewed as one stage in the manipulation or analysis of electromagnetic data obtained from remote sensing devices. This view is very limiting, and it ignores the ability of GIS to operate independently of remotely sensed data, as well as the unique analytical capabilities that give it the power to, for example, analyze networks for transportation studies. Instead, GIS and image processing software should be thought of as complementary technologies, with image processing software operating primarily (though not exclusively) on electromagnetic data and GIS as a more integrative tool, using a wide array of data types and data sources.

The utility of digital and other forms of remote sensing as data sources for GIS is unquestioned, especially for tasks such as rapid data updating and building temporal databases for large areas. Most satellite remote sensing data, as we saw in Chapter 3, are based on a raster format, with each grid cell (pixel) recording electromagnetic radiation as a number of radiometric values. The number is dependent on the type of system used. For example, LANDSAT data have a radiometric resolution of 256 grayscale levels, whereas AVHRR data derived from the NOAA satellite have 1024 radiometric levels. In either case, input to a raster GIS is easily accomplished because of the similarity of data structure. Even so, the raster data structure of satellite remotely sensed data should not drive the choice of a raster GIS data model over a vector model. Rather, the choice should be based on the use for which the database is being built. In addition, when any kind of remotely sensed data are being considered as input to the GIS, the data must be evaluated for cost, utility, and accuracy compared to data from alternate sources. Remember input rule three: avoid exotic data whenever possible. A familiarity with remotely sensed data and its spatial, spectral, and radiometric characteristics may make the choice of remotely sensed data advantageous. Let's briefly examine the sources of remotely sensed data inputs and review some of the characteristics of the data discussed in Chapter 2.

Aerial photography has long been a primary source of base map data for many common products. The USGS topographic maps are largely compiled and revised from viewing stereo pairs of aerial photographs, for example, and we have mentioned the use of aerial photography as a base document for soil maps. Because many maps come from aerial photography, it is wise to find out whether maps derived from such photographs already exist before choosing to use this mode as direct input. The choice may also reflect the data that were mapped from the aerial photography as well as the classification system used. The two primary stumbling blocks to the use of aerial photography as direct input to the GIS are the relationship between the classification needed and an ability to obtain those classes from the photography and, of course, the problems of rectification and lack of reference grid.

You should consult a text on general remote sensing or a specific text on aerial photography before beginning the GIS process. Prior to input, you need to determine how much distortion your project can tolerate and what specific categories of data you will need to perform your analysis. When these categories are known, they can easily be matched to the scale and spectral sensitivity of your photography.

A special type of aerial photograph deserves mention because the images do not contain the scale, relief, and tilt distortions normally characteristic of aerial photographs. These products, called orthophotographs (or orthophotoquads if they are based on the areas occupied by topographic quadrangle maps), are photographic images of the earth that resemble maps in that they have a single scale. Orthophotographs are subjected to a process called **differential rectification**, which involves point-by-point correction of the scale and relief displacements normally caused by differences in elevation between the aircraft and the topography over which it flies. Although a detailed description is beyond the scope of this book, it is important to note the availability of these products (both in analog and now in digital forms) as sources of input. The reader is referred to Lillesand and Kiefer (1995) for more information on rectification. If digital orthophotographs are not available, the analog versions can serve as an excellent source of manual data input to the GIS.

In our discussion of digital remote sensing in Chapter 2, we noted that there are generally two major products derived for input to the GIS: digitally enhanced imagery (designed to highlight certain features for analysis, such as edges) and classified images (obtained through complex computer manipulations designed to replace the visual analyst as a classifier of features). As an input to GIS, these classified images will most likely be used to update and/or compare their classification to classified data already inside the GIS. Even when two sets of classified data are obtained through digital remotely sensed image sources, the comparison is difficult. Remote sensing scientists often prefer that these comparisons be made on the raw, unclassified images because of the potential for confusion in classification (Haddad, 1992). Indeed, direct comparisons of classified digital data to maps classified from aerial photography or historical data are very difficult. Let's look at some of the technical difficulties associated with remotely sensed data input to the GIS as defined by Marble (1981) and Marble and Peuquet (1983).

Satellite data require preprocessing to remove geometric and radiometric flaws resulting from the interaction of two moving bodies (the satellite and the earth), sensor drift as the satellite systems age, and differences in atmospheric conditions. Techniques for correcting radiometric difficulties are readily available with most digital image processing software, and the necessary equations are quite easily obtained. For GIS input, the major problem related to preprocessing is a need to obtain geometrically correct ground positions for the imagery. This geometric correction requires a number of **ground control points (GCPs)** within the image, to place it in a correct spherical coordinate space on the surface of the earth. There should be a reasonable number (the more the better) of GCPs, and they should be evenly spaced. In some areas, it is reasonably easy to use devices such as GPS units to obtain very precise locations, but the ground points must also be observable on the imagery.

Obtaining adequate GCPs may be quite difficult, especially in areas like the tropics where the forest canopy or overstory is so dense that the GPS field unit has no direct line of sight to the satellite. In addition, even if line of sight exists, it is often difficult in such areas to find features that are distinguishable from the imagery. Care should be taken when using imagery lacking in GCPs because their absence degrades the coordinate accuracy severely, especially at the margins of classified areas. The issue of GCP accuracy should indicate the importance of coordinate accuracy to the overall functioning of a GIS. A proper geodetic framework will always improve the utility of a GIS to perform its measurement and other analytical functions.

We have already touched on the second major technical problem with using digital remotely sensed data, that of classification. If your project does not use direct biophysical inputs to the GIS, you most likely will have to convert from interval or ratio data to a nominal classification scheme. Remember that the classification process used in image processing software is often based on the ease with which it can be obtained. That is, the supervised and unsupervised classification methods are designed to improve the ability to obtain an adequate classification on the basis of what is in the image; the utility of the image in an integrated GIS database is not always addressed. Supervised classification, requiring human intervention, is generally an improvement over unsupervised classification, since the process can be more easily manipulated to conform to user needs, rather than being based strictly on the statistical characteristics of the data. Nevertheless, the question of correspondence to classifications for comparable themes remains a major consideration.

The classification of satellite imagery also implies that the results are accurate, not just compatible with existing map layers. Lillesand and Kiefer (1995) have shown that the ability of image processing software to produce classifications far exceeds our ability to assess the accuracy of the classification. This is especially true of comparisons of multiple-date imagery, where the accuracy of each data set must also be compared to the accuracy of the amount of change between the two dates. It is often best to use ancillary data to produce the classes from digital imagery for input to a GIS. The insertion of topographic data, preclassified data, rule sets, and other techniques will generally result in a much improved classification, and one that more easily conforms to the corresponding GIS map layers.

The final major set of technical problems associated with using digital remotely sensed data as an input source for GIS could more aptly be called institutional problems in that they hamper the process rather than produce error. Lauer et al. (1991) indicated six basic institutional issues that negatively influence the use of remotely sensed data. These institutional issues have been seen as more problematic than the technical problems in the integration of this data source within a GIS. We will look at the most important of these issues.

The first institutional problem is the general lack of availability of remotely sensed data. Although there are several major sources, the acquisition of remotely sensed data often requires the user to be familiar with the process of obtaining the data in the first place. Once the procedures are understood, the problem is one of obtaining imagery for a particular date within a given study area at a moment of minimum cloud cover. Mechanisms exist to have satellites view particular regions, at a cost to the user, and with a specified minimum cloud cover. In areas that experience persistent and substantial cloud cover, it is often necessary to piece together multiple sets of imagery, taken at different times, to obtain a cloud-free image. This process presents additional technical difficulties because of the differences in atmospheric conditions, and even changes on the ground resulting from the flowering or dying off of vegetation. In other cases, two or more contiguous satellite images may have to be pieced together to cover a large study area completely. If, however, the reflectances obtained from these two images are substantially dissimilar, there will be an obvious line between them, and the classification process will be degraded along that margin. Finally, the unavailability in historical archives of imagery covering earlier dates may result in temporal gaps affecting the performance of spatiotemporal analysis. A primary reason for many of these institutional problems is that the data and in most cases the satellite systems themselves were originally designed as experimental rather than operational systems.

With the commercialization of the U.S. land satellite program, a different institutional problem arose. Rather than the government's bearing the cost of satellite operation, as well as the cost of data dissemination, the customer now absorbs these costs through increased charges for imagery. This limits the user community to organizations whose budget can withstand $5000–$6000 price tags for LANDSAT thematic mapper data, for example. As a result, fewer GIS operations are likely to consider obtaining and using large quantities of these data in their day-to-day operations.

A third institutional problem with remotely sensed data input also involves money. Until recently, the hardware and software costs just to manipulate the data were too high for many would-be users. The ready availability of lower-priced image processing software running on standard computing hardware has improved this situation considerably. However, the costs of manipulating remotely sensed imagery are also linked to the availability of technically trained remote sensing professionals, especially those capable of linking the raw data to other GIS themes. Every attempt should be made to integrate the classified data with other available GIS themes, rather than placing the burden on the GIS user. This further illustrates the need for students of GIS to become familiar with the techniques and data available to the remote sensing community.

The foregoing considerations lead directly to education and training, a problem for both GIS and remote sensing. Today, technical institutions and image processing and GIS vendor companies offer programs that provide extensive, hands-on experience with the use of a particular remote sensing or GIS system. Although those who complete the programs may develop an in-depth knowledge of how particular systems operate and how to perform each individual function, it is shortsighted to reduce the complexity of the discipline to one of system operations. People trained and experienced on one system are capable of producing reasonable results from complex models as long as the conceptual nature of the model is explained in enough detail. However, technicians will seldom have the conceptual knowledge necessary to formulate the solution to a problem if they don't also know how the model should work and what the GIS or image processing system is capable of doing. As you might guess, communications between, for example, an environmental scientist, who knows nothing about system requirements and a technician who knows nothing about the environment lead to inadequate, and often quite incorrect, results from analysis. Although many job advertisements today list knowledge of specific systems, a person who knows the concepts and can find a way to implement them is generally going to prove more valuable. As a result, arguments over technical versus conceptual training are likely to continue in the remote sensing and GIS communities, as evidenced by another of the six institutional problems—that of professional certification. Much of this discussion entails turf conflict over who should be certified and who should not. It is of little concern to us here.

A final institutional problem of remotely sensed data input features organizational infrastructure. Neither GIS nor remote sensing has a clearly defined, well-organized, adequately funded sponsor within the federal establishment. This was clearly evident when, during the 1980s, the U.S. government attempted to discontinue the LANDSAT satellite program. Without a flood of letters from individual users and scientists, the future of the program would likely have been very short. Thus there is a need for a champion for remote sensing and GIS to allow mutual technical improvements that will enhance the utility of both technologies.

GPS DATA INPUT

An increasingly important source of primary data input to the GIS database is GPS data. Not only does the modern GPS allow for the development of a highly accurate geodetic framework (see Chapter 2), but it is often linked explicitly to the input of other primary data sources such as field data. Even in the early years of GPS availability, software packages such as Penmap® allowed for the simultaneous input of GPS locational information with field notes in a system they still describe as a surveyor's notebook.

With the advent of new handheld computer technology in the form of PDAs and iPAQs with full color displays and software packages like ArcPad®, we are now able to display digital map layers in full color on handheld units, to link those same layers to GPS coordinates, and to include a complete set of field notes. The field notes can include anything from descriptions of conditions at any given location to digital voice recordings, digital pictures, and even geometry. By taking multiple readings, for example, one could easily describe the accurate area and perimeter of area objects (such as forest patches or subdivision boundaries for urban subdivisions). This is especially useful as a total mobile solution in areas that are not well surveyed or for field checking satellite imagery for environmental surveys. The computer and GPS technologies have already been integrated to allow out-of-the-box single-unit solutions (Figure 5-10).

Although such mobile solutions are available, they are not commonplace because of the often high price tag. As the price for technology continues to drop, they will become much more common. The obvious advantages of mobile GIS/GPS solutions include reductions in time and costs in development of geographic databases, the simultaneous coregistration of datasets, rapid GIS and remote sensing data verification, incorporation of ancillary data and descriptions, and the easy transfer of these data to full GIS databases composed of both primary and secondary data.

Figure 5.10 Integrated iPaq and GPS technology illustrated by this Trimble® GeoXT GPS unit with built-in Hewlitt Packard iPAQ® and ESRI ArcPad© software.

SECONDARY DATA

An efficient method of building a GIS database is to limit the amount of time and costs necessary to develop databases in the first place. Fortunately, digital databases are becoming increasingly available. Digital elevation models, digital orthophotoquads, and digital line graphs are available from the USGS as well as third-party vendors (see Appendix A). The U.S. Bureau of the Census has TIGER and Dual independent map encoding (DIME) files, as does its Canadian counterpart (see Appendix A). The U.S. Department of Agriculture makes soil maps available in digital form. There are, of course, many more examples. In fact, members of an increasingly active and visible group of digital data entrepreneurs promise to provide much needed data to GIS analysts. These organizations, whose ads appear in such commercial GIS magazines as *GIS World* and *GeoInfo Systems,* are filling a significant gap in the GIS infrastructure.

But the availability of databases also introduces other problems—including some you will also encounter as you input data to your GIS. We will examine these technical problems with a view toward how we, as potential database providers, might avoid them. The first is the physical format of the media. Countless hours can be spent trying to obtain digital data in the proper format. As we will see in the next chapter, there are many formats from 9-inch tape formats to 8-millimeter tape formats to CD-ROM technology, and the like. You must be able to get the data in a format compatible with your retrieval equipment. Although this may seem obvious, a vendor may provide data in a standard (unusable) format unless you are quite specific about your needs. Even if the media are the same, data can be provided for numerous formats. This need not lead us to a detailed discussion of data formats and data exchange standards (Moellering, 1992), but it should suggest that you be aware of the format your current system requires or produces. You will learn the technical details as you gain experience with your system.

A more insidious problem with external databases deals with the quality of the data. Although some third-party data vendors may provide easier access for data than could be obtained from government bodies, you need to be aware that the service may not supply data in the original format. Some data, no matter the source, will be filled with easily viewable errors, some systematic and correctable, some not. You need to be aware of the quality control procedures used by each vendor. In addition, you need to know what your options are concerning return of poor-quality data. Ask where the data were obtained. Were they created in-house by qualified professionals, or were they obtained from digital sweatshops, frequently operated by severely underpaid and untrained individuals? All these questions are vital to the utility of the data. Unfortunately, current standards and practices are generally quite low. Some vendors are unwilling to define their quality control procedures; others are unable to do so because such procedures are lacking. Even descriptions of the data themselves are often inadequate or wrong. As in any purchase, you should demand a complete accounting of what you are receiving. Details should include the specific format provided, the quality control procedures under which it was developed, the quality you can expect, the return policies of the vendor, and any other pertinent information that will help guarantee successful integration of the data into your GIS. Vendors that fail to comply with these requests should be considered suspect.

A major problem that is often encountered with the use of an external database is one you should also take to heart as you prepare your own databases. Databases require information about their own content; such **data dictionary** material amounts to a catalog of your data content. There are two general forms, active and passive. Passive data dictionaries might include scale, resolution, the names of the data fields in the database, the codes used, and what they mean. Imagine the GIS applications person obtaining from a vendor a database that includes a category called "wetlands." The definition may seem self-evident to you, but you need to know more about the vendor's criteria for establishing this category. Remember, one person's wetland may be another person's watered lawn. The data dictionary should provide enough detail to ensure that any analysis based on it will be valid. This, of course, should remind you to keep a clear and concise record of your operations in a form that will enable someone unfamiliar with your original database input procedures to recreate them.

Active data dictionaries operate on the GIS database by performing checks for correctly coded inquiries. For example, if your vector GIS database management system is set up to allow only a four-digit code for a particular entity, the active data dictionary could check each inquiry to determine whether this four-digit limitation is uniformly met. Such checks are quite useful for allowing proper functioning of the system and for preventing erroneous results from incorrect input requests.

Beyond their technical problems, external databases are accompanied by certain fundamental legal and institutional problems. We will look at these problems in more detail later on, but they need to be mentioned for completeness here. A major institutional problem is that external databases are often hard to find, especially if they were produced by government agencies that may be tasked with dissemination but not promotion and advertising. There are presently no major efforts to consolidate GIS database catalogs to facilitate searching, although more electronic newsnets and home pages are being established all the time. This is being done on a piecemeal basis, however, and ignorance of available databases often results in costly data redundancy. The cost of the data is also an institutional problem that may limit access. It is not the overall cost per unit that is prohibitive but the frequent practice of providing data in large units that cover far more than a user needs.

Among the thornier issues facing the GIS user today is the fairness of having to pay for data produced through public funds. Tied to this, of course, are the twofold problems of data access and data security (Dando, 1991; Davies, 1982; Rhind, 1992). Although many believe that public data should be readily accessible by the general public, the problem of sensitive data, such as the specific locations of endangered species or military storage facilities, makes the issue much less simple. Even when data can be obtained through the U.S. Freedom of Information Act, the time required to complete the legal documentation may exceed the life cycle of the GIS project for which the data are needed. These problems will not be solved easily, but you will encounter them as you continue on your journey in automated exploration.

METADATA AND METADATA STANDARDS

Although data dictionaries are useful, they are often not enough for the growing set of GIS users, particularly where data are to be shared by many users. Many federal, regional, and local government organizations are now adopting the need for a rigorous set of standards that provide a detailed record of the datasets that they use and share with others. Such **metadata**, defined as data about data, are a major extension of simple data dictionaries and are becoming a major component of the GIS data input process. In the United States, for example, the Federal Geographic Data Committee (FGDC) (1998) has adopted a set of metadata standards that extend to a wide array of possible data types and require those who intend to share or provide data to the federal government that they be included.

Metadata describe in some detail the content, quality, condition, and other important characteristics of the data, thus helping others to both locate and understand them. Major uses of metadata are as follows.

1. Organize and maintain an organization's investment in the data.

    In this case, as personnel change and time passes, the integrity of the data is maintained because the content and appropriate uses of the data are cataloged.

2. Provide information to data catalogs and clearinghouses.

    Among the most difficult tasks of the GIS professionals is to identify what data are available for their particular needs. This need has prompted the development of both private sector and government organizations that keep catalogs of metadata, essentially acting as an index of available datasets. The Federal Geographic Data Committee (FGDC), for example, is developing a network of geographic data sharing called the National Geospatial Data Clearinghouse through which the metadata can be shared.

This allows potential users to become aware of what data are available and how to obtain them.

3. Provide information to aid in data transfer.

Any time the actual datasets are to be shared with others it is vital that the metadata be shared as well. This allows the users to process and interpret data correctly and to match the data with its own data holdings.

Metadata describe many different aspects of the geospatial data and may require the use of software for their creation. Most of these software packages are either free or relatively expensive and can easily be found by searching the web under the keyword "metadata." As a GIS neophyte, one might easily ask, "What can be so difficult about metadata that I would need software to help me compile it?" To answer this question, let's first take a look at seven basic categories of metadata that should be included and then parse out the more general individual data that should be included in each (Federal Geographic Data Committee, 2001).

1. Identification

a. What is the database called?

b. Who developed it?

c. What geographic area is covered?

d. What categories or themes of data are included?

e. When were the data collected, and how current are they?

f. What, if any, restrictions are there for data access?

2. Data Quality

a. How good are the data? (This can be a very complex question.)

b. How do the users know if the data are applicable for their use?

c. What is the positional and attribute accuracy level?

d. How complete are the data?

e. Was data consistency verified?

f. What were the source data from which the existing data were created?

3. Spatial Data Organization

a. What data model was used to encode the spatial data (raster? vector?)?

b. How many spatial objects are there?

c. Are methods other than coordinates (e.g., street addresses) used to encode locations?

4. Spatial Reference

a. Are coordinate locations encoded using latitude/longitude?

b. What, if any, projection was used?

c. What geographic datums (both vertical and horizontal) were employed?

d. What parameters should be used to convert the data from one coordinate system to another to preserve integrity?

5. Entity and Attribute Information

a. What specific attribute information is included?

b. How is the information encoded?

c. Were codes used?

d. What do the codes mean?

6. Distribution

a. Where and from whom do I obtain the data?

b. What formats are available?

c. What media types are available?

d. Can I obtain the data online?

e. How much does it cost?

7. Metadata Reference

a. When were the data compiled?

b. Who compiled the data?

As you can see, each of these questions may comprise multiple individual questions themselves, thus making the task of developing metadata a much larger one than you might have originally anticipated. As it turns out, three supporting categories of metadata will also prove useful: (1) citation information with details of the originator (source) of the data and publication dates, (2) time period information indicating whether the data represent a single date or multiple dates, and what the range of dates is, and (3) more detailed contact information, including primary contact person (or organization), addresses, and phone numbers. This latter information is particularly important for the clearinghouses that are attempting to make GIS data accessible to others.

Terms

data dictionary	information theory	percent occurrence
differential rectification	inverse map projection	presence/absence method
dominant type method	linearity	primary data
fuzzy tolerance	metadata	acquisition
ground control points (GCPs)	most important type method	registration points
		repeatability

resolution	scale-dependent error	splines
root mean square error (RMS)	secondary data acquisition	stability
	skew	tic marks
rotation	spatial information product	translation
scale change		

Review Questions

1. What four characteristics are shared by every method of GIS data input?

2. What are the five factors, beyond cost, that should be examined in deciding on a digitizer? Define each.

3. What are the fundamental differences between line-following scanners and drum scanners? What potential problems relating to the map document itself act as sources of error for line-following scanners?

4. Converting between raster and vector data structures can produce results that degrade the quality of the initial input. What primary problem might occur during conversion from vector to raster? Where is this most likely to occur? What problem often arises during conversion from raster to vector?

5. Describe the transformation processes involved in moving from the digitized map to three-dimensional coordinates and finally to an output map. What is an inverse map projection? How does it relate to the map projection process?

6. Illustrate the processes of translation, rotation, and scale change? Why are these important to the input subsystem of a GIS?

7. Why is map preparation important to GIS data input? What are registration points or tick marks used for? Why are they needed?

8. Why should you mark your map before digitizing? What kinds of information should you include on your prepared map document? Why do you need to provide information about projection and grid system when the digitizer is set up for planar or Cartesian coordinates?

9. What is fuzzy tolerance? Why is it important? What difficulties can arise if it is set too low? too high?

10. What is the potential impact of the cartographic medium on the digitizing process? How does this relate to temperature and humidity? What can be done to reduce errors due to media distortion?

11. How do you decide what to input to the GIS? What is a spatial information product, and how does it relate to GIS input?

12. List and explain the seven rules determining what should be input to the GIS.

13. What is a good rule of thumb to help you determine how much to input? What do we mean when we say that digitizing is a sampling procedure?

14. What does data input have to do with information theory? What general rule does information theory suggest for data input? Suggest situations in which this general rule is less useful than one might suspect.

15. What is scale-dependent error? How does it relate to the input subsystem of a GIS?

16. Why do you need to read the software manuals in regard to vector data input? Isn't this just a simple matter of pointing to a point and pushing a button? What are the numbered buttons on the digitizer puck used for? What can you do while you are digitizing to reduce editing time later on?

17. What is so critical about attribute data input in the vector domain? What is the primary problem that can arise if it is not done carefully? Why is this situation so hard to detect later?

18. What are the four basic methods of raster data input? How do they differ? What are the advantages and disadvantages of each? Which is (are) best for point and line data? Which results in data explosion?

19. What are the technical problems involved in using aerial photography for GIS input? How about digital satellite data? What are ground control points, and why are they important in using digital satellite data? What institutional problems are involved in using remotely sensed data as GIS input?

20. What positive impact are external databases likely to have on the growth of the GIS industry? What are some of the major technical and institutional problems with using external databases? Why is the data dictionary or metadata so important? What is the difference between active and passive data dictionaries?

References

Berry, J.K., and C. D. Tomlin, 1984. *Geographic Information Analysis Workshop.* New Haven, CT: Yale School of Forestry.

Calkins, H., 1975. "Creating Large Digital Files from Mapped Data." In *Proceedings of the UNESCO Conference on Computer Mapping of Natural Resources,* Mexico City.

Cameron, E.A., 1982. "Manual Digitizing Systems." Paper presented to the ACSM/ASP National Meeting.

Chrisman, N.R., 1987. "Efficient Digitizing Through the Combination of Appropriate Hardware and Software for Error Detection and Editing." *International Journal of Geographical Information Systems,* 1:265–277.

Dando, L.P., 1991. "Open Records Law, GIS, and Copyright Protection: Life After Feist." In *URISA Proceedings,* pp. 1–17.

Davies, J. 1982. "Copyright and the Electronic Map." *Cartographic Journal,* 19:135–136.

DeMers, M.N., 1992. "Resolution Tolerance in an Automated Forest Land Evaluation Model." *Computers, Environment and Urban Systems,* 16:389–401.

DeMers, M.N., and P.F. Fisher, 1991. "Comparative Evolution of Statewide Geographic Information Systems in Ohio." *International Journal of Geographical Information Systems,* 5(4):469–485.

Federal Geographic Data Committee, 1998. Content Standard for Digital Geospatial Metadata (revised June 1998). Federal Geographic Data Committee, Washington, DC. [http://fgdc.gov/metadata/constant.html] page 85.

Federal Geographic Data Committee, 2001. Content Standard for Digital Geospatial Metadata Workbook Version 2.0. Federal Geographic Data Committee, Washington, DC.

Fisher, P.F., and M.N. DeMers, 1990. "The Institutional Context of GIS: A Model for Development." In *Proceedings of AUTOCARTO 9.* Falls Church, VA: ASPRS, pp. 775–780.

Haddad, K.D., 1992. "CoastWatch Change Analysis Program (C-CAP) Remote Sensing and GIS Protocols." In *Global Change and Education,* Vol. 1, ASPRS/ACSM 92 Technical Papers, Bethesda, MD, pp. 58–69.

Lauer, D.T., J.E. Estes, J.R. Jensen, and D.D. Greenlee, 1991. "Instititional Issues Affecting the Integration and Use of Remotely Sensed Data and Geographic Information Systems." *Photogrammetric Engineering and Remote Sensing,* 57(6):647–654.

Lillesand, T.M. and R.W. Kiefer, 1995. *Remote Sensing and Image Interpretation.* New York: John Wiley & Sons.

Marble, D., 1981. "Some Problems in the Integration of Remote Sensing and Geographic Information Systems." In *Proceedings of the LANDSAT '81 Conference,* Canberra, Australia.

Marble, D., J.P. Lauzon, and M. McGranaghan, 1990. "Development of a Conceptual Model of the Manual Digitizing Process." In *Introductory Readings in Geographic Information Systems,* D.J. Peuquet and D.F. Marble, Eds. London: Taylor & Francis, pp. 341–352.

Marble, D., and D. Peuquet, 1983. "Geographic Information Systems and Remote Sensing." In *The Manual of Remote Sensing,* 2nd ed. Falls Church, VA: American Society of Photogrammetry, Chapter 22.

Moellering, H., Ed., 1992. *Spatial Data Transfer Standards: Current International Status.* London: Elsevier Applied Science.

Rhind, D., 1992. "Data Access, Charging and Copyright and Their Implications for GIS." *International Journal of Geographical Information Systems,* 6(1):13–30.

Shannon, C.E., 1948. "The Mathematical Theory of Communication." *Bell System Technical Journal,* 27:379–423, 623–656.

Walker, R., Ed. 1993. *AGI Standards Committee GIS Dictionary.* Association for Geographic Information.

CHAPTER 6

Data Storage and Editing

The GIS storage and editing subsystem provides a variety of tools for storing and maintaining the digital representation of a study area. It also provides tools for examining each theme for mistakes that may have crept into its preparations. Before we can successfully use these tools, we need to know what these possible mistakes are and how they can be discovered and corrected. If we have been careful in our input, we should encounter relatively few errors. As we have seen, however, even the selection of an improper fuzzy tolerance level can produce errors. Many of these errors will not appear until the GIS has completed organizing complete themes or grids. In raster, for example, we may need to display each grid to isolate illogical or out-of-place grid cells as we compare them to the input documents. In vector systems, we may have to build in topology after the initial data input, so that we can pinpoint any polygons that don't completely connect, lines that end in the wrong place, or points that occur where they should not. In the case of entity–attribute agreement, we may need to output sample portions of our map for comparison against the original input material.

There are many aspects of error detection and correction. While you read through this chapter, you will encounter a set of terminology that will be useful not only for building your geographic language filter but also for identifying the appropriate techniques with which to find errors and make corrections. When you encounter each type of error, try to relate it back to Chapter 5 on input. Ask yourself how such errors could be avoided in the first place through proper planning and pre-input preparations. Compile a list of special techniques that will facilitate error detection and subsequent identification for each type of error you encounter. These will sometimes be standard approaches, or they may be tailored to your own needs, the types of data you most often input, and the mistakes you tend to make systematically. Perhaps you consistently tend to digitize your lines just short of where they should intersect with other entities, when you are working in vector. Or you may make errors when you input attribute codes because you type more slowly than you read. It is a good idea to practice your input methods on sample databases and keep track of the kinds of errors you make. This simple precaution can eliminate many errors before they occur because your awareness of your tendencies will make you more careful when you digitize.

When you are finished with this chapter you should be able to:

1. Know what tiling is and what its purpose is.

2. Understand the three basic types of error that can occur in GIS, how these errors are edited, and the importance of editing in GIS databases.

3. List six areas of entity error with specific examples and suggest how they might occur, how they are edited, and how they can be avoided.

4. Describe the types of attribute error in both vector and raster, suggesting how they might occur and how they are detected and corrected.

5. Describe the process of converting projections using a vector GIS.

6. Understand the idea of edge matching and why it is needed.

7. Describe the process of conflation, explain why it is needed, and discuss conceptually how it is executed.

8. Describe the process and the purpose of templating.

STORAGE OF GIS DATABASES

An analysis of the precise computer methods for storing GIS databases is well beyond the scope of this text, as are the ever increasing types of hardware technology used to record the data. The methods themselves are also highly dependent on the data model used in your system (refer to Chapter 4). Still, the storage portion of the storage and editing subsystem is worth mentioning, at least as it relates to the necessity of editing and updating databases.

In raster systems, the attribute values for the grid cells are the primary data stored in the computer, usually on a hard drive, whether you are using a UNIX-based workstation or a personal computer. The locations of each grid cell are catalogued by their positions relative to the order in which they are placed in columns and rows. In other words, their positions are relative to the locations of the other grid cells. For this reason, editing is primarily concerned with the correct relative positions of each grid cell. Some raster systems, as we saw in Chapter 4, use compact methods of storage such as run-length codes, block codes, raster chain codes, and quadtrees. To effectively examine the relative positions of individual grid cells, most often you must be able to retrieve the data from storage for display in a manner that allows each individual grid cell to be identified separately by column and row position as well as by attribute code.

If your raster system allows a linkage to a database management system, the matter becomes somewhat more complicated in that each grid cell has attached to it a number of different attribute codes. Depending on how this is done with your particular GIS, you may have to display and analyze each set of coverage attributes as a separate map. Others may provide you with the ability to list the attribute codes for each grid cell as you examine it. You need to become aware of your system's editing capabilities and approaches.

In vector, the entities and attributes are either stored as individual tables within a single database or as separate databases, linked by a series of pointers. The separation of entities and attributes requires you to look at the editing procedures applied to entities, attributes, and databases. You can retrieve the graphic entities separately and display them to identify missing objects, incomplete links, and polygons. By retrieving the attribute tables, you will be able to examine them apart from their linked entities to determine whether you have typos, incorrect code sequences, or even the wrong attributes in the wrong columns of your table. Finally, you will be able to retrieve part or all of your database (i.e., parts of the graphics and/or parts of the database) and to examine both the entities and the attributes for agreement. You will most often be able to isolate individual entities and display, on the same screen, the attribute values you desire.

Many vector GIS systems also allow you to separately store portions of your database as large, predefined subsections for archival purposes. This process, called **tiling**, is most often used to reduce the volume of data needed for analysis of extremely large databases (Figure 6.1). Say, for example, that you are creating a detailed database for an entire county. You may wish to divide the database into smaller tiles based perhaps on the coordinates of the individual maps (such as topographic sheets) that you used for input. Although tiling does not require that you use such a formal framework, many find it useful for the sake of keeping better track of the tiles that exist. In addition, some analyses may require only that you select a particular portion of the database to work on, upon entering the GIS. Tiling the portions individually means that you retrieve just the portion of the overall database with which you are going to work, thus reducing your computational overhead and increasing system response. Another important purpose of tiling is to allow a system administrator to have final control over the editing and updating process by permitting only certain sections of the database to be operated on when needed. Even when small portions are released for editing and updating, the system maintains an original copy of the

Original database   Tiled database

Figure 6.1 Database tiling. Selecting portions of the database to be saved as individual units saves space and allows for easier retrieval later on.

pre-edited database until the system administrator is satisfied that the updates and edits are correct. Thus, by limiting access to those who are qualified to make changes, corruption of an entire database can be prevented.

Not only does tiling provide needed data security and reduce the amount of data but also acts as a screening mechanism for the entire editing process. Most often the database is completely cleaned and edited prior to tiling and archiving, and as a rule access is gained for updating and analysis. This is not always the case, however, and you will need to select the appropriate tiles to do your editing. In some cases, you may need to perform the process called **edge matching**—operating on more than one tile at a time to ensure a correct match between the two tiles for entities that extend across the tile boundaries.

In general, today's GIS software, whether raster, vector, or quadtree, provides a visual display mechanism that will enhance your ability to visualize the errors. The exact methods will depend on the data model you are using and the sophistication of your system. Since most systems allow interactive editing within this visualization subsystem, it should be possible to correct errors as you detect each one. This is a far cry from the old days of CAC, in which a hard-copy output of each entity's coordinates had to be produced and compared against the written coordinates typed in originally. Still, despite the sophistication of modern GIS software and its ability to find some obvious errors, the process is not completely automatic. You will need to interact closely with the software to both detect and edit the coverages. This is all the more reason for ensuring that your map preparations prior to input are complete. The prepared documents will often be used as a form of truth set against which to evaluate the digital database.

THE IMPORTANCE OF EDITING THE GIS DATABASE

Although some errors might occur as a result of computational miscalculations and rounding error in the GIS software, most database errors result from improper input. Even with the most meticulous map preparation procedures, the finest equipment, and the best-trained input technicians, mistakes will happen. Causes include simply pushing the wrong button on the digitizer puck, a hand made shaky by fatigue, typing errors during attributes input, and even registration difficulties. Actually, the potential sources for error are numerous and include problems with the input documents themselves (Laurini, 1994). But the most annoying aspect of input errors is not their source; rather, since such mistakes are generally very small and extremely difficult to find even with the best GIS software, correcting them is time-consuming and costly. It is not unlikely that more time will be spent correcting even a small number of errors than was used to prepare your maps and input them to the system. Your instructor may provide an exercise in editing to acquaint you with the frustration of trying for an hour to find a single entity error, which can then be corrected. Such a lesson is a good reminder to prepare thoroughly before you input because editing is even more tedious than the input step.

We have discussed three general types of error. Let us repeat them now, so they will be prominent in your mind. The first type of error you will encounter deals primarily with vector systems and is called **entity error** (positional error). Entity error can take three different forms: missing entities, incorrectly placed entities, and disordered entities. We will discuss these in more detail later on. The second type of error is **attribute error**. Attribute error occurs in both vector and raster systems, often with the same frequency. Most often attributes are typed in, and the sheer volume of typing required for large databases often constitutes a major source of error. In vector systems, attribute errors include using the wrong code for an attribute, as well as misspellings, which make an attribute impossible to retrieve if a query uses the correct spelling. In raster, the input most often consists of attributes, so the result of typing a wrong code number or placing it in the wrong grid location is a map that displays these incorrectly coded grid cells in the wrong place. Such incorrectly placed attribute data comprise the third kind of error, **entity–attribute agreement error** (logical consistency), which also occurs in vector systems when correctly typed codes are attached to the wrong entities.

Of the three basic types of error found in GIS databases, the last two, both relating in some degree to attributes, are the most difficult to find. Mistyped attributes placed in correct locations (e.g., in a particular place in an external database) might be found if an active data dictionary is part of the system. This feature generally is helpful if you have violated a rule already established for the data dictionary, such as putting in numbers in locations requiring letters, or putting 80 percent of a five-digit code in a four-digit slot. However, misspelled attributes may not be found until you actually perform an analysis. Entity–attribute agreement error is often even more difficult to find than misspellings or incorrect codes. In raster, the only way you can observe problems of this type is to display the map to identify misplaced grid cells. With vector, you will most often be able to point to entities and display their attributes on your monitor. However, the GIS is not likely to be able to tell you that you have the wrong attribute attached to a particular entity. Instead, you will need to have a copy of your input map beside you as you display or highlight each entity.

If you created a very complex database, you may have to spend months evaluating each of the thousands of entities and making comparisons to your input document. It is far better to check these errors in small groups, as you input the database. For one thing, you will be more familiar with the data when you are inputting them than you will be if you go back later. In addition, the input document is already there in front of you. For this reason, some software vendors allow you to use the editing portion of the GIS as a method of input, rather than the input subsystem. Some vendors take an alternate approach and build the editing capabilities into the input subsystem. In either case, you can examine your map for entity as well as attribute errors and determine agreement problems as they occur. Although these steps will slow the input process, once again remember that it is much better to do it right the first time than to spend hours correcting mistakes after an entire coverage has been encoded.

Although you've heard this before, it won't hurt to hear it again. Error-prone data will contribute to error-prone analysis. And although simple errors may seem quite innocuous, even the simplest can produce results that are grossly incorrect. As a simple example, imagine a database containing over 8000 polygons, some of which illustrate the locations of highly toxic materials; a single polygon (let's say it's polygon number 2003) bears an incorrect attribute code indicating that the location contains no toxic material. In your analysis, you are searching for the polygons that correspond to the highest rates of cancer mortality in an area. As it turns out, the highest rate of cancer mortality on one coverage corresponds to polygon 2003 in your toxic substance coverage. Thus, although common sense would lead you to suspect a direct spatial correspondence between the highest cancer mortality and very high toxic substances, your analysis fails to demonstrate this. In the short term, your GIS analysis has produced incorrect results. In the long term, decisions concerning the cleanup of toxic substances for health reasons will likely be incorrect. Such incorrect decisions are among the topics covered in an ongoing discussion about the legal liability of GIS databases as tools for decision making (Epstein, 1989; Seipel, 1989). Although this hypothetical example may seem extreme, it should point up the constant possibility that minor errors in data can produce major errors from analysis. Fear of litigation may be as good a reason as any to spend the time necessary to ensure database integrity and accuracy.

DETECTING AND EDITING ERRORS OF DIFFERENT TYPES

As we have seen, a GIS database is subject to errors involving entity, attribute, and entity–attribute agreement. Although these three errors are distinctly different, in the following discussion we will look at entity errors first, considering attribute errors and entity–attribute agreement errors together in a second group. Most often attribute errors are detected and identified because the entities and attributes fail to agree. Although this is not always the case, the detection of pure attribute errors is most commonly performed by producing tabular listings of the attributes or their tables. Although this is certainly part of the process of error correction, a complete description is unnecessary. Examples for each type of GIS would take up a large portion of the book, so we will see how errors are detected for a common system or two, and you can modify your precise procedures on the basis of the GIS you use.

Entity Errors: Vector

Upon completion of the digitization, GIS systems require you to perform an operation that builds topology, unless this was part of the digitizing procedure itself. In either case, the topology, by providing explicit information about the relationships of the entities in your database, should permit you to identify the types of entity error in your digital maps. Some of these errors will be pointed out through text-based error flags telling you that you have a problem. Others must be interpreted by looking at database statistics concerning the numbers and types of entities or by inspecting the graphics displayed on the screen for error the GIS is not designed to detect. You will be looking for six general types of errors represented by the negative case of the following statements (Environmental Systems Research Institute, 1992):

1. All entities that should have been entered are present.
2. No extra entities have been digitized.

3. The entities are in the right place and are of the correct shape and size.

4. All entities that are supposed to be connected to each other are.

5. All polygons have only a single label point to identify them.

6. All entities are within the outside boundary identified with registration marks.

A major commercial GIS should be able to provide these general spatial relationships, and you can use them to identify errors. A useful procedure for comparing the entities you digitized and the original map document is to produce a monitor display or even a hard-copy plot of the pre-edited database. The latter will allow you to physically overlay the two maps using a standard light table. In addition, many GIS systems provide a number of symbols that indicate some errors. To save time, you should become familiar with these before you begin to edit. Let's now go through and identify specific types of error we can find related to the six general types.

As you remember from our discussion of vector data models, nodes are special points that indicate a link between lines composed of individual line segments. In such vector data models as POLYVRT and DIME, for example, the nodes are often described as "to nodes and from nodes," indicating the overall extent of the line feature. Nodes are not just points between line segments that show directional changes in the line; in addition, they have specific topological meaning. Nodes may be used to identify the existence of an intersection between two streets or a connection between a stream and a lake, but they should not occur at every line segment along a line or a polygon. Thus the first type of error that can be detected entails false nodes called **pseudo nodes**, which occur where a line connects with itself (an **island pseudo node**: Figure 6.2a) or where two lines intersect along a parallel path rather than crossing. Your GIS should be able to flag the existence of pseudo nodes by means of an easily identifiable graphic symbol. When you build your first GIS database, you may be perplexed by an abundance of pseudo nodes showing up on your coverage. Before you panic, you should be aware that some pseudo nodes are not errors but are merely flags indicating the presence of potential problems. A pseudo node connecting a line with itself may simply be the beginning and ending of an island polygon (sometimes called a **spa-**

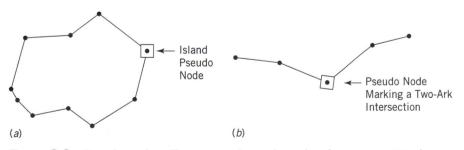

(a)   (b)

Figure 6.2   Pseudo nodes. Two types of pseudo nodes that are permitted are those that indicate the presence of an island, called island pseudo nodes (a), and those that indicate attribute changes along a line, such as a speed limit change (b). If, in (a) there isn't an island, this indicates an error and must be corrected. In (b), if there is no change in attribute along this line, then this pseudo node condition also needs to be corrected.

tial pseudo node), in which case its flag can be ignored. Or, for two line segments with an intervening node (also known as an **attribute pseudo node**), the node may indicate something as simple as a speed limit change (Figure 6.2*b*).

Spatial pseudo nodes that are not the result of your efforts to produce an island polygon (i.e., a polygon within a polygon) are most often due to a misplaced data point or to pushing the wrong button on the digitizing puck. In other words, either you were trying to create a nonclosing structure but placed the puck in the wrong place or you were trying to create a polygon that connected to other polygons (i.e., had neighbors that share line segments) but pushed the wrong button. As a means of avoiding improper spatial nodes, you can number your points when preparing your map or use a special code or symbol to indicate where your nodes are going to be. A good practice is to use a numerical code that is identical to the numbers on your digitizing puck that correspond to nodes. This procedure will also reduce the likelihood of producing an erroneous pseudo node for line entities.

If your software indicates that your coverage has one or more pseudo nodes, the prepared map document can be used to help you correct any mistakes. First, you need to determine whether the pseudo nodes are, in fact, errors. Correct pseudo nodes (i.e., those present for a particular purpose) can be ignored. Incorrect nodes can be corrected by either selecting each individually and deleting when necessary or by adding nodes where needed to convert to a polygon an island that is attached to other polygons. The process is generally easy to perform with commercial software.

Another common node error, called the **dangling node**, can be defined as a single node connected to a single line entity (Figure 6.3). In some GIS packages, you often need to have a From node and a To node rather than just a single

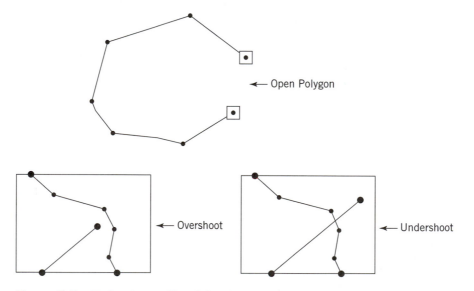

Figure 6.3 Node errors. Illegal dangling nodes come in three basic types: those that are a result of a failed polygon closure, those classified as undershoots because a node falls short of an object to which it is meant to be connected, and overshoots, where a node lies beyond an object to which it is supposed to be connected.   *Source:* Figure derived from Environmental Systems Research Institute, Inc. (ESRI) drawings.

node. Dangling nodes, sometimes just called dangles, can result from three possible mistakes: failure to close a polygon, failure to connect the node to the object it was supposed to be connected to (called an **undershoot**), or going beyond the entity you were supposed to connect to (called an **overshoot**). In some cases the problem is a result of incorrect placement of the digitizing puck, whereas in others the fault lies in a fuzzy tolerance distance not set large enough to account for the level of digitizing accuracy you are normally capable of producing. Setting the fuzzy tolerance properly is one way of avoiding this problem, and map preparation is another. It is generally easier to find overshoots than undershoots. If you tend to produce dangling nodes, a good practice is to overshoot rather than undershoot the line you are trying to connect. Although accurate digitizing would be even better, this method has proven useful for some who produce this error frequently.

Dangling node errors generally are identified by a graphic symbol different from the one used for pseudo nodes. In addition, if your dangles indicate an open polygon, the GIS will alert you by telling you the number of complete polygons in the database: if it differs from the count you had prepared prior to digitizing, you know you need to look for these dangles as incomplete polygons. Corrections are again quite simple. For undershoots, the node is identified and is moved or "snapped" to the object to which it should have been connected. Overshoot errors are corrected by identifying the intended line intersection point and "clipping" the line so that it connects where it is supposed to. In the case of an open polygon, you merely move one of the nodes to connect with the other. Most often the GIS will eliminate one of the nodes when this has been done.

As with pseudo nodes, some dangles are intentionally input to the GIS for a particular purpose. Most often these nodes serve as indicators of something important at the end of a line or arc. For example, you might use nodes to indicate the locations of residential cul-de-sacs (Figure 6.4). In even more unusual circumstances, a line used to indicate the location of a multistory building may contain numerous nodes, each one indicating the location of a separate floor. Although this abstraction deviates from the normal cartographic form, it illustrates the potential for dangling nodes to be legitimate objects within the database, rather than existing as errors.

Although you are digitizing polygons, you will need to indicate a point inside each polygon that will act as a locator for a label on which you will display text information about the polygon. You need only one label point per polygon, but

Cul-de-sacs

Figure 6.4   Acceptable dangling nodes. Legal dangling nodes created to indicate the existence of cul-de-sacs along a residential street.   *Source:* Figure derived from Environmental Systems Research Institute, Inc. (ESRI) drawings.

you do need it. Two types of errors can occur relating to label points in polygons: **missing labels** and **too many labels** (Figure 6.5). Both of these errors are most often caused by failure to keep track of the digitizing process. Although good map preparation will reduce the occurrence of label error, most often the problem is caused by confusion, disruption in the digitizing process, or fatigue. Fortunately, such errors are very easy to find. They are, once again, indicated by a graphic device distinguishing them from other error types. Editing is simply a matter of adding label points where necessary and deleting them where superfluous.

Another type of digitizing error most commonly occurs when the software uses a vector data model that treats each polygon as a separate entity. In such cases you are required to digitize the adjacent lines between polygons more than once. Failure to place the digitizing puck at exactly the correct location for each point along that line will often result in a series of tiny graphic polygons called **sliver polygons** (Figure 6.6). Sliver polygons can also occur as a result of overlay operations (see Chapter 12) or when each of two adjacent maps is input from a separate projection (see later section in this chapter). We will limit our current discussion to the sliver polygons produced through the input process.

The easiest way to avoid sliver polygons on input is to use a GIS that does not require digitizing the same line twice; in fact, this requirement is becoming less common. At times, however, you will accidentally digitize the same line twice. The result is the same: sliver polygons. The method of finding sliver polygons depends somewhat on whether you actually completed the adjacent line with nodes that are effectively placed on top of each other. If you digitized the same line twice by accident, you may also have a dangling node hanging around because an unneeded line has been created. In this case the line can be removed, eliminating the problem.

Finding slivers in the absence of a dangling node is more difficult. One way is to compare the number of polygons produced in your digital coverage with that of the original input map. It is often very difficult to locate slivers, however, even though you know they are there, somewhere. Most often you have to move through your image, searching for suspect polygon boundaries, then zoom in to

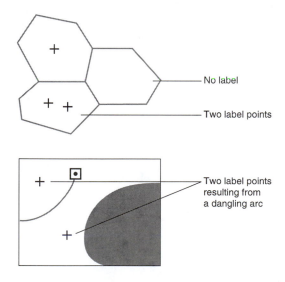

No label

Two label points

Two label points resulting from a dangling arc

Figure 6.5   Label errors. Inside each polygon there must be a single point, somewhere, to which attributes can be attached. Errors result when polygons have no labels or more than one label.

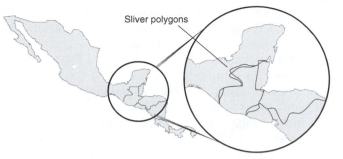

Figure 6.6 Sliver polygons. These result from poor digitizing along common boundaries where the line must be digitized more than once. Highly irregular national boundaries, as in Central America, are particularly vulnerable to such digitizing fuzziness.

see the slivers. In some cases you may have to zoom in several times to see them. You often don't know until you see the slivers whether you are zooming in on a single line or on sliver polygons. Sometimes when you have a series of very tiny polygons but no dangling node, it is possible to adjust (coarsen) the fuzzy tolerance you used during data input. If this measure is possible within the editing subsystem, the software will remove the slivers automatically.

A separate problem related to polygons is the production of **weird polygons**, which are defined as polygons with missing nodes (Figure 6.7). In this case, the polygon is a graphic artifact that appears to be a true polygon but is missing one or more nodes. Generally, this occurs when two or more lines cross over, producing the semblance of a polygon. The most frequent cause of this error is a point digitized in the wrong place or in the wrong order. For example, let's say you have a rectangular polygon that requires only four points to define it (Figure 6.7). You would want to start at the upper left-hand node, move to the upper right, then lower right, and lower left, and end at the upper left where you began. However, perhaps because you did not number these points during your map preparation procedures, you instead go from upper left to upper right, and then mistakenly go to the lower left and then the lower right (in a sort of figure-eight arrangement), finishing at the upper left where you had begun. Although

Figure 6.7 Weird polygons. Example of the creation of a weird polygon by digitizing points out of sequence. Although graphically we seem to have two polygons, the point at which the lines cross in the middle does not have a node.

you digitized all the points correctly, your polygon looks more like an hourglass than a rectangle. In effect, it appears like two triangular polygons connected at a central point. However, the central point has no node.

As you can surmise from the example of Figure 6.7, a simple way of avoiding this problem is to number the input points. Even if you don't, however, you can avoid the problem by establishing a set pattern for digitizing polygons. For example, you may decide always to digitize polygons in a clockwise fashion. Such a systematic procedure for digitizing will most often keep you from omitting essential nodes. As an aside, many users also take this systematic approach in deciding which portions of a map to digitize when multiple sessions will be required to complete the procedure. It is a good habit to adopt.

Detecting weird polygons is difficult but is not impossible. One straightforward method is to highlight the nodes and display them as part of the polygon coverage. In this way, the areas that should have nodes but don't will stand out from those that are properly digitized. Editing the error involves moving the lines to the correct locations, thereby placing the nodes in the correct sequence. Sometimes it is easier to simply remove the offending lines and use the editing subsystem to redigitize the points in the correct sequence.

The errors we have seen thus far are among the easiest vector mistakes to find; as a rule, you can make the necessary adjustments without plotting out your map to identify the errors. A more annoying group of entity problems is identified by numbers 1, 2, 3, and 6 in the preceding list. Problems of missing objects, additional objects, and displaced or misshapen objects are easiest to detect by plotting the digitized coverage at the same scale used for input (Figure 6.8). If you overlay the original document and the plot on a light table, you can identify problem areas. Nearly all these errors are due to lack of preparation, although the problems of disruptions and fatigue will always play a part in their occurrence. Correcting them requires you to mark the problem areas on the map document, preferably by noting the exact nature of the problem and how you will correct it. If you are missing an object, mark it by indicating, in order, the points, lines, and polygons that need to be digitized, including any other information concerning locations of nodes and other topological informa-

Missing arc

Figure 6.8   Missing arc. Illustration of the use of source map comparison to identify common graphical errors. Notice the missing arc in the digitized version of this map of Canada.

tion that may be needed. Additional objects should be marked for removal. For objects that extend beyond the map extent borders defined by your registration marks, the points need to be removed and redigitized. Objects that are misshapen or incorrectly placed but still within the map extent can most often be individually selected and moved, without being redigitized.

In all these cases, consult your GIS software manual to determine the exact commands needed to perform the operations. Because you are now aware of the typical graphical problems that can occur, you can easily find the appropriate instructions in your documentation. But before we leave the vector entity error procedures, there are a couple of caveats. First, remember that you have made changes to your entities, and the computer needs to understand that these changes may modify the spatial relationships you originally input. Most GIS software will require you to enter one or more commands to indicate that the changes have been made. The software will most likely also have to be instructed to rebuild the topological structures on the basis of the new entities. Finally, you must save and store your new map. Forgetting to save changes to a coverage representing hours of editing is a mistake you will not want to repeat.

Attribute Errors: Raster and Vector

As we have seen, attribute errors, including entity–attribute agreement, are among the most difficult to detect. This is primarily because the GIS does not know which attributes are correct and which are not. Because vector entity and raster grid attributes (and their assumed entity locations) differ significantly from application to application, and because there is no attribute equivalent of topology, there are no rules against which the GIS can check your accuracy. That is, there are no explicit rules to indicate that a particular attribute occurs with a particular, stable pattern with respect to its neighbors. If there were such rules, much of what we do in analytical operations of the GIS would be redundant. In fact, it is the search for these consistent patterns that often drives the analysis in the first place. Perhaps, with a few decades of research, we may be able to define some of these patterns but in the interim we are forced to compare the attributes in our digital database to the original map document to determine most of the attribute errors that might occur.

Missing attributes are perhaps the only attribute errors that are detectable without direct comparison to the input document. In raster these tend to occur as maps that are missing whole rows or columns or portions of rows or columns. These are detectable because familiarity with the original map shape will alert us to the absence of rows or columns of grid cells that substantially alter the overall shape (Figure 6.9). Missing rows and columns in raster are most often caused by missing one's place while typing the grid cell values; seldom will this problem occur when a digitizer is used. If you are typing in grid cell values and miss several of the attribute values, quite often an unusual value will appear in your map, one that has little correspondence to the rest of the data. This data value is probably a row indicator value that tells the GIS which row number you are on as you input the grid cell attribute values. From that point on, the grid cell values will be in the wrong place. A simple way to avoid this type of error is to display the text equivalent of the attributes prior to complet-

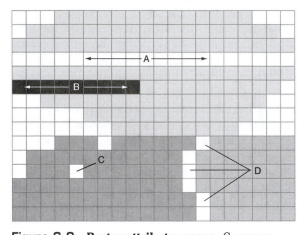

Figure 6.9   Raster attribute errors. Common raster attribute errors that are identified on the basis of how they distort the coverage: A, missing row; B, incorrect or misplaced attributes (appearing as one or more rows of vastly different values); C, incorrect attributes occurring singly; D, attributes errors occurring along area margins (most often caused by digitizing problems).

ing the input. This is very easy if you are using a word processor for creating the database, and you will be able to make the necessary corrections prior to database development. Alternatively, if you are using the software to input the values by typing them in, the GIS most likely will indicate the row numbers as you enter the data row by row. If you are typing in the codes without the assistance of the software, the potential for missing values is increased. By having someone read the data to you as you enter the values from the keyboard, however, you can focus on your typing, minimizing this and other types of attribute error as well.

In raster, if you miss a complete row of data, or more than one row of data, your map will appear shorter than it should. Visually, this discrepancy is more difficult to spot, but the software most likely will be able to tell you how many rows, columns, and even total grid cells the database contains. This response is easily checked against a known value for the complete database. The specific locations of such missing rows are relatively easy to detect if naturally contiguous patterns seem to stop abruptly along a given row. Most often there will appear to be a distinct line of demarcation between one portion of the map and another. Correcting the error is often a simple matter of downloading your grid data to a text processor and typing in the missing values, renumbering the rows as you go. Alternatively, you might be able to use the editing capabilities of the GIS to locate each grid cell and recode it. If the error occurs near the beginning or even the middle of the map, however, you will have to spend too much time editing to warrant using this method. It is best to use this method for correcting small numbers of grid cell values, or for correcting for missing rows that occur near the bottom of the map.

Missing attribute values in vector GIS are commonly caused by simply not including anything in the attribute tables for individual points, lines, or polygons.

These can be identified by listing the tabular information and identifying missing attribute values in the tables, or by outputting the entities and their attributes as a video display or plot. The missing attribute values simply will not be included next to their respective entities. Commonly, if you use the second method, you can edit each entity by selecting it from the rest and inputting the appropriate attribute values. And, of course, you need to remember to save your edited work.

Incorrect attribute values are very difficult to detect, both in raster and in vector. In raster GIS, when they occur as individual cases, or short row or column segments, they are most often caused by typing errors when that form of input is used. When wholly incorrect attributes occur for large areas, they are most likely a result of inputting the wrong attribute value, either through digitizing or a block encoding method. In raster, an incorrectly coded grid cell is most likely to be identifiable as one individual that seems "out of place" among the surrounding grid cells. Unusually high or low attribute values show up easily. They normally appear as out-of-place grid cells or groups of grid cells that disrupt the natural organization of the map. If they occur as continuous strips of incorrect grid cell attributes, most software will allow you to use a run-length encoding strategy to edit these interactively. Individual grid cell errors can be selected individually and edited.

In raster maps that have few contiguous areas (e.g., maps of raw topographic data), incorrect values will not easily be seen on a two-dimensional display of the data because there are no large area patterns to disrupt. In such cases, a three-dimensional display of the raw data may show unusually high spikes or very low dips in the surface. Although these abnormal features may be errors, be sure to check them out—they may be real attribute anomalies. Generally, such incorrect values occur as individual grid cells and can be selected and edited interactively as before.

Incorrect attributes in raster systems also may be found along the margins of the areal patterns. In such cases, the typical culprit is either unreliable digitizing algorithms that do not use one of the four basic grid cell input methods developed by Berry and Tomlin (1984) or carelessness in determining the correct attribute codes along these margins. The pitfall here is that the incorrect values are most often identical to the neighboring area, thereby giving the impression that they are correct. You will need to compare the shapes of area patterns on the raster map against the original shape of the input map. Correcting such a problem usually means reevaluating each grid cell as to its correct attribute. In other words, which of the two adjacent areas does this grid cell really belong to? Once the answer has been determined, each can be selectively edited as before.

Incorrect attributes are often more difficult to identify in vector than in raster because the vector case generally calls for intimate familiarity with the source map and its attributes and attribute patterns before these can be compared to the digital version. If you are using a form of coding strategy that replaces the actual names of the items by, for example, a numerical value, there are many chances to enter an incorrect number. In such cases, the codes will sometimes not correspond to linked tabular information included in other portions of your database or even in the data dictionary. The software should be able to flag such inconsistencies. Take, for example, a numerical coding strategy devised to represent the names of individual plant species for point entities in your coverage. Although this releases the user from having to spell out the species names exactly during query operations, it increases the likelihood of input errors. If you have established an appropriate set of rules in your data dictionary—especially if it is an active data dictionary—it should be possible to flag any code you enter that does not represent a real species name stored in the tables. Many times, however, the code sequences become ingrained in your subconscious, resulting in plant codes that are real. They just aren't the correct ones. The only way to prevent such errors is to double-check each code as you enter it. Identifying this type of error requires you to display the coverage and compare it to the original. The work is tedious, but you should be able to detect most of the errors. By selecting the offending entities, you can easily edit their attribute values as before, interactively.

A common source of the foregoing type of attribute error, however, is failure to keep track of the attributes as you are typing them in: most of the time the problem is not simply that the wrong attributes are typed in but that attributes are misplaced—that is, matched with the wrong entities. This mismatch is also a result of unconsciously using the wrong code, but we will treat it separately here, simply because it is distinctly different from placing a meaningless code in your tables. In many cases, these misplaced codes occur predictably. For example, you may tend to lose track of your place as you are typing codes in. This knowledge may provide a clue to finding these errors, particularly when the tables are printed and compared with the input data. Finding them with the use of graphical output is again a function of comparison to the input map. If the attributes were input using the label points on the entities themselves, this would be the best way to identify the errors. Alternatively, if your attribute tables were created separately and later linked to the attribute codes, systematic errors would be more easily detected by comparing tabular data from the database with the original data from which the tables were produced. The systematic errors that are most likely to occur involve short periods during which incorrect codes are entered (again primarily by reason of losing your place), and then, suddenly they begin to assume the correct order of entry. As before, a good way to avoid this problem is to get assistance while typing in the data. If your software permits you to input your attribute data interactively, you are likely to produce not large systematic attribute misplacements but only occasional ones.

DEALING WITH PROJECTION CHANGES

Although a major function of the storage and editing subsystem is the correction of entity and attribute errors, it is also used to convert between the Cartesian (digitizer) coordinates and real-world coordinates based on a reference globe. Quite often the software will require you to identify the projection of the map upon input. But some programs, especially in the case of many raster systems, do not. In any event, the conversion to a set of projected map coordinates is a necessity for vector systems for any analysis requiring real-world measurements. Moreover, because not all your input maps will have the same type of projection, it will be necessary to standardize them to permit comparison of the different maps.

If you had to input the reference points using Cartesian as well as projected coordinates, you would use these as reference points for the coordinate transformations necessary to project your map. If your system operates on real-world coordinate systems but has you input the data strictly in a Cartesian coordinate structure, you must define the reference points as latitude/longitude coordinates to make the projection possible. Another important factor is the method by which the GIS stores and manipulates these coordinates. Some use latitude/longitude coordinates recorded in **degrees**, **minutes**, **and seconds (DMS)**, whereas others will require you to convert these to **decimal degrees (DD)**. The formula for converting from DMS to DD, which gives the numbers in degrees or fractions of degrees (Environmental Systems Research Institute, 1992), is

$$degrees + minutes/60 + seconds/3600$$

In many cases you will operate first on the reference coordinates by creating a separate map layer with these values only, recorded in digitizer inches. The values can be read directly from the associated tables, then edited by typing in their longitude and latitude equivalents. The reference points are now in geographic coordinates. Upon completion of these edits and after double-checking for accuracy, you will save the data and link them to the geographic coordinate system (latitude/longitude). This tells the computer that the numbers you edited are defined by a real-world (geographic) projection. That is, you are telling the computer that you have changed the Cartesian coordinates to the geographic projection.

As you remember, the map from which you input your data will likely have been produced through map projection. You must get your coordinates, which are now in a geographic projection, into the same type of projection as the original input document. Usually only a few commands, or even a single command, will be necessary to set the GIS about the transformation procedures of scale change, rotation, and translation that will produce the projection required for your reference points. Thus far we have been working with the reference points as a separate coverage, rather than with subsequent thematic maps. Since, in addition, these transformations require mathematical manipulation of your original geographic coordinates, errors will always be part of the procedure. No projection procedure is without error. Many software packages provide some measure of this error, and you should look at these numbers. High amounts of error resulting from this transformation are often a result of inaccurate digitizing, map sheet distortion, or incorrect recording of original reference coordinates (Environmental Systems Research Institute, 1992).

You now have a map of reference points, or ticks, that represent the projection of the original map input. Upon completion, the reference points can easily be linked to the input coverage from which you extracted the reference points to create the reference coverage. Whenever you use the original coverage, it will be connected to the reference map, which also will have correctly transformed coordinates. Thus measurements can be produced in real-world units rather than in digitizer inches.

JOINING ADJACENT MAPS: EDGE MATCHING

Now that you have created a coverage that contains all the pertinent coordinate transformation information, you must consider a closely related topic in the storage and editing subsystem—edge matching. In edge matching, two adjacent maps, usually of the same theme, are physically linked to permit the analysis of a larger study area (Figure 6.10). There are two sources of difficulty when two adjacent maps are input.

First, two maps that were input with the same projection but because they were put in separately are likely to display entity errors that are somewhat different. Keep in mind that the maps are registered to the digitizing tablet separately, the tick marks or reference points were input separately, and all the entities were input during a separate digitizing session. Therefore, although each map may be reasonably accurate with respect to itself, the differences in input errors between the two will most likely cause mismatches—sometimes subtle, sometimes obvious—between and among individual objects. You will need to link all the line and polygon entities that are supposed to be connected. For example, if a road digitized on one map sheet is supposed to be a straight line that runs across the two sheets, make sure that when the sheets are connected, you don't have a jagged edge or a slightly offset road. Both portions of the road will have to be connected so that they exist as a straight line.

The second situation that is likely to cause problems requiring edge matching arises either when two adjacent maps are input from different projections (or with the same projection but based on different baselines or other starting points) or when the projection is applied specifically to that map layer without regard to its possible effects on the neighboring map layer.

Try taping any two adjacent USGS 7.5-minute quadrangle sheets together so that all line features are lined up. If they line up perfectly (a very unusual case), you are very fortunate, and you will only have to worry about errors introduced through digitizing. These errors will be minor if you prepared your map well and were careful during input. If the sheets do not line up perfectly, it is because the projection was applied to a single sheet at a time. You might still ask why the edges don't match up. The answer is simply that the mathematical procedures of map projection are imprecise, both because they are approximations of the three-dimensional reference globe and because of rounding error inside the computer.

Let's take a more obvious example. A match between two maps of the same area but produced with different projections is impossible. As any GIS vendor will tell you, however, the software can project the data into any projection you want. That is another statement that ignores the two sources of imprecision:

(a)   (b)

Figure 6.10 Edge matching. Two adjacent parts of a map (a) before and (b) after edge matching.

projection errors and computational rounding. Often you will find that the two maps, even when projected to the identical projection inside the GIS, will not completely line up. You may even see the Frankenstein effect mentioned in connection with the digitizing of adjacent polygons. After all, if the two maps have polygons, they too will be offset from each other.

Edge-matching problems can also occur in raster systems, at least for those that operate on projected surfaces, rather than simply in flat, Cartesian space. A common example of edge matching in raster systems is found in the use of remotely sensed data products, such as LANDSAT MSS (80-meter) images. Because horizontally adjacent scenes are sensed at different times, often days apart (Lillesand and Kiefer, 1995), the satellite may not be located at exactly the same latitudinal coordinates. This often results in a skew between the two images of one to several pixels (grid cells). Most often this is corrected by moving one of the datasets until the pixels match. Finding these errors is not unlike our example for finding missing lines from raster data input. Even very complex satellite images usually contain areas that produce relatively similar responses that, when classified, would often correspond to a single area class. When this group of pixels is offset, even before classification, the condition is relatively easy to detect because the regular patterns themselves show the incongruity along this line. In addition, geographic coordinates are most often provided with such data, allowing the user to match the edges by matching the coordinates along both adjacent images.

CONFLATIONS

There are frequent circumstances, particularly in vector, when some content of two digital source maps needs to be combined. This may be required to make a composite map that is superior to the originals. You may, for example, have a map showing hydrology that is extremely accurate, but you also have a map of vegetation where you wish to compare the vegetation along the river to the river. This requires you to combine both maps to create a map of riparian vegetation. You may also wish to compare multiple maps when you have multiple dates for the same region, but each map was digitized from a different set of aerial photographs. Other circumstances include the consolidation of multiple thematic datasets within a single organization, or the addition of features to an existing map. All of these require a complex process called **conflation** (sometimes called **rubber sheeting**, although this terminology is most often applied to the edge mapping problem) to combine these datasets with an acceptable level of accuracy.

The process of conflation involves identifying features within one reference map that most reasonably correspond to accurate locations of real earth objects that are to be combined with one or more target maps. Next, one selects control points within the maps, and their locations are reconciled with the selected objects. Some portions of the map will move a great deal, whereas others may not move much or at all (Figure 6.11). In many cases the process may have to be repeated to guarantee that the "best" possible outcome is obtained. Each iteration may require the selection of additional control points, removal of others, and an overall evaluation by the user. This suggests that the process has not yet been fully automated, nor is it likely to become so in the near future.

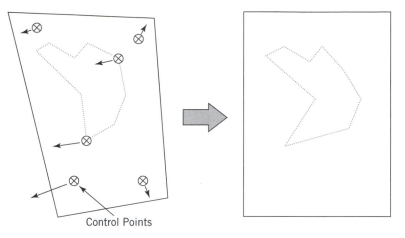

Control Points

Figure 6.11   Conflation. The use of conflation to move objects in a map: some entities are tacked down to prevent them from moving, whereas the others are shifted to adjust to the same coordinates as the base data layer.

TEMPLATING

In the preceding section, we mentioned the use of a single coverage as the most accurate among several for the same area. When you look at multiple maps of the same theme for different time frames, you notice a number of graphical discrepancies, including one we have thus far neglected. Viewing the same maps simultaneously, you also notice that, despite all your efforts to prevent it, the outside dimensions of all four study areas seem to differ slightly in shape. When you input these maps, you chose certain data points as reference points and assigned them to true geographic coordinates, yet the four maps are not identical. Probably, then, the differences in location of these reference points from coverage to coverage, combined with the nuances of the projection algorithms and computer rounding error, have produced slightly different results for each coverage.

If you must later perform an overlay of these four maps, there will be numerous areas along the margins of some maps that will not have associated areas for other maps. Once again, you must select the coverage you trust most to be representative and use it as a **template** (some call it a cookie cutter). If the boundary of the template is within the boundaries of all the other maps, you simply use this pattern to cut out the study area of the rest. However, if any of its boundaries extend beyond those of another map, you will need to select coordinates somewhat inside the margins of the template to ensure that all subsequent maps will occupy all areas within the template. Once applied, all the coverages should have the same shape, coordinates, and size.

A note here about the statistics reported from multiple themes is important. It is entirely possible that once you have applied a template to all your themes, then produced output statistics, there will be slight differences in total area for each of the four. This was the result in a study of land use change in which every attempt to remove the discrepancy failed (Simpson et al., 1994). Much of the incongruence can be attributed to a combination of computational rounding and the algorithmic methods by which the GIS calculates area. For the most part, if the error is a minor percentage of the overall database, it should simply be accepted. No amount of editing is going to remove it. Your task is complete, and you are now ready to begin analysis.

Terms

attribute error	entity error	spatial pseudo node
attribute pseudo node	incorrect attribute values	template
conflation	island pseudo node	tiling
dangling node	missing attribute	too many labels
decimal degrees (DD)	missing labels	undershoot
degrees, minutes, and seconds (DMS)	overshoot	weird polygons
edge matching	pseudo nodes	
entity–attribute agreement error	rubber sheeting	
	sliver polygons	

Review Questions

1. What is the purpose of tiling in the storage and editing subsystem?

2. What is so important about editing a database? What problems can occur as a result of even a simple error in the database? Give an example other than one from the text.

3. What are the three basic types of error that need to be edited? Describe each. Which of these is (are) the most difficult to find? Why?

4. What are the six major areas of entity error that need to be addressed? Give an example of each.

5. What are pseudo nodes? How are they produced? How can they be avoided? Give an example or two of a pseudo node that is not an error.

6. What are dangling nodes? overshoots? undershoots? How are these caused? What can you do to avoid them?

7. Why is a missing polygon label point a problem? Why do we sometimes end up with missing or multiple label points? How can this result be avoided?

8. What are sliver polygons? weird polygons? Describe methods of avoiding and correcting these conditions.

9. How do you find missing attributes in vector? raster? How can you avoid creating a map that is incomplete in this respect?

10. How do you find incorrect or displaced attributes in vector? raster? What are the principal causes of each?

11. Describe the process of converting projections using a vector GIS.

12. What is edge matching? Why is it needed?

13. What is rubber sheeting (conflation)? Why is it needed? How is it done?

14. What is templating? What is it used for? How do you decide which coverage to use as the template?

References

Berry, J.K., and C.D. Tomlin, 1984. *Geographic Information Analysis Workshop Workbook.* New Haven, CT: Yale School of Forestry.

Environmental Systems Research Institute, 1992. *Understanding GIS: The ARC/INFO Method.* ESR I: Redlands, CA.

Epstein, E.E., 1989. "Development of Spatial Information Systems in Public Agencies." *Computers, Environment and Urban Systems,* 13(3):141–154.

Laurini, R., 1994. "Multi-Source Updating and Fusion of Geographic Databases." *Computers, Environment and Urban Systems,* 18(4):243–256.

Lillesand, T.M., and R.W. Kiefer, 1995. *Remote Sensing and Image Interpretation.* New York: John Wiley & Sons.

Seipel, P., 1989. "Legal Aspects of Information Technology." *Computers, Environment and Urban Systems,* 13(3):201–205.

Simpson, J.W., R.E.J. Boerner, M.N. DeMers, L.A. Berns, F.J. Artigas, and A. Silva, 1994. "Forty-eight Years of Landscape Change on Two Contiguous Ohio Landscapes." *Landscape Ecology,* 9(4):261–270.

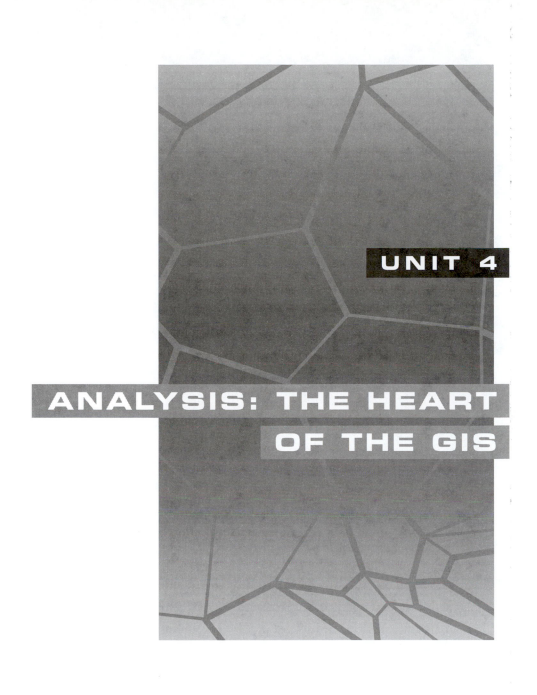

UNIT 4

**ANALYSIS: THE HEART
OF THE GIS**

CHAPTER 7

Elementary Spatial Analysis

INTRODUCTION TO GIS SPATIAL ANALYSIS

The analysis subsystem is the heart of the GIS: it is what the GIS is all about. It is also the most abused subsystem of the GIS. The abuses range from attempts to compare noncomparable nominal spatial data with highly accurate ratio spatial data to statements about the causative nature of spatially corresponding phenomena made without testing for alternative causes. Much of the abuse of the analytical subsystem results from a lack of understanding the nature of the spatial data contained in the system. For example, some data are purposely ranked at the ordinal level, primarily on the basis of their perceived importance for a particular problem. Then, the same ordinal data are used for other, nonrelated analyses for which the ranking no longer applies beyond the original spectrum of values from which they were produced. This is not unlike correlating history grades to calculus grades to compare a student's intellectual capabilities. In addition, it is not uncommon for the numerical values stored in a grid-based GIS, and meant to represent nominal categories such as land use type, to be multiplied or divided by ordinal, interval, or ratio data, yielding numerical values that essentially have no meaning. Yet the results of such analysis are frequently used for decision making. The list of possible offenses is quite large, mainly because the inherent power of the GIS is too great to be effectively applied without a firm understanding of fundamental geographic, numerical, and statistical concepts.

Beyond the problems of GIS abuse, there is a common belief that GIS is the panacea for all geographic problem solving. Although the tool is quite powerful, GIS is generally driven more by market demands than by computational power. In short, GIS is an incomplete set of spatial analytical tools. In many cases, the user is obliged to combine GIS tools with other computational tools in order to accomplish his modeling tasks. Fortunately, there are many types of GIS software, each with its own strengths and weaknesses. The limitations of the GIS you choose should be weighed against the kinds of analysis you most often perform. You should therefore be aware of the myriad possibilities available in the larger field of digital geography rather than being versed in a particular package with no analysis background.

171

Most GIS packages are heavily oriented toward the ability to overlay two or more maps to analyze corresponding patterns or to operate on digital data obtained through remote sensing systems. Although these two tasks are a small portion of the potential of a GIS, they do illustrate the market-driven nature of the GIS analysis capabilities. But the line between these two types of systems is becoming less well defined through time. It is up to you as a GIS analyst to recognize the potential capabilities because such awareness will provide you with the conceptual framework to operate on the largest "superset" of geographic analysis capabilities you have at hand.

In this and the following six chapters the framework of geographic ideas is limited to those that deal fundamentally and explicitly with maps. Although there are far more techniques that allow analysis of mapped data implicitly, these are more appropriately studied in courses in geographic analysis, statistical analysis, systems modeling, and geostatistics.

LEARNING OBJECTIVES

When you are finished with this chapter you should be able to:

1. Create and describe a simple GIS flowchart.

2. Discuss why the analysis subsystem of the GIS is often the one most abused and why it is considered an incomplete set of geographic analysis tools.

3. Understand the importance of the analysis subsystem even for those who will primarily be building databases for others.

4. Explain the process and the importance of isolating, identifying, counting, and separately tabulating and displaying individual items.

5. Describe the differences in process for isolating, counting, and identifying the different object types (points, lines, areas) in a database.

6. Explain the use of measurable attributes in searching for lines and areas.

7. Define *higher-level objects,* giving examples for points, lines, and areas, and describing their utility.

8. Describe the approach to identifying higher-level objects.

9. Be able to define and describe the different types of region.

10. Suggest how collections of points, lines, and areas could be considered to be communities.

PRELIMINARY NOTES ABOUT FLOWCHARTING

Before we describe the possible things a GIS can do, it is important to provide you with a preliminary introduction to flowcharting. We will develop the flowchart idea more fully in Chapter 13, but the ideas of GIS analysis and GIS model flowcharting are so central to the analysis subsystem that many software pack-

Figure 7.1   GIS flowchart. This is the general form of most GIS flowcharts. Although the specifics and symbols will change, virtually every GIS model begins with input, performs some operation (with or without conditions), and produces some input.

ages now incorporate some form of flowcharting utility (cf. ESRI's Model Builder, ERDAS's GIS Modeler, and the flowcharting capabilities of IDRISI). Given the emphasis that the GIS software vendors are placing on flowcharting, and their recognition of its ability to systematize our thinking and documenting our procedures, we should begin thinking about them immediately. In this way we will get a chance to flowchart simple analysis tasks and become comfortable with the idea.

All GIS tasks, no matter how simple, require us to perform at least one identifiable operation, typically on at least one thematic layer as a model input to produce some form of output. This provides us with the general form of a GIS model flowchart (Figure 7.1). As you can see from this simple diagram, we have an input (typically a map), an operation (a search, a measurement, or other GIS task), and an output of some kind. For the sake of simplicity it is assumed that the output is a map in the figure. Notice how the flowchart shows both the objects and operations and a flow direction. This type of flowchart is called an **implementation flowchart** because it shows how the model actually works. As our modeling expertise grows, we will need to create a flowchart that flows in the other direction first so we can move from expected outcomes to the basic data elements on which the model is built. This type of flowchart is called a **formulation flowchart** because it is used to design the modeling tasks. As you will see in Chapter 13, it is frequently useful to begin modeling with this type of flowchart to help us organize our thinking. It is usually easier in complex modeling to decide what our desired outcome is before we know what the elements that we need to obtain that output are. For the time being, you will see examples of implementation flowcharts to illustrate how the portions of a fully developed GIS model flowchart are created. Examples of simple implementation flowcharts will be included for selected analytical tasks so that you can begin flowcharting now. The general form described above will be used for these examples, but the exact terminology will change based on the command names and modeling approaches of your software employs. It would be useful to try to modify this general flowchart form to conform to whatever software you are using.

GIS DATA QUERY

Each digital database typically contains many different maps, each possessing a separate theme. Within each theme, as within each focused field study, you are likely to be trying to isolate numerous features for study. You will often want to know which of the selected features occur most often, how often they occur, and where they are located. Let us say, for example, that you are interested in the number of individual trees (point features) located inside your coverage. Because these are point data, you are going to be counting objects that conceptually occupy no spatial dimension. A raster GIS offers numerous ways to find these items, but the simplest is to create a new coverage that eliminates all unnecessary data. To do this you mask all other portions of the database through a simple reclassification process (see Chapter 9), classifying all the trees of the selected species as your target points and reclassifying everything else in your coverage as background. Most raster GIS packages will then allow you to output the results in a table that allows you to count the target points (grid cells) and the background points (grid cells) (Figure 7.2, Table 7.1). Generally, this will also allow you to output percentages directly as well, making possible comparisons of amounts of the coverage occupied by the selected tree species. The trees are only located within the grid cell, not wholly occupying it, so the percentage of coverage occupied by these target cells is rather skewed.

Identification of the absolute locations of point data in a raster GIS is, if you will forgive the pun, a pointless task because the grid cells divide up or quantize space into spatially uniform packets. Still, you can easily identify the grid cell locations, most often by using an onscreen cursor device that allows you to point to each grid cell individually. This usually results in a readout giving the row and column numbers, as well as the attribute codes for the item selected or pointed to. Even in raster, knowledge of the location of individual points is important. For example, you might be comparing the locations of trees to grid locations from remotely sensed data as a means of testing the sensor's ability to recognize the presence of trees at that location based on the electromagnetic properties returned to the sensor from that cell.

In raster, GIS lines are merely collections of grid cells that touch one another either along each side or diagonally in Cartesian space. Polygons are groups of grid cells that are connected in much the same way, or share attribute values (called **regions**). Because of the way attributes are linked in the simplest raster

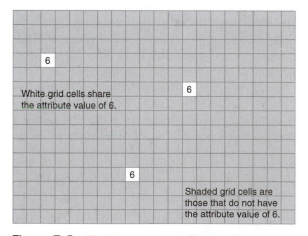

Figure 7.2 Finding raster attributes. Raster geographic information system (GIS) showing the isolation of points by attribute. Each grid cell is identified by a color or shading pattern that represents its unique attribute values.

**TABLE 7.1   Listing of Grid Cells
Chosen by Attribute Value Codes**

Category	Attribute Value	Cells[a]
Not coded	0	225
Prime land	1	642
Urban	2	201
Federal land	3	188
Grazing	4	981
Water	5	64

[a]Number of cells of each type found in the database.

GIS software, to identify them as entities you must identify them by attribute, just as we did earlier when we isolated point data. And, as before, you can use a cursor to identify column and row locations and even attributes for each location. Tabulation of the results will again show the amount of each category and the percentages. Again, you need to keep in mind that for lines, percentages of the total database occupied by the various objects are most likely meaningless. For areas, the accuracy of the percentages is largely determined by the method used to encode the grid cell categories.

In vector systems, the process of finding points is performed by displaying all points in the cartographic portion of the database. If you need to find points of a particular kind (e.g., points that have attributes indicating that these are telephone poles, bird nesting sites, etc.), you will need to access the attribute database and its tables (Table 7.2). Most often you will perform a search that identifies all table locations that exhibit the appropriate code for the objects you wish to see. Because these are linked to the entities, you will be able to selectively isolate these specific types of point objects. Because vector data structures contain explicit spatial information, you can easily obtain the exact coordinates in projected space. These can be produced as tabular output or viewed directly on the screen by means of a pointing operation. In some cases you will be able to point to individual items, and in others you can surround several with a graphic window that will give the desired information for all items inside it. As with raster systems, you can also obtain attribute and coordinate information.

Because explicit coordinates are used in vector data structures, you can also obtain entity information about lines and polygons by selecting them individually. For example, you can identify all points in the coverage by simply displaying only points, or you can obtain a listing by accessing the attribute tables specific for points. This is true for lines of all kinds as well as for polygons. It is no difficult matter to identify all the polygons in a map, plus their attributes. To identify the absolute coordinates for one- and two-dimensional objects (i.e., for lines and areas), you will need multiple sets of coordinate pairs for each of the points that are used to identify all the line segments with which they are defined. Again, this can be done by pointing to individual objects or by surrounding them with a graphic window, depending on the software. And, as with point

TABLE 7.2   Listing of Entities Chosen by Attribute Value in a Vector System[a]

Landuse Code	Name	Number of Polygons
100	Row crops	21
110	Fallow	18
120	Grains	65
130	Grazing	3
200	Residential	982
210	Commercial	124
220	Light industry	192
230	Heavy industry	54

Landuse Code	Name	Polygon ID
130	Grazing	25
130	Grazing	28
130	Grazing	29

[a]The lower half of this table shows a list of polygons selected from the top half. Notice that three polygons matched landuse code 130.

data, any line or polygonal object can also be selected on the basis of a database query of the attribute tables.

Locating and Identifying Spatial Objects

Although nearly any GIS can be subjected to find or locate queries, before we go any further we might ask why we need to be able to find and locate objects in our GIS database. The twofold answer is blatantly obvious (because it is a fundamental process comparable to traditional map reading) and yet subtly obscure (because of its relationship to more complex calculations). In the first case, it is important to be able to isolate, count, and locate objects because these activities give us an understanding of the overall complexity of our maps. The objects in the map represent features on the ground. On the basis of our traditional communication paradigm, one primary purpose of the map is to display these objects, to facilitate the identification of the spatial relationships between and among themselves as well as to other objects on the landscape. The more objects of a particular type occur in the coverage, the greater their density in the real world. It is, for example, important to know whether there are many houses or just a few in a particular region. These numbers often relate directly to population densities, which may, in turn, be vitally important in determining the risk of a population to hazardous materials spills, flood events, earthquakes, and other disasters, to cite just a few examples. Similarly, the number of plants or animals in a theme may very well be related to the environmental health of the region. Or the number of road networks might provide information important for routing sales staff or truck traffic. By extension, knowledge of the relative absence of these objects or features may be as useful under given circumstances. In other words, a quantitative measure of the amounts of these objects will allow us to make direct, analytical comparisons to other coverages or other variables in the same coverage. The actual existence of objects and their locations and distributions is often very important. Comparisons to other features on the landscape might be used to determine causes for distributional patterns, or at least to suggest strong spatial relationships among cartographic variables. For example, an enumeration of the number of houses may be useful for examining their relationship to the amount of land available for new housing. Areas on a map with low numbers of housing units can be compared to a map layer of available land for house construction through a process called **overlay**, which we will cover in Chapter 12.

But the simple enumeration of objects and their locations also lets you examine their relationships to more prominent objects in the same map. Landscape ecologists for example, are interested in the relationships between small, isolated polygons, which they call patches, to what they call the matrix, or overall background of the study area (Figure 7.3). By knowing how many patches occur in a study area and how much area they occupy in the database landscape, ecologists obtain a measure of patchiness or fragmentation that is taken to be one indicator of overall landscape health. There are applications in economics as well. The locations of shopping centers, for example, may be highly fragmented, requiring additional roads and public utility services to afford access to isolated polygons of retail activity.

As implied by the last two examples, among the most important aspects of being able to find and locate objects on the map is the ability to make further measurements and comparisons. From absolute numbers we can make extensions to relative numbers. For example, we can move from the total number of houses, roads, or polygons of a particular land use to relative numbers of houses per unit area, the relative number of road miles per square mile, or the relative number of polygons of land use type 1 to that of land use type 2. Each set of measurements or comparisons provides additional insights into our environment. But all require the initial ability to isolate, count, and locate the individual entities within the database.

Figure 7.3   Elements of a river landscape showing upland regions and a patch and matrix configuration within the stream banks. Modern geographic information system (GIS) software can evaluate such maps to derive fragmentation, patchiness, and connectivity measures. To do this, the patches must be able to be isolated from the rest of the map elements.

DEFINING SPATIAL CHARACTERISTICS

Simply being able to find point, line, and area entities on a map is of little value. Most of the objects analysts find, count, and locate in coordinate space are selected not on the basis of entity type alone but rather on what they represent in the real world. Just as we would not simply record points as points, lines as lines, and areas as areas in a field survey of an area, we likewise do not keep them stored as sterile entities in our GIS database. Instead, as we would in the field, we make a note of the types of point, line, and area objects, the amounts or values of each, and their categories, because our greatest interest is in their attributes or descriptors. For the same reason, we most often search for objects, count their numbers, and note their locations on the basis of these attributes. We will examine individually the three entity types (points, lines, and areas) on the basis of how we would search a database for their attributes and the insights we might gain from doing so. We will examine surfaces separately (see Chapter 10) because they are stored in the GIS in a fundamentally different way.

Point Attributes

Point objects, like all entities, differ not only in their locations but also in their attribute characteristics. Trees are different from houses, which are different from cars, which are different from disease locations, and so on. These differences provide us with different but often related spatial patterns of each group of objects. Point objects can differ by nominal type; for example, we can isolate patients in a region that share diseases related to specific organ groups (Figure 7.4). These types may also be ranked within type. For example, digestive diseases could be ranked as mild, moderate, or severe. This would allow the user

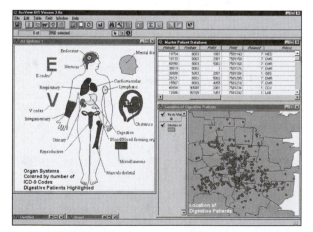

Figure 7.4 **Body Viewer is a classic example of how point objects can be selected by type.** In this case, selection of a graphic image of a human organ or system allows display of a linked table of disease types associated with those organs.

to identify patterns of distribution of such phenomena and to determine whether there is some reasonable cause for this pattern. We will look at pattern in more detail in Chapter 11.

If we can classify point objects by nominal and ordinal categories, we can also separate out these objects by value in interval or ratio scales as well as by type. For example, houses could be selected on the basis of market value: houses that are below $50,000, between $50,001 and $100,000, and so on. These values would have to be stored in the database to permit retrieval of the corresponding points. Other examples might include the annual amount of sales for individual businesses, the amount of nitrogen fixation for selected plants, and the shoulder heights of deer located through telemetry. These quantitative measures will allow us to select a wide range of groups or classes of each object, depending on our needs.

It is vitally important that a functioning GIS be able to define each group or category separately and tabulate the results of a search. In addition, we need to be able to produce graphic coverage of the categories or values of point variables of interest, and only those categories or values. In other words, we need to be able to isolate these groups of point objects from those that are not important to our output or analysis objectives. Because GIS operates in a holistic cartographic paradigm, we should be able to retrieve all the raw data and then group the results in any fashion that suits our purpose.

Beyond simply isolating categories of data, the GIS must allow us to locate each item individually for each class of data and compare it with others of its own kind. That is, we need to be able to show the spatial relationships between individual items of a selected point object in a class of objects so that we can later perform analytical operations to quantify those relationships. Are all houses costing under $50,000 located within a given distance of one another? Are they distributed evenly or randomly? Are they clustered or dispersed in their distribution?

If we need to show spatial relationships between like-kind or like-value objects, it stands to reason that eventually we will need to compare them to other similar point objects of a different type or value. We will need to be able to show and later quantify the relationships, for example, between houses selling for $50,000 or less to those costing $50,001 and over. But even more, we are likely to want to show the relationships among point objects of different types, say houses and streetlights. It follows that if we need to be able to compare the numbers and spatial locations of one group of point objects to another, we will soon need to show the relationships among entities of different types. For example, there may come a time when we will need to know the relationships between houses and available paved roads, sewerage lines, water and electric utilities, parks, and shopping centers. As we will soon see, each of these other entity types will also need to be separable on the basis of type or magnitude.

Before we leave our discussion of points, it is important to note that although most GIS software will allow us to isolate, select, quantify, and locate point objects, many, perhaps most, GIS packages are not specifically designed for determining the spatial relationships among these objects. Still, even though the software may not allow us to perform advanced spatial analytical techniques on these objects, we should be able to perform at least these basic functions; then we can use alternative analytical software to learn more about the spatial relationships that exist. Your GIS, at a minimum, should allow you to separate out the items you wish, and only those items, accompanied by their attributes and locational coordinates. Once these have been selected, you should also be able to produce a separate coverage of what you have chosen, to permit further analysis on them either within the GIS or using separate software.

Line Attributes

Recall that line objects are one-dimensional entities defined by two or more points with corresponding coordinate pairs. The line object can also contain nodes that are points specifically indicating either the beginning or the ending of a line, or some change in attribute along the line. As with point objects, line entities can be identified by attributes with varying levels of data measurement (i.e., nominal, ordinal, interval, or ratio). Examples of line item types include railroads, streets, fault lines, fencerows, or streams. Each of these is separable from the others because all differ in kind (Figure 7.5). In other words, each different object should be able to be identified, retrieved, and located separately. And they should also be able to be tabulated and displayed separately to identify the unique pattern each exhibits on the landscape.

Like point objects, lines should be separable based on ordinal rankings or some measure of magnitude. Highway types such as single lane, double lane,

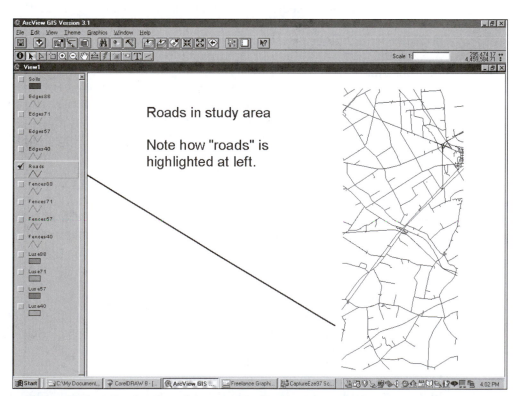

Figure 7.5   Selecting line objects on the basis of their attributes. Note how only the theme with fencerows, rather than the edges theme, is activated.

three lane, and interstate freeway are examples of line objects organized by ordinal ranks. These distinctly different road types can be compared only across
the single spectrum of highway types; quantitative comparison to other non-
highway line features is not permitted. Line items that exemplify a measurable
difference in magnitude might be exemplified by lines representing the actual
stream flow of different tributaries of a river system, probably measured in cubic feet per second or cubic meters per second.

In some cases, a single line may experience a change in attribute type, rank,
or magnitude along its overall length. For example, a road may change from single to double lane, a stream's flow may change to a higher value because of inputs from tributaries, or the measured traffic flow along a city street may
change along a portion of its length where it connects to another street. By using nodes to indicate the changes and by storing each segment between nodes
with the appropriate attributes, we identify each segment as a separate, identifiable entity. The attributes can therefore be used to retrieve either whole lines
or portions of a line.

Another attribute characteristic involves not just the attributes of the line itself but also a comparison of what falls on either side. For example, we may
want to define a hedgerow or fencerow not on the basis of type of vegetation
but rather a comparison of the land covers that fall on either side of the entity.
We might want to be able to identify, for example, all hedgerows that have forest on one side and farm fields on the other, or all hedgerows that have only
farm fields on both sides (DeMers et al., 1995). In raster systems, chores like
this may be difficult, requiring us to perform elaborate categorical manipulations called **neighborhood functions** (which we will discuss in Chapter 9). In
vector systems that employ a topological data model, a search could be performed to identify each attribute for each bordering polygon. Since, however,
the relationships between the line and its neighboring polygons are explicitly
encoded into the database during input, a simpler method for both raster and
vector is to encode the attributes for the line that relate to the neighboring
polygons. With raster this would mean either creating separate coverage for
the line entities to permit the attributes to be coded or, if your raster system
employs a database management system, storing the attributes there as separate identifiers. In vector, the latter approach is most often used. Once again,
you can see the importance of thoughtful planning before data entry—in this
case with a view to easily answering questions you might ask.

To locate the lines, it is necessary to identify all the coordinate pairs that
make up the line in vector, or all the grid cell column and row values in raster.
This adds three other factors that might be used to retrieve the line—its length,
its azimuthal direction, and its shape. Line entities may be straight with single
orientations, or they may be jagged or sinuous with multiple orientations, like
street or road patterns or meandering streams, where each straight-line segment indicates a unique orientation. Some line entities are simple, with a single
line; others are complex, with a branching network that may form a hierarchy
of lines, as in a branching stream network. All these are difficult for most raster
GIS software to calculate, but vector GIS packages can easily perform length
(see Chapter 8) and azimuthal calculations. Length calculations are merely a
matter of performing the distance-between-points calculations using the
Pythagorean theorem, if the data are in Cartesian coordinates, or great circle
calculations for data that are projected (Robinson et al., 1995). Azimuthal calcu-

lations use the standard formulas for spherical trigonometry between each set of points (Robinson et al., 1995). Shape calculations for line entities are most often measures of **sinuosity**, or some ratio of the amount of overall length of a curvilinear line to its straight-line length. This merely requires two separate sets of calculations, one for the total length and one for the straight-line length. More complex analyses involving combinations of these calculations are possible but call for numerous sets of operations (Mark and Xia, 1994).

These three principal types of linear measure—length, azimuthal direction, and shape—can be defined as individual attributes for each line entity, or for each separable portion of a line entity. For example, you may want to find all hedgerows that are longer than 80 meters, or you might want to separate out all roads that have a north-south orientation, or you may want to isolate highly sinuous streams. To allow the lines or line portions to be selected, tabulated, and displayed separately, then, each should be separable with your GIS. In some cases, the line entities may extend beyond the map boundaries rather than being wholly contained inside the map. If you want to be able to eliminate these, it might be useful to include an attribute indicating that the line is incomplete. Analyses that are not able to separate out incomplete line entities are likely to be inaccurate even if the data are aggregated. Most of these operations are designed to obtain overall or aggregate statistics for your map, such as the average length of a certain line entity in a map or the overall average azimuthal direction or overall average sinuosity. At times, however, it may be necessary to combine operations to create regions of similarity in a map where the lengths, azimuths, and sinuosities vary from portion to portion on the map.

Area Attributes

As with point and line entities, area or polygon entities can be defined, separated, and retrieved on the basis of category, class, or magnitude (Figure 7.6, 7.7). As before, each of these attributes needs to be explicitly stored in the database, whether as grid cell attributes or as vector polygon attributes. Isolation and retrieval are performed in exactly the same manner as for points and lines (Figure 7.7). But, like lines, areas have added dimensionality that permits more attributes to be assigned to them on the basis of measurements of their dimensions.

Among the more useful attributes for polygon entities would be a measure of their shapes. This shape could be strictly Euclidean in that it could be measured as a deviation from some known geometric shape with predefined properties, such as a circle or a square. Another group of shape measures uses a set of mathematics called fractal geometry, in which the irregularity of the outside of a polygon is measured. Related to shape is a measure of a polygon's elongation, or the ratio between its long and short axes. Although not generally a built-in function of vector GIS, measuring the long and short axes and expressing these values as a ratio are relatively simple to do. The coordinates for each polygon point can be used to identify the points that are the farthest in a particular direction. However, this requires the use of a higher-level object, the **centroid**, against which to measure the rest of the polygon edges (see next section). Of course, elongation implies that one might also wish to determine the

Selecting land use polygons based on land use codes. Code 206 represents agriculture and shows as the lighter color on the map.

Figure 7.6   Selecting area objects on the basis of their attributes. In this case, the areas are selected by land use code.

particular orientation of elongated polygons with respect to cardinal directions. Although measures of shape are not frequently invoked by the GIS community, there is growing interest in the relationships between shapes of both human and natural area features and their functioning (Forman and Godron, 1987). At a minimum, some simple measures of shape would be useful to facilitate the isolation, selection, tabulation, and display of polygons of a given category of shape and orientation.

Another somewhat more commonly applied measurable attribute of polygons is their size. In raster, size is determined by counting the number of grid cells of a given category. These categories do not need to be restricted to a single polygon; rather, they simply indicate the number of grid cells with a given value. Often, a reclassification process (see Chapter 9) will be necessary to isolate these polygons, and then their area can more readily be found by a simple count of grid cells. Measures of size in raster are not going to be particularly accurate because of the quantized nature of the grid cell. In vector, the perimeter

Figure 7.7   Attribute selection flowchart. A simple flowchart illustrating the use of land use codes to retrieve selected types of area attributes.

of the polygonal lines is easily calculated by adding the line lengths, whereas the area is calculated much as would be done manually, by a series of length-times-width calculations for portions of each polygon. The more complex the polygon, the more calculations must be performed. But in general, any commercial GIS package should be able to provide reasonably rapid response times for area calculations. To isolate polygons on the basis of area or perimeter, one need only select a set of category criteria, place the appropriately sized polygons in each, and perform a simple retrieval on those selected. Polygon size, like line length, is most often (but not necessarily) performed on each polygon for each category and is then averaged for tabulation. This indicates to the user the relationship of polygon size to total area of the database occupied by that category of polygon.

Two other measures regarding areas need to be mentioned as well, although they will be discussed in more detail in Chapter 8. The first is **contiguity**, a measure of the wholeness or amount of perforation of a polygon (Berry and Tomlin, 1984). A large polygon that contains many holes (small polygons contained within the larger polygon) demonstrates much less contiguity than one with only a few holes or none at all. Analyses may require the amount of contiguity for animal habitats or fire hazards for forest regions. Most raster GIS software contains some measures of contiguity, but vector GIS is less adept at performing this analysis. By determining the amount of contiguity, then classing the results into groups, the analyst can easily retrieve and display these groupings or use them for further modeling.

The second additional measurable attribute that could prove useful for polygonal features is the **homogeneity** of an area, not necessarily defined as a single polygon (Berry and Tomlin, 1984). Homegeneity is a measure of how much area of a given portion of a map is directly in contact with polygonal features sharing the same attributes. For example, two polygons that have identical attributes may touch each other only along a small portion. In this respect, homogeneity is very similar to a simple measure of size. However, homogeneity can also be defined to include the amount of internal **heterogeneity** that an area contains. One can, for example, group a number of polygons on the basis not of the similarity of their attributes but of some heterogeneous mix of attributes. Other homogeneity measures of attributes include minimum values, maximum values, averages, totals, and even diversity of attributes. Each of these can be grouped, with the result that the polygons contain a new set of identifiable attributes that describe them. For example, we may have a forest polygon that shows, when separated out on the basis of species of trees, that there are 12 different species of trees contained within its borders. We could isolate all polygons that demonstrate 12 or more trees as a measure of the amount of diversity (in this case called species diversity) within them. Thus, we could map all areas that showed high diversity of tree species. Alternatively, we might be interested in polygons that, for example, show a high degree of human ethnic similarity—a lack of diversity in residential areas. We could then select all the polygons that contained fewer than three or four ethnic groups, to perhaps indicate some measure of ethnic segregation in a city.

Both raster and vector GIS typically have some capability to perform this type of analysis, although in some systems a direct function is lacking. Instead, a number of commands might be used to achieve the desired results. As always, these measured attributes should allow us to isolate, retrieve, and output the results individually. When we begin looking at classification inside the GIS, we will return to all these techniques as methods of reclassifying our existing coverage attributes.

WORKING WITH HIGHER-LEVEL OBJECTS

Thus far, we have worked with points, lines, and areas, based either on readily available descriptive characteristics or on attributes that are measurable through sometimes simple, sometimes complex methods. All of these entities, however, have certain attributes that set them apart from the rest. Some attributes are a result of the methods of encoding—for example, nodes that are encoded during the digitizing process. Others may have to be determined—for example, centroids indicating the center of an area. We call these "higher-level objects" for lack of a better term, simply because of their uniqueness and usefulness for later GIS analysis. We will separate these higher-level objects into points, lines, and areas and examine each individually.

Higher-Level Point Objects

Two primary types of higher-level point objects are centroids and nodes. A centroid is most commonly defined as the point that occurs at the exact geographic center of an area or polygon (Figure 7.8a). Its calculation is simple for simple polygonal shapes such as rectangles; when the polygons become quite

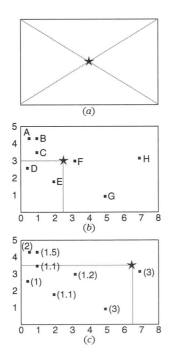

Figure 7.8   Types of centroid. *(a)* Simple centroid, *(b)* center-of-gravity centroid (mean center), and *(c)* weighted mean center, where numbers in parentheses indicate assigned weights, *f* values in Table 7.4 (note how high weights for points *G* and *H* of *(b)* pull the mean center to the right).

complex, the complexity of the calculations needed to perform the task increases proportionately. Raster GIS is not well suited for this procedure. In many cases, even vector GIS does not include this calculation as a stand-alone function. Simple or geographic centroids in vector are calculated by a rule called the **trapezoidal rule**, which separates the polygon into a number of overlapping trapezoids. Then each trapezoid's centroid, or central coordinate, is calculated and its weighted average calculated (Figure 7.9). The centroid may be important if you are attempting to produce a surface map from samples taken from within areas. For example, if you were to produce an **isoline** (defined as a line of equal value) or fishnet map of the population of the United States, but the data were sampled at the county level, you would have to place a centroid in each of the approximately 300 counties. The centroid would act as a point location, as if the data were actually calculated at this point. Then, through interpolation, the isolines or surfaces could be calculated on the basis of these point locations.

Centroids can also be placed at the center of a distribution of some phenomena, instead of at the absolute geographic center of a polygon (Clarke, 1990). For example, to continue our analogy of using county-level population data to produce an isoline map, we may find that for some counties, most of the population is clustered to one side or another. Although we could still use the centroid of the county as the estimate of its population center, it would be more accurate if we placed it closer to the center of the distribution. This location, called the **mean center** or the **center of gravity**, requires us to separately average the X and the Y coordinates for all points in the map (McGrew and Monroe, 1993; Muehrcke and Muehrcke, 1992) (Table 7.3, Figure 7.8b). The final result would be a single pair of values representing the central point of this distribution of points. In addition, if these points had additional weights—for example, if the points indicated both the locations and the amount of shopping done at each store—we could further place our center on the basis of this additional weighting factor. The procedure, called the **weighted mean center**, simply requires us to multiply each X and each Y by a weighting factor (amount of shopping in this case), then sum them up and divide by the number of points (Mc-

Figure 7.9 Trapezoid rule. Diagram of the trapezoids defined in an irregular polygon. Each overlapping trapezoid has a calculated mean center from which the weighted mean center can further be determined. This weighted mean center is the centroid of the polygon.

TABLE 7.3 Table of Values
Set Up to Allow Calculation
on the Center of Gravity[a]

Point	X	Y
A	0.5	4.5
B	1.0	4.5
C	1.0	3.5
D	0.5	2.5
E	2.0	2.0
F	3.0	3.0
G	5.0	1.0
H	7.0	3.5
Total	20.0	24.5
Mean center	2.5	3.0625

[a]"Centers of gravity" (mean centers) are
found by dividing the total values for X
and Y by the number of points.

Grew and Monroe, 1993) (Table 7.4, Figure 7.8c). The result is a single pair of X
and Y coordinates indicating the mean center of the distribution, modified by
the weighting factor.

Such calculations are often applied in large-scale market analysis and economic placement analysis to define the center of the market. Having done this,
the market analyst might choose to select areas near the market center for lo-

TABLE 7.4 Table of Values Set Up to Allow Calculations of the
Weighted Mean Center on the Basis of the Weights Assigned to Each
Set of X and Y Coordinates

Point	X	Y	f	fX	fY
A	4.5	2.0	0.5	2.25	1.0
B	1.0	1.5	4.5	6.75	1.5
C	1.0	3.5	1.1	1.1	3.85
D	0.5	2.5	1.0	0.5	2.5
E	2.0	2.0	1.1	2.2	2.2
F	3.0	3.0	1.2	3.6	3.6
G	5.0	1.0	3.0	15.0	3.0
H	7.0	3.5	3.0	21.0	0.5
Total				52.4	28.5
Weighted mean center				6.55	3.512875

cating a new shopping center or other type of business. There are many other reasons for selecting either simple centroids or center-of-gravity centers; examples include studies of animal foraging activity (Koeppl et al., 1985) and analyses of the change in movement of population centers through time as a measure of large-scale migration (McGrew and Monroe, 1993). As mentioned before, many raster GIS systems do not have this capability built in, but the majority of commercial vector systems will allow the calculation of centroids and weighted mean centers.

Our second type of higher-level point object, the node, was mentioned in Chapter 4. In this case the points are significant not as individuals but as specific locators along line and area entities. Nodes do not occur as specific objects in raster GIS, so there is no need to comment on their calculation in those systems. Because nodes carry with them an indicator of a change in attribute, the ability to identify them is vital to many attribute selection and manipulation procedures. Nodes are generally encoded explicitly during input and should be easily separated or identified through simple search procedures in the GIS. Difficulty will arise only when a node has been improperly coded as a simple point rather than as a node. In such circumstances, the ability to isolate line segments defined by nodes will be compromised. This is yet another illustration of the importance of good planning and careful execution during the input phase of the GIS.

One other situation that is covered in later chapters needs brief mention here: namely, use of the distribution of points to develop areas. For example, where large numbers of points are found to exist, whatever they might represent, their area is distinctly different from places where few points exist. You

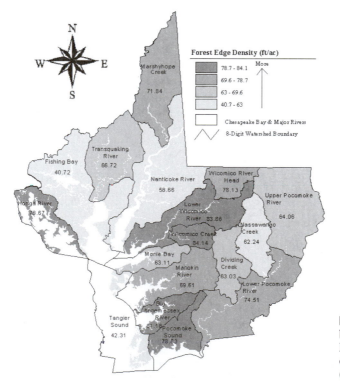

Figure 7.10 Communities based on object density. Map illustrating different communities of points (trees) based on density.

may, for example, find a great many weeds in particular parts of agricultural fields, indicating that these areas are different. What remains to be determined is the source of the difference: a former disturbance, a lack of pesticide, or something inherently different about the soil or how it is manipulated.

Other pattern characteristics such as uniformity of distribution of point objects, or randomness, or another identifiable pattern can also be used to define areas as specific **communities**, or areas sharing distributional patterns (Figure 7.10). Defining areas on the basis of point distributional patterns is not a strong part of most GIS systems, but it can often be calculated in both raster and vector systems. We'll return to this topic later in this chapter.

Higher-Level Line Objects

Three different types of lines are particularly important and justify calling their objects higher level. We touched on the first type when we examined the relationships between line attributes and associated attributes for the adjacent polygons. These lines are most often called **borders**, and a major change in attribute value or even collections of attributes is recognized or assumed as one moves across them. In other words, borders are particularly important because of their locations relative to adjacent polygons. As in our hedgerow example, borders need to have associated with them a set of attributes that easily allow portions of the map to be separated out because all the objects on one side are substantially different from all those on the other. Take a simple example of a national border between the United States and Canada. The line serving as the border between these two countries should allow one to identify all the states in the United States as fundamentally separate from all the provinces and territories of Canada on the north. Although this may seem obvious, if the line separating these two areas on the map is not provided with attribute information clearly indicating its status as an international border, you may have to force the GIS to separate the United States from Canada, rather than simply performing a search on the basis of the neighbors of the borderline.

Lines also can become higher-level objects when they are located in a particular way relative to other lines rather than to polygons, as in our last example. In many cases, lines are not simply indications of locations of linear objects or of boundaries between polygons but rather are connected by nodes to form **networks** (Figure 7.11). Networks can most appropriately be defined as a set of interconnected line entities whose attributes share some common theme primarily related to flow. As we will see in Chapter 11, networks allow us to perform modeling of a multitude of flow types from automobile and rail traffic to the flow of particular commodities, or even the movement of animal species along corridors. In all these cases, we need to be able to operate on networks, which means that the lines must share the attributes necessary for analyzing these flows (speed limits, frictions, etc.). Raster GIS is not particularly efficient at handling networks because there is no way to define them explicitly except to assign specific attribute values to the grid cells. Still, this can be done, although most GIS professionals will defer to vector GIS for working with networks.

The three major forms of networks are straight lines, as one might find in a major interstate highway (Figure 7.11*a*); branching trees, as one might expect

(a)   *(b)*   *(c)*

Figure 7.11 Types of networks. Networks composed of line segments with attribute codes indicating flow. The three major types are shown: *(a)* a straight-line network, *(b)* a branching network, and *(c)* a circuit.

when looking at stream networks (Figure 7.11*b*); and **circuits** (Figure 7.11*c*), as one might find in street patterns whose lines lead directly back to the starting point (Muehrcke and Muehrcke, 1992). In addition, all these network types can be defined as directed or **undirected**. In a **directed network**, the flows are allowed to move in a single direction only (Figure 7.12*a*). Streams, for example, will flow downslope and will not, under normal circumstances, flow in the opposite direction. Similarly, one-way streets limit the flow of traffic to a single direction. In the event that one link in the network intersects another at an angle, a change in flow direction may occur, or it may be necessary to restrict the locations at which turns can be made in passing from one link to another. At the intersection of a two-way street and a one-way street, for example, you will not be allowed to turn from the two-way street into the oncoming flow of one-way traffic. But in networks called undirected networks (Figure 7.12*b*), the flows can go back and forth along the network in either direction.

Because networks have the capability of modeling flows in either a directed or an undirected fashion and because some network links will be connected to certain links but not to others (e.g., a road overpassing another road), all these

Figure 7.12 Directed versus undirected networks. Directed networks *(a)* restrict the flow to a single direction, whereas undirected networks *(b)* allow flow in both directions.

characteristic attributes must be explicitly encoded during either data entry or later (as edited attributes). Nearly all vector GIS software has the capability of storing these attributes and modeling flows with them. Some software has been specifically designed for work of this kind (see Chapter 11). A lack of attribute data for networks severely limits the use of linear features as higher-level network objects. Lines connected to each other without attributes indicating that they provide a common pathway for flows offer no basis for network modeling.

A final characteristic related to lines is not unlike the last one mentioned for higher-level point objects: that is, collections of lines can also act as identifiers for area patterns or communities. There may be areas of high-density road networks or areas low in hedgerows. Because of the density or lack of density of the line objects, we can define these areas as communities sharing these characteristics. As with point entities, additional patterning such as regularity, or randomness, can be used to identify communities of these line features. Most GIS software, however, does not contain algorithms specifically designed for this process. Most often you will need to use alternative software or modify the existing software through the use of onboard macro languages and other tool kits that may be obtainable from the vendor under certain conditions.

Higher-Level Area Objects

As with points and lines, areas can also occur as higher-level objects. The polygons themselves can be used to define regions of similar geographic attributes. Among the more important aspects of geographic research today, as in the past, is the definition of regions: areas of the earth that exemplify some unity of describable attributes. For example, political regions are defined by national boundaries, ethnic regions by a similarity of origin, and biogeographic regions on the basis of organismic similarities. Inside the GIS, these regional definitions can be based on the attributes defining each polygon or set of polygons. We might, for example, be able to define a region within our GIS by selecting all the polygons that show forest as the major vegetation component. This will give us a "forest" region. We will have to know beforehand which regions we are going to be looking for and how they are to be defined. Because defining regions is a major endeavor in itself, the chances are good that simply selecting the appropriate polygons or sets of grid cells will not suffice to produce definitions. Instead, we will most likely have to combine several sets of attributes from several different coverages to define our regions. In fact, the power to define regions on the basis of a large variety of characteristic attributes put together in as many ways as we need is one of the nicest features of the GIS. For this reason, we can define regions on the basis of what we intend to do with the data. This leads us to another classic problem in geography, that of classification itself, and we will cover this in some detail in Chapter 9. Meanwhile, the selection of regions can be thought of as isolating homogeneous sets or homogeneous combinations of factors. In some cases, regions can also be regarded as areas containing a similar heterogeneous mix of attributes rather than sharing only a few common attributes.

Regions differ not only in their attributes and in the way the attributes are manipulated to define them but also in the way they are configured in space.

There are three basic types, based on spatial configuration: **contiguous regions**, **fragmented regions**, and **perforated regions** (Figure 7.13). Contiguous regions are wholly contained in a single polygon. Although a contiguous region is wholly contained, its attributes can be defined as homogeneous or heterogeneous but with some similarity of heterogeneous mix, as we have seen. Fragmented regions, on the other hand, share some commonality of attributes. Again, they are either a homogeneous or a heterogeneous mix containing identifiable similarities but are composed of more than one polygonal form separated by intervening space that does not share the same mix of attributes. For example, one could define as a forest region a number of polygons, scattered throughout the coverage but having the same mix of trees or tree types. There are no limitations to the distance between polygons as long as the similarity of attributes is maintained. Similarly, for perforated regions, the defining criterion for the region remains the same—homogeneity of attributes or of the attribute mix. In perforated regions, however, the uniform polygon is interspersed with smaller polygons that do not share a mix of attributes with the surrounding polygon. As such, the region is defined as the surrounding matrix, whereas the smaller internal polygons are said to be the perforations. Clearly, there is a possible relationship between perforated and fragmented regions. If the smaller polygons contained in the perforated region are found to share common attributes, they too may be considered to be a region and can easily be separated out from the background region.

All the objects we have identified, whether simple or higher order, must be identifiable by the user so that they can be operated on for further analysis. Each must be able to be isolated, separately tabulated, and displayed. The ability to perform analysis is strongly linked to the ability of the GIS to do this. In vector data models, and in raster data models that are linked to a database management system, most often these objects are selected through a formal search of the attribute tables. For simple raster model systems, the separation is once again relegated to some form of reclassification process—the topic of Chapter 9. But first, Chapter 8 will take us through some important measurements that also are essential to more advanced analysis.

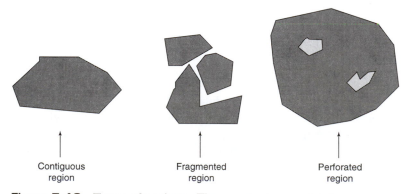

Contiguous   Fragmented   Perforated
region   region   region

Figure 7.13 Types of regions. Three major types of regions: contiguous, fragmented, and perforated.

Terms

borders
center of gravity
centroid
circuits
communities
contiguity
contiguous regions
directed network
formulation flowchart

fragmented regions
heterogeneity
homogeneity
implementation flowchart
isoline
mean center
neighborhood functions
networks
overlay

perforated regions
regions
sinuosity
trapezoidal rule
undirected networks
weighted mean center

Review Questions

1. Explain why the analysis subsystem of the GIS is the most abused subsystem and give examples to support your explanation.

2. Explain what we mean when we say that a GIS is an incomplete set of geographic analysis tools. What accounts for this?

3. If your career goals revolve around database creation rather than GIS analysis, why is it important to know about the analytical capabilities and limitations of GIS?

4. Why is it important to be able to isolate, identify, count, and separately tabulate and display individual items?

5. Describe the process of isolating, counting, and identifying point objects on the basis of their attributes within a raster and a vector GIS. Why do we need to use attributes to find these objects?

6. What are the differences between finding point objects and finding line and area objects?

7. What measurable attributes can be used to search for lines? areas?

8. What are higher-level objects? How do they differ from simple objects?

9. Give some examples of higher-level objects for points. Do the same for lines and areas.

10. What are centroids? How are they found in a vector GIS? What are the different types of centroid? How do they differ in how they are found in a vector GIS?

11. What are networks? How do they differ from simple lines? What forms can networks take? What needs to be done to identify networks, as opposed to simple lines, in a GIS?

12. What is the difference between a directed and an undirected network?

13. What is a region? What are the differences and similarities among contiguous, fragmented, and perforated regions?

14. Give some examples of how collections of points, lines, and areas might be considered to be communities.

References

Berry, J.K., and C.D. Tomlin, 1984. *Geographic Information Analysis Workshop Workbook.* New Haven, CT: Yale School of Forestry.

Clarke, K.C., 1990. *Analytical and Computer Cartography.* Englewood Cliffs, NJ: Prentice Hall.

DeMers, M.N., R.E.J. Boerner, J.W. Simpson, A. Silva, F. Artigas, and L.A. Berns, 1995. "Fencerows, Edges, and Implications of Changing Connectivity Illustrated by Two Contiguous Ohio Landscapes." *Conservation Biology,* 9(5):1159–1168.

Forman, R.T.T., and M. Godron, 1987. *Landscape Ecology.* New York: John Wiley & Sons.

Koeppl, J.W., G.W. Korch, N.A. Slade, and J.P. Airoldi, 1985. "Robust Statistics for Spatial Analysis: The Center of Activity." *Occasional Papers,* 115:1–14, Museum of Natural History, University of Kansas, Lawrence.

Mark, D.M., and F.F. Xia, 1994. "Determining Spatial Relations Between Lines and Regions in ARC/INFO Using the 9-Intersection Model." In *Proceedings of the 14th Annual ESRI User Conference,* pp. 1207–1213.

McGrew, J.C., and C.B. Monroe, 1993. *Statistical Problem Solving in Geography.* Dubuque, IA: Wm. C. Brown Publishers.

Muehrcke, P.C., and J.O. Muehrcke, 1992. *Map Use: Reading, Analysis, Interpretation.* Madison, WI: J.P. Publications.

Robinson, A.H., J.L. Morrison, P.C. Muehrcke, A.J. Kimerling, and S.C. Guptill, 1995. *Elements of Cartography,* 6th ed. New York: John Wiley & Sons.

CHAPTER 8

Measurement

Although locating and counting objects is the first step in many GIS operations, it is often just that—a first step. In many instances we will want to know how large they are, how far apart, and the distances among them. This sets the stage for more advanced analysis by providing a set of quantifiable feature attributes for each theme that can later be compared within a single theme or to those of other themes. Distances between objects will be measured as simple Euclidean distances; in other cases we will define our distances incrementally, adding the distance of one leg of our journey to another to produce a total distance; and in still other cases we will search for better paths and less arduous terrain, seeking always the easiest, least-cost distances. These least-cost distances may be from point to point or from a single point to all other points in our terrain. In some instances we will measure distance based on a need to traverse or to avoid barriers, and we will include some measure of the impact of the relative difficulty of traversing surfaces.

LEARNING OBJECTIVES

When you are finished with this chapter you should be able to:

1. Describe the process of measuring both straight and sinuous lines in vector and in raster and discuss the advantages and disadvantages of each process.

2. Explain the process and the purpose for measuring polygons to determine long and short axes, perimeter, and areas.

3. Explain some of the simpler measures of shape for lines and areas and discuss why they might be important.

4. Understand the concepts of spatial integrity and boundary configuration, describe how these might be measured, and state what they tell us about the real-world objects we are analyzing.

5. Explain the process of measuring both isotropic and functional distances among objects, through friction surfaces, and around barriers.

6. Explain the methods and potential problems associated with assigning friction and **impedance values** for modeling functional distance.

MEASURING LENGTH OF LINEAR OBJECTS

Aside from surfaces there are three basic cartographic object types. Points are assumed to have zero dimensions; lines, a single dimension of length; and areas, two dimensions—length and width. Because points have zero dimensionality, there is no appropriate measure for them aside from magnitude attribute values, which are assigned to them and stored as grid cell attributes or saved in an attribute table linked to the entities themselves. Although these point magnitude values are important for analysis, they require no formal calculations, so we can eliminate them from the remaining discussion.

Lines, with a single dimension of length, will need to have this dimension measured or calculated, again assuming that other magnitude attributes formally linked to the line entities can be simply retrieved from the database. But as we have seen, line lengths can be used as search criteria. In vector when locational data are recorded on a globe, we use trigonometric formulas to calculate great circle distance. Within a planar reference system, the traditional distance equation, based on the Pythagorean theorem, takes advantage of the explicit X and Y coordinate systems for vector measurements of distance. Through the digitizing process, each line segment is a straight line, and its distance can therefore be calculated on the basis of the distance equation. By adding each of these distance measures, we can easily determine the total length of any line.

One often overlooked problem with calculating line distance is that of lines whose lengths extend over a range of elevational distances. For example, a map may show that the road distance from one point to another is somewhat less than what an automobile odometer might show. This typically happens when the vehicle is traveling up and down topographic slopes, thus adding road log miles to the overall distance (Figure 8.1). If the change in elevation is essentially linear, the additional mileage can be calculated with simple trigonometry. Because most topographic surfaces change in a nonlinear fashion, the calculation becomes more complicated but is easily within the capabilities of most GIS software.

Calculating line lengths in raster is a matter of adding the number of grid cells together to achieve a total. Let us begin by working with a straight-line entity in grid format that occurs as a set of vertical or horizontal grid cells. Knowing the resolution of each grid cell, generally assumed to be from side to side **(orthogonal)**, rather than on the diagonal, we need only add up the number of grid cells and multiply that value by the grid cell resolution value. If a vertical line is composed of 15 grid cells, each with a resolution of 50 meters, our line length is calculated by multiplying 15 grid cells by 50 meters to obtain a total line length of 750 meters. The identical process would be performed for a horizontal line.

But suppose that a line is perfectly diagonal. What do we do when the grid cells in a line are connected to each other along a corner? A simple, but inaccurate, approach would be to assume that the resolution is the same for the diagonal measure as for a side-to-side (orthogonal) measure and add the number of

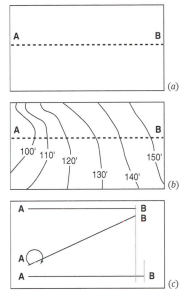

Figure 8.1 Relationship between map distance *(a)* and road log distance *(c)* as influenced by changes in elevation *(b)*.

grid cells and multiply by the resolution value. Most modern raster systems are able to calculate the diagonal distance between each cell in a diagonal line of grid cells through the use of simple trigonometry (Environmental Systems Research Institute, 1993). Because a diagonal through a square grid cell produces a right triangle, the right-angle sides of which are identical to the grid cell resolution, the hypotenuse is calculated with the Pythagorean theorem (see Chapter 2). You should note that the calculation of the hypotenuse will always give you a value approximating 1.414 times that of the horizontal or vertical distances. Therefore, it is a simple task to multiply the resolution of each diagonally connected grid cell by 1.414 to obtain the correct value for distance. Depending on the resolution and the irregularity of the line entity's path, whole loops can be represented by a single grid cell, in which case the length of the object will be underrepresented (Figure 8.2). For this reason, it is best, whenever possible, to use a vector data structure if your analysis relies heavily on measurements of linear objects.

The calculation of length along a vector line entity is computationally more exhaustive than simply adding the number of grid cells, as in raster, but the results are decidedly more accurate, as is the representation of the line itself. For each straight-line segment in the linear object, the software will have stored a

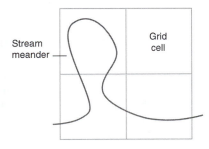

Figure 8.2 Subresolution line object. A highly sinuous linear object, such as a meandering stream, may have whole meanders self-contained within a single grid cell. This indicates the limitation of measuring length with raster data structures.

set of coordinate pairs. The distance between members of each coordinate pair can be calculated through the Pythagorean theorem. Simply adding line-segment lengths will produce a relatively accurate measure of the total or accumulated line length. Just remember that the vector representation of linear entities also employs a form of sampling, where changes in direction are represented as separate straight-line segments: the more line segments, the more accurately the linear object will be represented by the data structure, and the more accurate the measures of total line distance will be. This again reminds us of the importance of good map preparation and careful data input.

MEASURING POLYGONS

Polygonal features exhibit two dimensions—length and width. Because of the added dimensionality of polygons, we have more measurements that we can make upon them. We can, for example, measure the length of the short or long axis of a polygon, calculate the lengths of its perimeter, or measure its area. As we learned in Chapter 7, these measures can be used as search criteria for individual polygons, which can then be used for subsequent analysis.

Calculating Polygon Lengths

Because the orientation of the polygon is often linked to an underlying process, the orientation of polygonal features will at times be important to the GIS user (Muehrcke and Muehrcke, 1992). For example, asymmetrical forest patches oriented in a particular manner may be easily observed by birds as they fly over (Baker, 1989; Forman and Godron, 1981; LaGro, 1991; O'Neill et al., 1988). Or a scientist attempting to determine the movement history of glaciers may wish to know whether polygons representing certain glacial features have a particular direction. Conceptually, the idea of polygon orientation is quite simple. It is merely a matter of determining the direction of the longest axis for the feature. Most raster GIS software, however, lacks techniques for easily ascertaining the long axis. And because raster uses relative locations of grid cells, it is no easy task to determine orientation either.

In vector, the solution is one of calculating the lengths of each pair of opposing polygon vertices, much as we did with line entities. A comparison of these vertices will then show which line is the longest. That line is the long axis of the polygon, and its angular direction is calculated with spherical trigonometry (Robinson et al., 1995).

At times, however, the GIS analyst will be searching not for orientation but rather for the relationship between the major and minor axes. The latter gives a simple measure of shape that can also be used as subsequent search criteria (Figure 8.3). Suppose an ecologist wants to study small mammalian species that prefer to locate in long, skinny forest areas next to agricultural areas (Turner, 1991). In any event, it is obvious that being able to calculate the long-to short-axis ratio is a useful GIS function. Its initial calculation in vector GIS is performed as before, but first distances of all the opposing vertices must be

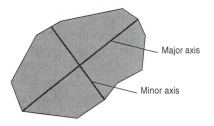

Major axis

Minor axis

Figure 8.3   The ratio of the long (major) axis and the short (minor) axis provides a simple measure of shape as defined by elongation.

found. Then, by comparing their lengths, we can identify the shortest and the longest axes.

When one is calculating long and short axes for polygons, it is notably easier to work with convex polygons. If there are concave portions, or if the polygons are highly irregular, the calculations become considerably more difficult to describe, and the results are less useful. One approach, called the **least convex hull**, allows us to sort of shrinkwrap the polygon so that it essentially resembles a totally convex polygon. These calculations, though useful, are beyond the intent of this introductory work; you may consult any good book on computer graphics, or seek more detailed algorithmic descriptions provided by your GIS vendor.

Calculating Perimeter

Perimeters around polygons are created by connecting line segments in a **closed cartographic form**, where the beginning coordinates of the first line segment are identical to the ending coordinates of the last line segment. Therefore, calculating the perimeter is a matter of calculating the distance of each line segment using the distance formula and then adding these individual values. Once again, raster data structures are not particularly well suited for this task, but, like measuring distances themselves, they can be made to produce a perimeter. To do so, we must separately identify and reclassify each grid cell located at the outside perimeter of the collection of grid cells that make up a polygon to produce a map of just these perimeter-forming grid cells. Quite often this task calls for extensive interaction with the software. Finally, however, the perimeter is defined by the sum total of grid cells, multiplied by the grid cell resolution (Figure 8.4). As before, the more complex the polygon, the more grid cells will be at a diagonal to their neighbors and the less accurate your results will be.

each grid cell = 2 km/side

perimeter =
28 grid cells x 2 km = 56 km

Figure 8.4   Finding perimeter in raster. By adding the perimeter grid cells and multiplying that number by the resolution, we obtain a reasonable estimate of perimeter.

Calculating Areas of Polygonal Features

A frequently used measure in GIS is that of polygonal area. Area gives us some quantitative measure of the amount of each classified areal feature. Finding area for a region, as defined in Chapter 7, is calculated in raster by selecting the grid cells that share common attributes (our region) and counting the number of grid cells that the region occupies. Actually, you don't even need to perform a calculation. Just tabulate the data in your grid layer and read the number of grid cells sharing the assigned attributes from the resulting table (Figure 8.5). The table will also give you a percentage of grid cells in the database for that region, compared to the overall database. Other ratios can be performed by comparing the numbers of grid cells of one region to those of another.

If, on the other hand, you do not want to measure the grid cells of fragmented regions, the problem is a little more difficult. If you want to know only the area of a particular polygonal area, whether or not it constitutes a contiguous region, you need to isolate each polygon separately by selecting its grid cells and reclassifying them so that their attribute values are unique to the grid layer. As we will see in the next chapter, typically functions are available in raster GIS that allow you to isolate such areas by size, providing an easy way to select the polygon you want to measure. If you have many polygons of approximately the same size within your grid layer, however, you will need to isolate them either by their row and column locations or by individually recoding the grid cells of your target area using available editing functions. In any event, the process is not difficult; it simply requires forethought and planning.

In vector, the target polygon or polygons are selected on the basis of either coordinate locations or attribute values in the attribute tables. Then the area of the polygon is calculated. For simple cartographic forms, such as rectangles, triangles, circles, parallelograms, and trapezoids, the calculation is no different from what we learned in our youth. It is only when the polygons take very complex forms that the computations become more complex. The most common solution is to divide complex polygons into shapes that can be easily measured with readily available formulas (Clarke, 1990) (Figure 8.6). In many vector GIS, the areas of polygons are calculated during the digitizing process itself. As the operator digitizes a polygon, a code is normally given to indicate that a polygon, not a line, is being digitized. Therefore, as the units in each set of line segments are digitized, the software determines the simple geometric shapes produced as the process continues and then calculates the areas of these shapes. Finally, the totals are computed to obtain the final polygonal area. Because the polygonal areas can be computed individually, their values are passed to the at-

each grid cell = 2 km/side

12 grid cells × 2 km/side
=
24 square km area

Figure 8.5   Finding area in raster. By adding the grid cells and multiplying that number by the area of each grid cell, we obtain a reasonable estimate of area.

Right triangle subset of
the polygon

Original polygon

Figure 8.6 The area of odd-shaped polygons can be calculated by dividing them into smaller objects whose areas are easily calculated. Shown here is a right triangle as the first portion of the polygon whose area is to be calculated.

tribute tables immediately upon data input. Therefore, the area of a polygon can be obtained by making a simple selection from the attribute tables.

Frequently, there is a need to combine the values of area and perimeter in a ratio. This relationship, called the **perimeter/area ratio**, provides a measure of the complexity of each polygon. Thus, a circular object will have the smallest perimeter/area ratio. By contrast, long, skinny polygons tend to have larger perimeter/area ratios because their perimeters are so large and their areas so small. Although the calculation of such a ratio might seem to be pointless, numerous tasks, especially in natural resource-related operations, require it. For example, the smaller the perimeter/area ratio of a forest stand, the more likely you are to have living within that forest stand animals that prefer interior environments for habitat. If you are a wildlife specialist, you probably have some idea of the requirements of the animals you are trying to protect. Alternatively, if you are trying to develop lakefront property, the higher the perimeter/area ratio, the more beach area you can offer to prospective lot buyers. Thus the higher the perimeter/area ratio, the more money you will make developing beachfront property.

MEASURING SHAPE

A close relationship exists between shape and such measures as perimeter and area for polygons and even length for linear objects. The shapes of polygons or some measure of the sinuosity of linear features will often provide insights into the relationships between objects and their environment (Boyce and Clark, 1964; Lee and Sallee, 1970). For example, the sinuosity of a stream is related to such functions as stream sediment load, slope, and the amount of water flowing through the stream. In turn, these functional relationships have much to do with whether a stream is aggrading (building up sediment load), at grade (in balance with stream inputs and outputs), or degrading (downcutting). Hydrologists, geomorphologists, and other environmental scientists find these values useful in overall analyses of conditions in a region. In polygons, the relationship between perimeter and area has much to do with the functioning of anthropogenic and natural features. Therefore, it is important to have at least a basic understanding of what these simple shape measures are and how they might be performed in a GIS.

Few measures of shape are readily available in GIS, although some relatively sophisticated ones have been reported in the literature (Moellering and Rayner, 1982). Traditional Euclidean geometry limits us to a rather small list of identifiable shapes whose dimensions have been documented and quantified. There is an emerging interest in nontraditional geometries, such as fractal geometry, which uses imaginary numbers to define shapes. But this approach, too, is atypical, so you or your instructor may choose to ignore it. Interestingly, however, at least two major efforts have been made to incorporate such measures, as well as additional measures of shape and object interactions, into both raster and vector GIS (Baker and Cai, 1992; McGarigal et al., 2002). For the time being, we will limit ourselves to simple measures of shape that are commonly available in most GIS software.

Measuring Sinuosity

Two simple measures of sinuosity can be used as a measure of linear shape. The first, which we have already examined to some degree, is the ratio of the straight-line distance (in the numerator) and the actual linear distance (in the denominator) (Figure 8.7). The closer this value is to 1, the less **sinuous** the line. However, there is often a need to know more about the shape of a curve along a sinuous linear object. Very sharp curves in roads, for example, are likely to cause accidents. And in streams, sharp curves lead to a high degree of erosion along the outside of the formation and more sedimentation along the inside. For this reason, it is useful to be able to find the radius of each curve in the line. To do this, we assume that the curves are essentially circular, although this may not always be the case (Strahler, 1975). Then we fit a semicircle to each curve and measure the radii. When streams are being represented as polygonal shapes rather than as lines, a ratio of the radius to the width of the stream gives yet another useful measure of shape.

Measuring Polygon Shape

There are two fundamental aspects of measuring the shape of individual polygons. The first, based on the idea of perforated and fragmented regions, is given the general name of **spatial integrity**. A second measure is based on the boundary configuration, more closely related to our earlier discussion of perimeter and area relationships. The second measure is often used in conjunction with other commands that allow each individual contiguous polygon to be separated from polygons with the same attribute values. In other words, we can measure each polygon's boundary configuration in isolation from the rest of a possible fragmented region, rather than simply noting the overall size of the region made up of individual polygons.

Figure 8.7 | **One simple measure of sinuosity is a ratio of the straight-line distance to the true distance.** If the stream is perfectly straight (no sinuously), its ratio would be 1.0. In this case, the ratio is 18.5/23.5, or approximately 0.787.

Spatial integrity is a measure of the amount of perforation in a perforated region. The most common measure of spatial integrity is called the Euler (pronounced "oiler") function and named after the Swiss mathematician (Berry, 1993). Many organisms—for example, many bird species—prefer large, uninterrupted patches or polygons of a particular type of land cover, whereas others, like deer, seek out large forest areas broken up by intervening smaller patches of grassland or other land cover types. A patch of forest that is unbroken is said to be a contiguous region; if it completely surrounds smaller polygons, it is said to be perforated; and if it is completely separated from similar patches by some intervening type, we call it fragmented. The **Euler function** is a numerical measure of the degree of fragmentation as well as the amount of perforation. Let's take a conceptual look.

Figure 8.8 shows three possible configurations of polygons. The Euler function describes these functions with a single number, called the **Euler number**. Numerically, the Euler number is defined by the following simple equation:

$$\text{Euler number} = (\#holes) - (\#fragments - 1)$$

where

#holes is the number of self-contained polygon perforations in the outside polygon

#fragments is the number of polygons in the fragmented region

In Figure 8.8*a*, there are four holes or perforations in a single contiguous region. By inserting these numbers into the equation, we get:

$$\text{Euler number} = (4) - (1 - 1)$$
$$= 4 - 0$$
$$= 4$$

In the second example (Figure 8.8*b*), we find two fragments in our region, each of which has associated with it two perforations or holes. Again substituting in our formula, we find that our Euler number is $4 - (2 - 1)$, or $4 - 1 = 3$. The third example (Figure 8.8*c*) shows that with three fragments and a total of five holes, our equation gives $5 - (3 - 1)$, or 3, exactly the same as in Figure 8.8*b*.

(*a*) Euler = 4   (*b*) Euler = 3   (*c*) Euler = 3

Figure 8.8   Example of Euler numbers. Three different land configurations and their associated Euler numbers: (*a*) four holes in a single contiguous region; (*b*) two fragments with two holes in each; (*c*) three fragments, two each with a pair of perforations and the third with a single perforation. It is important to note that the Euler numbers for (*b*) and (*c*) are identical even though the configurations are different.

This shows that, although the Euler number gives a measure of spatial integrity, care must be taken in explaining the results. You might want to examine a number of cases to see what results you get for, say, an area with three fragments but no holes: $0 - (3 - 1) = 22$. On the basis of the last example, you should see that at times the numbers will not be positive. In addition, if you tabulate some of your results, you may begin to see the relationships between Euler numbers for different combinations of holes and fragments (Table 8.1). This table will prove useful in determining the twin meanings of any given Euler number.

The second group of polygonal shape measures, those related to boundary configuration, are quite numerous. Among the more prevalent are those based on the axial ratio, those based solely on perimeter, those based on perimeter and area, those based only on areas, those based on areas and areal lengths, and others measuring circularity mean side and variance of side. The formulas for these, which can be found in Chapter 5 of a useful text on geology (Davis, 1986), can be referred to as your interests in shape analysis develop and your GIS skills improve. Some of the more esoteric measures are available in commercial GIS, whereas others are available in geostatistical software or as macros for commercial GIS (Baker and Cai, 1992; McGarigal and Marks, 1994).

Most of these measures are strongly related to the ratio of perimeter to area. As you will soon see, the perimeter/area ratio itself can be considered to be a measure of polygonal shape. However, this doesn't necessarily describe the actual geometric shape of the object. Instead of a simple ratio, it would be desirable if we had a measure that was more shapelike. For this we most often compare the polygonal shapes we encounter to more familiar shapes we can easily describe (Muehrcke and Muehrcke, 1992). We might, for example, want to compare the polygonal shapes to such geometric forms as parallelograms, trapezoids, and triangles. However, these shapes come in a wide array of forms, each one of which would have to be described separately. And, of course, the sim-

TABLE 8-1   Matrix of Euler Numbers for Different Hole–Fragment Combinations[a]

					#fragments				
#holes	1	2	3	4	5	6	7	8	9
1	1	0	−1	−2	−3	−4	−5	−6	−7
2	2	1	0	−1	−2	−3	−4	−5	−6
3	3	2	1	0	−1	−2	−3	−4	−5
4	4	3	2	1	0	−1	−2	−3	−4
5	5	4	3	2	1	0	−1	−2	−3
6	6	5	4	3	2	1	0	−1	−2
7	7	6	5	4	3	2	1	0	−1
8	8	7	6	5	4	3	2	1	0
9	9	8	7	6	5	4	3	2	1

[a]Note the number of configurations that can produce the same Euler number. Note also the mirroring of Euler numbers on either side of the diagonal made with the zero Euler values.

plest, most compact, and most easily defined shape is the circle. For this reason, the basic method of measuring shape is to compare it to the circle.

Because we use the circle as the comparative shape, we can say that this measure of shape is also a measure of the **convexity** or **concavity** of the polygon. A circle has perfect convexity because no portion of its surface is concave or indented. This is why the circle is the most compact geometric shape and why we use it to measure the shapes of other objects. As it turns out, all other geometric shapes have more perimeter than a circle. To compare our polygonal geometric forms to that of a circle is essentially the same as examining the amount of convexity of the polygon versus that of the circle. The general form of the convexity formula in vector GIS is

$$CI = \frac{kP}{A}$$

where

CI = convexity index

k = a constant

P = perimeter

A = area

So we have a perimeter/area ratio for each object, multiplied by the constant. The constant is based partly on the size of the circle that would **inscribe** the irregular polygon. In addition, it is designed to provide a range of positive values from 1 to 99, where 100 indicates 100 percent similarity to a circle. Stated differently, a 1 is as far from a circle as can be measured, and a 99 is as close to a circle as a shape can get without actually being a circle. A perfect circle thus has a value of 100.

In raster, the formula is based on the exact same concept, but the area is now recorded as the number of cells, and its square root is used to provide the same 1–99 range of similarity values. Therefore, the general form of the formula for calculating convexity in a raster GIS is

$$CI = \frac{P}{\sqrt{\#\,cells}}$$

where

CI = convexity index

P = perimeter

$\sqrt{\#\,cells}$ = area in raster format

As before, a value of 1 is as far from a circular or totally convex shape as can be measured, and 99 is as close to perfect convexity as a shape can get without being a circle. In raster it is physically impossible to get a perfect circle.

As before, the important question again arises. For what reason do we use measures of convexity anyway? Ask yourself this question: When the American Plains Indians were attacking a wagon train, why did the wagon master have the wagons form into a circle? why not a trapezoid? or a parallelogram? The answer is that it has the lowest amount of edge possible, given the area. Many creatures, including humans, like the security of a circular shape, with its readily defended perimeter (hence the term's use in the military). Other interior species take advantage of the lack of edge that offers protection from competition for needed resources like food and shelter. Yet some species—for example, some small rodents—really like edges. They tend to use them as base areas from which to forage in the open along field crops, while the forest remains accessible for escape and shelter.

This scenario suggests another, quite simple measure of boundary configuration called **edginess** (Berry, 1993) that uses a device called a **roving window**. Also known as a **filter**, a roving window is a matrix of numbers of a preselected size that can be moved across a raster GIS or digital remote sensing database. The idea is to move the window across the database either to examine what is there or to modify the grid cell values. In remote sensing, two fundamental tasks for which the filter is used are **edge enhancement** and **smoothing**. In edge enhancement, edges and lines are detected, and their values are increased so that these features become more prominent. In smoothing, highly different values are suppressed, and their numbers are made more like their surroundings through some form of averaging. Although these uses of filters could more closely be associated with classification, we will look at their use for edginess here as a measure of boundary configuration.

Let's assume that we are using a 3×3 filter (a nine-cell window), which we will move along the respective boundaries of two raster-based polygonal structures to examine the numbers covered (Figure 8.9). We assign digital values to all the cells inside the filter: 1 for cells that are the same attribute value as the edge we are interested in and 0 if they are attribute values for any other object in the coverage. The amount of edginess is obtained simply by counting the

Figure 8.9 Raster measure of edginess. A moving filter used to determine the edginess of a polygonal edge in raster. Note how much more edge there is as the number of neighboring polygons with a single attribute goes down. *Source:* Modified from J.K. Berry, *Beyond Mapping. Issues and Concepts in GIS.* Fort Collins, CO, GIS World, Inc., © 1993.

number of grid cells that are 1s, or share the same attribute values as the polygon. The more 1s we get, the less edge we have, and the more interior. Thus the value of 7 in Figure 8.9 illustrates very little edginess: in the first portion of the matrix covered, nearly all the grid cells are connected to each other. Similarly, a 9 would be total interior, with all grid cells connected and no edge. By contrast, the edginess value of 2 in Figure 8.9 indicates that only a small number of the grid cells exist in the second portion of the map over which we have fitted the filter; the remainder are essentially background or matrix grid cells (0s). The 2 indicates a small, skinny protrusion into the matrix. Or stated differently, the lower filter covers a great deal of edge.

MEASURING DISTANCE

The distance along and among features is important not only in later analyses of the relationships between and among them but also more immediately because it provides a means of estimating travel to, from, and around them (often called travelsheds). Distance can be measured in simple terms where its calculation is based on the absolute physical distance between places on the map. In addition, distance can be measured to include costs incurred while traveling rugged terrain, the cost of travel specifically along networks rather than off-road, or difficulties involving barriers that restrict or prevent movement. These latter approaches to distance are collectively known as **functional distance**.

Simple Distance

Calculating simple distance, known as **Euclidean distance**, is relatively easy in both raster and vector GIS. In raster, the method is to add the column and row distances in grid cell units, then multiply the number of grid cells by their grid cell resolutions—whether orthogonal or diagonal—to convert to standard distance measures. You might remember that the diagonal distance is 1.414 times the orthogonal distance (the hypotenuse of a right triangle for each grid cell).

Until now, we have been discussing distances between point locations in a thematic map or along the grid cells of a linear object. However, this limits our analysis of distance. We may, for example, want to know the distance between a single point location on a map and all other possible locations. In raster this is done by producing a series of concentric rings, one grid cell in diameter, around the starting grid cell. Each ring is essentially one grid cell away from the preceding ring. The result of such a distance measure, assuming the surface is completely uniform, is called an **isotropic surface** inasmuch as it is completely uniform in all directions (Figure 8.10a). If displayed in three dimensions, it gives the appearance of a cone whose center is the starting grid cell. An isotropic surface map from a single starting point to all other locations is essentially a travel map showing distances to every other location in the database. Its advantage over a simple pairwise calculation of point-to-point distance is apparent when there are many of these to calculate: the isotropic surface has already calculated them.

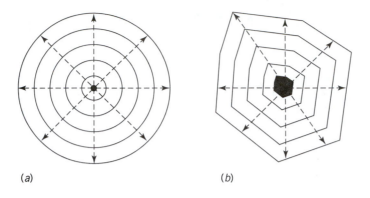

(a)   (b)

**Figure 8.10   Isotropic surfaces around a point (a) and a
polygon (b).**

It is not always necessary to measure distances across a surface starting
with a point as the origin. We may wish to measure the distance from a linear or
polygonal object to all other locations on the map. Suppose, for example, that
you want to know the distance from the edge of the city, whose boundaries are
represented by a large polygon, to all other locations on the map. The GIS operates as before but begins its measurement at the margins of the polygon and
moves outward in all locations. The result is another isotropic surface, but with
an apparent hole in the middle where the starting polygon was located (Figure
8.10b). Figure 8.11 illustrates a simple flowchart of the process of measuring
distance to produce a travelmap.

In vector, once again, calculation of distance is simplicity itself, at least for
calculating distances between points. By using the Pythagorean theorem for
each connected line segment and then adding, we obtain a total Euclidean
distance measure. The same operations are used to calculate distances between any two point objects that are not connected through line segments.
However, the points must actually exist in the map. Because the vector data
structure does not explicitly define the intervening spaces between existing
objects, no calculations can be performed in these spaces unless objects are
placed there.

The foregoing limitation will also apply to developing isotropic surfaces,
whether from points to all other locations or from polygons to all other locations. To perform such calculations requires a different vector data structure
specifically designed for modeling surfaces. The most common of these vector
surface data models, the TIN, was described to some degree in Chapter 4 on

Figure 8.11   Distance flowchart. Notice how the distance operation is modified
based on its source or starting place. This source might be a point, a line, or a polygonal object. The result of measuring from a starting place to all other places on the
map is called a travelmap.

cartographic and GIS data structures. We will return to the TIN model and describe its operation in more detail in Chapter 10. Distances can be calculated in the TIN data model, but at considerable computational expense. GIS users who work with surface models most often defer to the use of a raster system, especially when it is necessary to measure distances along surfaces.

Functional Distance

Although measuring Euclidean distance is essential, the chance that one will encounter a totally isotropic surface is practically nonexistent because it is a theoretical construct. Still, we often ignore the surface conditions to simplify our surface measurement. In many situations, however, surface conditions cannot be ignored. Such conditions typically involve the measurement of non-Euclidean distance based on some functional variable such as time, energy, or gas consumption—thus the name functional distance. The implication here is that movement across nonisotropic surfaces incurs a cost.

We need to consider two basic concepts when looking at functional distance: **friction surfaces** and **barriers**. Both involve some form of impedance or friction value, which is a measurable condition that imposes the cost of movement. In friction surfaces, the impedance or friction value is generally considered to exist to some degree throughout a map area. Topography, itself a continuous variable, may impose a friction value based both on surface roughness and the variable cost of moving up or down hill (Figure 8.12). This might be combined with still another variable such as vegetation (Figure 8.13) to create a more complex friction surface.

In other cases rather isolated conditions or discrete obstacles called barriers interfere with movement across a surface. Barriers come in two types, absolute and relative. **Absolute barriers** (Figure 8.14) (cliffs, fenced areas, lakes, etc.) prevent or deflect movement. Although one normally characterizes absolute

Figure 8.12   Topographic friction surface. Notice how the movement (dashed lines) get closer together as they move upslope and further apart as they move downslope. This shows how topography itself can impose a cost on movement.

Figure 8.13 Vegatative friction surface. Notice how the existence of forest vegetation slows the movement (closely spaced dashed lines), while the absence of forest vegetation results in faster movement.

barriers as impermeable (thus preventing movement) (Figure 8.14*a*), they may be permeable with points of access through which travel is permitted (Figure 8.14*b*), such as bridges over streams. In such cases, movement across the barrier is restricted to these access points, producing a bottleneck that results in a radial pattern of movement on its down-movement side. This is not unlike the pattern achieved as perpendicular onshore waves encounter the inlet to a bay, where the water that gets through produces a series of concentric waves inside the bay.

Relative barriers (Figure 8.15)—like their less discrete friction surface counterparts—incur a cost for movement. Such barriers might be exemplified by narrow ridges of hilly terrain, shallow streams that are passable for off-road ve-

Direction of Movement ⟶

(*a*) Absolute Barrier (impermeable)

Direction of Movement ⟶

(*b*) Absolute Barrier (permeable)

Figure 8.14 Absolute barriers. Notice how absolute barriers can be impermeable *(a)* where they stop or deflect movement and permeable, *(b)* which results in restricted access such as bridges across a river.

Relative Barrier

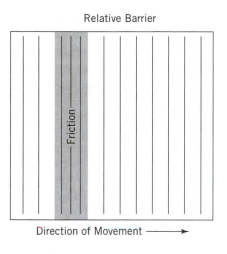

Friction

Direction of Movement ⟶

Figure 8.15   Relative barriers. Much like friction surfaces, except in their discrete spatial occurrence, relative barriers incur a cost for movement from one place to another.

hicles, or patches of forest that restrict but do not prevent completely the movement of animal herds. This results in either slower movement or the expenditure of additional energy. You might imagine the slower movement of large mammals through a swamp as opposed to rapid travel on an open prairie.

Whether you are employing friction surfaces or barriers, it is common practice to employ both an original map layer (without friction or barriers) and another layer (with friction or barriers). In such cases, you are modifying your measurement of Euclidean distance by the friction (or barrier) map layers. So movement across a topographic surface or through forests will modify the speed of movement. Figure 8.16 illustrates a simplified flowchart characterizing such a process. In all such cases, one is required to model the friction surfaces or barriers themselves through which, or over which, the measurements will be moved.

Raster modeling of barriers and friction surfaces is conceptually fairly simple. If, on an isotropic surface, the GIS adds one grid cell for each unit of travel, the resulting map will show that each additional ring of the isotropic surface is recorded as 1 + the row before it. So if we travel 10 units from our starting point or polygon, the outside ring will have the value 10, the next inner ring will have the value 9, and so on, since the GIS counts grid cells as it moves along the surface. Let us say, however, that we want to place a relative barrier across our

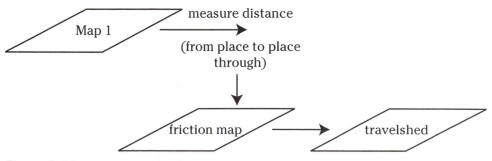

Figure 8.16   Flowchart illustrating distance measure through a friction map. The result of such a process is generally known as a travelshed.

map from top to bottom. We want to show travel distance as a functional distance from left to right across the map. To create the barrier, we place values along the vertical line and will not allow the coding of another cell beyond it until it has counted beyond the highest value coded for the barrier. For example, if we used the impedance value of 5 for our barrier and began counting from left to right, it would continue to increase the next grid cell by a value of 1, until it reached the barrier. It then must add three additional counts before it can continue. Such a barrier is considered to be a relative barrier because once it has counted five additional numbers, it will continue to move along, adding one number to each additional grid cell (Figure 8.17a). In many cases, such relative barriers or friction surfaces may vary in value and extend across the entire surface of the coverage. For example, we might want to move through a map layer of land cover containing five or six different land cover types, each with its own friction value for travel. Most often this is done by reclassifying the land use attributes to reflect their friction value and naming this new coverage as the friction coverage (Figure 8.17a).

To make the barrier an absolute barrier, we assign it a weighting factor sufficiently high that it will exceed either some maximum number designated in the GIS program or what should be the maximum number of grid cells in the GIS coverage. So if we have a raster coverage that is 100 grid cells wide, we might put a barrier value or impedance value of 1000 to ensure that the process of moving and counting the grid cells stops before it passes the barrier (Figure 8.17b).

Although the concept of assigning friction values and impedance values is a simple one, operationally it causes some difficulty. Just how much impedance does a forest cause to deer populations as they travel through it? How much more gasoline is used in moving up a 15-percent inclined surface than in moving over a flat surface? How much longer does it take a large mammal to swim a stream 100 meters wide than it would to walk across a bridge 100 meters long? These questions are real questions, many of which will have to be answered to some extent before we can create our barriers and friction surfaces. Most often the answers not only will not be available but will also depend on a variable as slippery as the agility of a deer, the gas efficiency of an automobile, or the swimming ability of a zebra. In other words, most often we lack absolute measures for impedance or friction values. Generally, we tend to use a more ordinal

Figure 8.17 Impedance. Simple accrued values of grid cell counts as movement encounters a relative barrier *(a)* and an absolute barrier *(b)*.

ranking system to assign impedance values, on the basis of a comparison of the relative difficulty of travel through each of the intervening friction surfaces or barriers. This is not always easy, and the results are not as precise as we would like. Caution is advisable in assigning these values to ensure that a reasonable explanation is given for the system used and that some justifiable comparison can be made among the friction surfaces. Above all, the results of analysis of distance based on barriers and friction surfaces should be viewed with some trepidation, especially if they are to be used for decision making.

Before we discuss the mechanics of raster approaches to non-Euclidean and functional distance, we must examine two additional characteristics of distance. Distance can be viewed not only as Euclidean or non-Euclidean or as isotropic or functional but also as **incremental distance** or **cumulative distance**. Incremental distance simply measures each unit, or increment, traveled, adding the second to the amount of distance traveled in the first, and so on. Each successive increment is added simply as a measure of distance, much like our procedure for working with the isotropic surface. In other words, the incremental distance measures the shortest path between two places without adding any friction or impedance values along the way. If incremental distance is measured for an entire surface, the result is a **shortest-path surface**, whereas if the technique is restricted to lines or arcs (or linear arrangements of grid cells), we have a shortest path, rather than a surface, as a linear product.

As we saw with friction and impedance, however, cost increases continuously as we encounter surfaces or obstacles. The purpose of calculating functional distance on a friction surface (e.g., a topographic surface) is to find the **least-cost distance**, or the shortest path between two points in a map layer. We may also want to calculate the **least-cost surface** for movement from one point location to all other locations. Let's take a look at these cases individually.

For this discussion, let us assume that our surface is a real topographic surface and that the cost is related to the change in elevation from one grid cell to another. Conceptually, we are placing a drop of water at the top of our topographic surface and watching as it streams or drains down the topographic surface to lower elevations, finding the path of least resistance. Indeed, many GIS software packages use such terms as drain or stream as commands. To create a least-cost distance (as opposed to a **shortest-path distance**) in 5×5 raster matrix, we begin at the top grid cell and perform a search of the eight nearest grid cells to evaluate which has the lowest friction or impedance value (in this case, the lowest elevation relative to the starting point). This grid cell, which is assigned a numerical value or a pointer to indicate that it has been selected, then becomes the starting point for the next iteration in a search of its eight neighbor cells. The process continues until the lowest elevation value in the database has been selected. What we have then is a selected route from the top of our hill to the bottom that incurs the least amount of effort.

Let's continue with our simple visual example. We want to find the easiest path down the mountain represented by the 5×5 matrix of Figure 8.18a. Starting at the top (column 5, row 5), we examine our neighbors (there are only three because we are at the corner of our coverage) to determine which of the connected grid cells is lowest relative to (5,5). We note that grid cell (4,4) has the lowest value, and we move there next. Proceeding to grid cell (3,3), we again perform a search of all neighbor cells, again choosing to move to the lowest available grid cell value (2,2). When we have reached the bottom of the hill

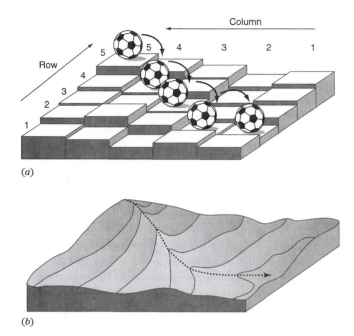

(a)

(b)

Figure 8.18   Least-cost path. *(a)* Calculation of a least-cost path requires the computer to compare the starting grid cell to its immediate neighbors. A soccer ball bouncing down the steps (grid cells) of a raster surface is used to illustrate the idea. *(b)* The same function as it might be performed by a stream flowing down a mountainside.

at location (1,1), the trail left behind is our least-cost path. An analog version of this search is shown in Figure 8.18*b*.

Although the least-cost route often suffices, it may prove useful to examine all possible routes from a particular place as we accumulate a cost for moving up or downhill through a friction surface. In this case, our friction values might be envisioned as topographic impedances, or friction impedances encountered during a trip through different types of land cover. Take a look at Figure 8.19. Here we have two grids—one indicating the friction or impedance factors and another indicating the calculated values of accumulated distance (weighted by the impedance factors). The process of computation is slightly more sophisticated than the shortest-path model we just examined. In this case, we see that

1	1	3	1
1	3	1	1
3	1	1	1
1	1	1	1

(a)

4.828	3.828	4.414	3.00
3.828	4.242	2.414	2.00
4.828	2.414	1.414	1.00
3.00	2.00	1.00	0.00

(b)

Figure 8.19   Accumulated distance. Accumulated cost surface shown by indicating *(a)* the friction coverage and *(b)* the calculated values. Note how the calculation of diagonal grid cells uses a value of 1.414 as the distance rather than 1.0 for orthogonal grid cells. The shaded grid cells are those whose friction value is 3. To accumulate distance, the previous value must be added to the calculated value.   *Source:* Modified from J.K. Berry, *Beyond Mapping. Issues and Concepts in GIS.* Fort Collins, CO, GIS World, Inc., © 1993.

the program doesn't simply search for the lowest value but rather calculates a value for each adjacent cell based on both Euclidean distance and the impedance values. This example uses rational (decimal) numbers; diagonal distances between cells are measured when necessary.

We begin the process in the grid cell at column 0, row 0. The program then identifies all bounding grid cells and notes the friction factors (if any) for each of these bounding cells. The distance is calculated in half-step increments by multiplying each occupied grid cell (including the starting grid cell) by its friction factor and then by a value indicating its width (1 for orthogonal grid cells and 1.414 for each diagonal grid cell). Because we are moving from the center of the starting grid cell to the center of the next, we multiply each by 0.5 to indicate the half-step increments. Thus our formula for each step of half a grid cell is (Berry, 1993)

$$0.5 \text{ (grid distance} \times \text{friction factor)}$$

We then must add this to the previous friction value. So for a movement starting at column 4, row 4, moving to grid cells (3,3), (2,2), and (3,4), our resulting calculations and values are

3,3	$0.5 (1.414 \times 1.00)$	=	0.707
	$0.5 (1.414 \times 1.00)$	=	0.707
	Subtotal	=	1.414
	+ Previous value	=	0
	Total	=	1.414
2,2	$0.5 (1.414 \times 1.00)$	=	0.707
	$0.5 (1.414 \times 3.00)$	=	2.121
	Subtotal	=	2.828
	+ Previous value	=	1.414
	Total	=	4.242
3,4	$0.5 (1.000 \times 1.00)$	=	0.500
	$0.5 (1.000 \times 1.00)$	=	0.500
	Subtotal	=	1.000
	+ Previous value	=	0
	Total	=	1.000

The next step is to choose the smallest accumulated distance for each of the adjacent cells. Then the process is repeated. The friction distances for all steps are added together to yield an accumulated distance. As you can see, the process can be quite tedious. That is why we use computers. The result is a least-cost surface rather than a least-cost path, and we can choose the least-cost path to and from anywhere on the coverage. This is done by simply selecting the adjacent cells with the lowest cost values.

In cases of vector coordinates describing points, lines, or areas available for calculating distances, a series of modifications of the standard Pythagorean theorem can be used to calculate the distances. We will examine these by first reminding ourselves of the original formula for Euclidean or straight-line distance between any two points:

$$d_{if} = \sqrt{(X_i - X_j)^2 + (Y_i - Y_j)^2}$$

In this equation, the distance between points i and j is the square root of the sum of the square of the differences between the X and the Y coordinates.

If, however, we are unable to perform a straight-line distance—in other words, if there is an obstruction requiring us to deviate from a straight line—we can generalize the formula to a non-Euclidean form:

$$d_{if} = [(X_i - X_j)^k + (Y_i - Y_j)^{k\,1/k}]$$

where the squares and the square root sign have all been replaced by k, a variable representing any of a range of possible values (McGrew and Monroe, 1993). Instead of squaring the differences between the X and the Y distances, we raise them to the kth power. And where the sums of the X and Y coordinate values were under the square root sign, we now raise them to the $1/k$th power (keeping in mind that the 1/2 power is the same as square root).

By inspecting the above equation, you will see that by replacing each variable k with 2, you return to the original equation for calculating Euclidean distance. So the generalized equation works just as well for straight-line distances

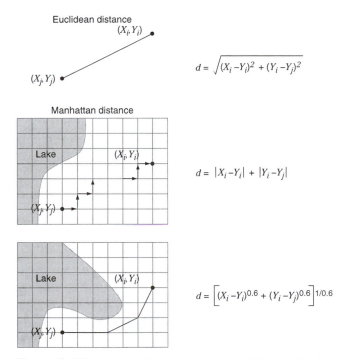

Figure 8.20   Vector distance measures. Two methods of calculating non-Euclidean distance in vector; each is a direct modification of the Pythagorean theorem formula. The exponent or k value is the only part of the equation that needs to be modified.

as it does for non-Euclidean distances. Let us assume, for example, that we are trying to find the distance between two points in the borough of Manhattan, where dozens of tall buildings and square blocks restrict our movement in square units. This measure, called the Manhattan distance, describes a rather restricted movement and changes our k factor to 1. Given that value, our formula now converts to the following:

$$d_{if} = |X_i - X_j| + |Y_i - Y_j|$$

As you can see, the calculation of any distance in a vector database can be calculated by simply modifying the k factor to fit our needs (Figure 8.20). A k factor of 1.5, for example, simulates the distance modified by a combination of Manhattan and Euclidean. Even friction surfaces can be estimated through the use of these metrics by modifying the k factor. Take, for example, the use of a k factor of 0.6, which allows us to find the shortest distance around a barrier, such as a lake.

Although these metrics are very useful, we will at times want to perform distance measures for networks. We have the capability to find shortest paths and least-cost paths and to perform operations employing frictional impedance values as well as barriers. However, these are more closely related to connectivity than to distance, so we will examine them in Chapter 11 when we look at comparisons of variables within a single map layer.

Terms

absolute barrier	Euler number	least-cost surface
barrier	filter	orthogonal
closed cartographic form	friction surface	perimeter/area ratio
concavity	functional distance	relative barrier
convexity	impedance value	roving window
cumulative distance	incremental distance	shortest-path distance
edge enhancement	inscribe	shortest-path surface
edginess	isotropic surface	sinuous
Euclidean distance	least convex hull	smoothing
Euler function	least-cost distance	spatial integrity

Review Questions

1. Describe the simplest possible method of measuring a line in raster format. How can a result so obtained be converted to an actual measure of length when the grid cells are orthogonal? What do you do when they are lined up diagonally?

2. What are the possible pitfalls for measuring length with raster, especially with regard to the way grid cells capture linear data?

3. How do we measure the length of a sinuous linear object in vector? What is the relationship between sampling when digitizing a line in vector and the length of the line as it is measured?

4. Why would we be interested in measuring the long and short axes of a polygon?

5. Why is the relationship between the perimeter and area of a polygon important? How might differences in this ratio be important in the natural world? in the anthropogenic world?

6. How is the perimeter of a polygon measured in vector? in raster?

7. How is the area of a polygon measured in vector? in raster?

8. Given the relationship between feature shapes and function, why are there so few operational methods of shape analysis in GIS?

9. Describe two methods of measuring sinuosity of linear objects. Give two examples of such measures that might be useful to the GIS user.

10. What is spatial integrity? How can we measure it? Why is a given Euler number not a definitive measure of the number of holes and the number of polygons?

11. Describe how polygon boundary configuration might be calculated. What is a measure of convexity? Why is the circle the most frequently used geometric object against which polygonal shape is compared?

12. How can a roving window be used as a measure of polygon boundary configuration in raster?

13. How is Euclidian distance from a point to all other points in a coverage measured in raster? What is an isotropic surface, and what is its significance?

14. What is a friction surface? a barrier? What are the two types of barriers? What is the relationship between a relative barrier and a friction surface?

15. Describe a simple raster method of measuring distance through a friction surface.

16. What is the difference between incremental and accumulated distance? How can we measure the path of least resistance (best path) through a raster coverage that includes surface data? What is a more accurate method of measuring accumulated distance (best path surface) through a raster GIS than simply adding up the friction numbers?

17. How can we modify the Pythagorean theorem to account for Manhattan distance, paths around barriers, and paths through friction surfaces in vector?

18. What are the potential problems of assigning friction or impedance values for modeling distance?

References

Baker, W.L., 1989. "A Review of Models of Landscape Change." *Landscape Ecology,* 2:111–133.

Baker, W.L., and Y. Cai, 1992. "The r. le Programs for Multiscale Analysis of Landscape Structure Using the GRASS Geographical Information System." *Landscape Ecology,* 7(4):291–301.

Berry, J.K., 1993. *Beyond Mapping: Concepts, Algorithms, and Issues in GIS.* Fort Collins, CO: GIS World.

Boyce, R.R., and W.A.V. Clark, 1964. "The Concept of Shape in Geography." *Geographical Review,* 54:561–572.

Clarke, K.C., 1990. *Analytical and Computer Cartography.* Englewood Cliffs, NJ: Prentice Hall.

Davis, J.C., 1986. *Statistics and Data Analysis in Geology,* 2nd ed. New York: John Wiley & Sons.

Environmental Systems Research Institute, 1993. *Learning GIS, The ARC/INFO Method.* Redlands, CA: ESRI.

Forman, R.T.T., and M. Godron, 1981. "Patches and Structural Components for Landscape Ecology." *BioScience,* 31:733–740.

LaGro, J.A., 1991. "Assessing Patch Shape in Landscape Mosaics." *Photogrammetric Engineering and Remote Sensing,* 57:285–293.

Lee, D.R., and G.T. Sallee, 1970. "A Method of Measuring Shape." *Geographical Review,* 60(4):555–563.

McGarigal, K., S.A. Cushman, M.C. Neel, and E. Ene, 2002. FRAGSTATS: Spatial Pattern Analysis Program for Categorical Maps. Computer software program produced by the authors at the University of Massachusetts, Amherst. Available at the following web site: www.umass.edu/landeco/research/fragstats/fragstats.html

McGrew, J.C., and C.B. Monroe, 1993. *Statistical Problem Solving in Geography.* Dubuque, IA: Wm C. Brown Publishers.

Moellering, H.M., and J.N. Rayner, 1982. "The Dual Axis Fourier Shape Analysis of Closed Cartographic Forms." *Cartographic Journal,* 19(1):53–59.

Muehrcke, P.C., and J.O. Muehrcke, 1992. *Map Use: Reading, Analysis, Interpretation.* Madison, WI: J.P. Publications.

O'Neill, R.V., et al., 1988. "Indices of Landscape Pattern." *Landscape Ecology,* 1(3):153–162.

Robinson, A.H., J.L. Morrison, P.C. Muehrcke, A.J. Kimerling, and S.C. Guptill, 1995. *Elements of Cartography,* 6th ed. New York: John Wiley & Sons.

Strahler, A.N., 1975. *Exercises in Physical Geography,* 2nd ed. New York: John Wiley & Sons.

Turner, M.G., 1991. "Landscape Ecology: The Effect of Pattern on Process." *Annual Review of Ecology and Systematics,* 20:171–197.

CHAPTER 9

Classification

Classification of the earth's features is an old problem. It has engaged the minds of natural scientists of many types, each looking at the earth through different eyes and using a different vocabulary and a different intellectual filter. Although the concept of classifying features dispersed throughout the earth's surface may seem quite simple, the practical aspects of the task are far from straightforward. How we classify what we see will be strongly reflected in how we analyze data and what our conclusions mean. Indeed, the way we classify data may even determine what our conclusions will be. The process cannot be taken lightly, for poor classification will often lead us astray.

Let's say we have a set of digital maps of an environment developed some time ago. Among the many maps is one called land cover, which includes a classification of some areas called wetlands. This is simple enough. Consider the following questions and statements concerning wetlands, however. Suppose you have a large backyard, say a full acre. You have just watered it, and it now has many large areas of standing water. Are they wetlands? Are wetlands the opposite of uplands? Some well-established classification systems say they are. How big does a wet area have to be before we call it a wetland? Is it a wetland if it has waterfowl? What if the waterfowl show up only during migration periods at the changes of seasons?

As the foregoing questions indicate, how we classify things has much to do with how we view them, the scale or resolution at which we view them, and how we use the data later on. Our mode of classification may also influence the instruments we choose to employ: whether ground based or deployed from aircraft (aerial photography) or a satellite. The classifications may be simple, based on a single observed criterion such as ground cover, or complex, based on multiple sets of criteria (elevation, rainfall, ecological functioning, etc.). Some classifications can be created by combining many measurements of many different themes. The ways we classify can change dramatically for a single set of themes, based entirely on what questions we are trying to answer.

In this chapter you will become acquainted with the general problem of land classification. We will take a brief look at some classification systems and some basic methods of classification that can be used in the GIS environment. Finally, we will see how classifications themselves can be decomposed or combined with data from other coverages to reclassify what we originally had. This material is broken down into relatively simple reclassification procedures and more complex classifications.

When you are finished with this chapter you should be able to:

1. Understand the enhanced importance of classification afforded by new technology, as well as the common themes for the many classification systems available.

2. Know how to aggregate data in both vector and raster to achieve new classification systems.

3. Be able to reclassify data based on the manipulation of nominal, ordinal, interval, or ratio data.

4. Understand the types of filter available and know how to use both static and roving window filters for reclassification.

5. Have a thorough understanding of what neighborhoods are, what types are available, and how total and targeted neighborhood analysis can achieve new classifications.

6. Be able to describe the many ways neighborhoods can be created in GIS for two-dimensional map data.

7. Understand the differences among focal, zonal, block, and global neighborhood operations.

8. Describe how buffers are used for creating neighborhoods and understand the four basic methods for determining the buffer distance.

CLASSIFICATION PRINCIPLES

Among the most natural activities people undertake is classifying things. We classify even the most mundane items we encounter every day. We wear different types of clothes, drive vehicles of different kinds, have different categories of employment. We are male or female; Republicans, Democrats, or independents; adults, teenagers, or children; and so on. The classifications we use are designed to put people, places, and things in understandable sets based on selected criteria. In this way, we both group the members of each class and separate them from other sets. This is the essence of classification.

The tendency for humans to want to impose a classification framework has long been extended to the analysis of earth surface features. The centuries-old history of cartography shows us how people have grouped features by physical type (land vs. water), by political subdivision, and by human endeavor (inhabited vs. noninhabited). For centuries, these classifications were placed on maps with little formal discussion about how they should be organized or what impact their meanings might have on decisions. Only since the importance of map classification as a means of decision making and scientific inquiry was recognized have the problems of areal classification been examined at length (Sauer, 1921). Although the interest in examining the impact of land classification began to spread in the early part of the 1900s among organizations needing to make decisions about resources and land planning, a greater urgency seems to have emerged with the advent of two simultaneous technologies: satellite remote sensing and digital cartography/GIS. Because these systems had the ability to provide or manipulate vast amounts of spatial data, the time was ripe for developing an interest in better classifying portions of the earth. Then, as now, it is vitally important that the sometimes mundane task of classification be viewed with a measured sense of cynicism because of the potential limitations these classification **filters** impose on our interpretations and decisions.

Land classification depends on the types of objects we are going to group. Separate classifications have been set up for vegetation (Küchler, 1956), soils (Soil Survey Staff, 1975), geological formations and wetlands (Cowardin et al., 1979; Klemas et al., 1993), and agriculture, land use, and land cover (Anderson et al., 1976; U.S. Geological Survey, 1992). These classifications may be simple, such as a classification of vegetation for mapping based strictly on the plants present, or they may be more functional, such as mapping ecosystems rather than vegetation alone (U.S. Department of Agriculture, 1993; Walter and Box, 1976). Variations in classification are dictated by scale when, for example, vegetation is mapped not for small regions but for the entire earth (United Nations Educational, Scientific, and Cultural Organization [UNESCO], 1973). Or the system might be specifically oriented toward the mandates of a particular organization (Klemas et al., 1993; Wilen, 1990). Still other cases of classification are dictated more by the technology used to obtain the raw data, such as with the use of satellite remotely sensed data (Anderson et al., 1976). Yet others are designed specifically to address decisions based on known or suspected factor interactions, as exemplified by maps of biophysical units, land capability analysis, or land suitability, which have strong similarities but reflect subtle differences in the manner in which the classification is to be applied. Biophysical units use a combination of spatially arrayed co-occurrences of biological and physical phenomena related, for example, to soils and topography to evaluate the viability of the land units to support natural systems. Land capability most often applies specifically to the soil's ability to support housing, septic systems, wildlife, agriculture, and other major categories. Land suitability maps areas on the basis not only of soils but also other biological and physical properties; they are used primarily to evaluate the utility of the respective regions to support a variety of human activities.

Collectively, all these classifications, and the thousands of others that have been and are still being produced, have one thing in common: they all have a target audience. The end users of some classified datasets, such as land suitability for a single land use type, will be very specific. In other instances, the maps will be usable by a much larger audience. The UNESCO (1973) vegetation classification system was designed to unify vegetation classifications around the world, to permit anyone interested in vegetation, anywhere on earth, to talk about plants using the same basic system. Although this broad approach to classification may be practical as a general standard, its usefulness as a decision tool is restricted. In introducing some of the specific problems with selected map types in Chapter 3, we discussed the rigid nature of traditional maps developed under the communication paradigm. With today's GIS, we can store raw data before they are classified and manipulate them within the GIS, depending on our needs. This is the most important point about classification in the GIS environment: the more closely your classification can be made to fit user needs, the more useful it becomes. A GIS provides a wide variety of ways to classify and reclassify stored attribute data to achieve this result. Its functionality might actually be called one of reclassification because the data that are input to the GIS are already classified. The GIS operator either displays the existing classification (a storage and retrieval operation) or manipulates the existing attributes to create a classification more appropriate to the questions being asked and the decisions to be made. In the remainder of the chapter, then, we will refer to the function as "reclassification."

ELEMENTS OF RECLASSIFICATION

On many occasions point, line, and area features, together with their attributes, are input to a GIS under a specific set of criteria. We have seen how points and lines can be reclassified by simply recoding the attributes in their tables or by recoding the grid cell values to produce new point or line themes. In this simple process, the users change the attributes themselves, nothing more. The process is much the same when working on polygonal features with raster systems. By selecting the attributes for the areas in which you are interested, you merely change the numerical codes or attribute names for the grid cells.

GIS reclassification is at its simplest within raster data models. Whether focusing on individual point locations or groups (neighborhoods) of cells, the process is essentially one of changing the attribute codes. In the past, the legend associated with the grid cells did not always accompany the changes in numerical grid cell values. This meant that the user had to keep track of the changes that were made and adjust the legend accordingly. In addition, these primitive systems often required the user to perform a series of such numerical operations rather than operating on the categories. This forced the user to pay particular attention to the sequencing of mathematical or logical processes. Such sequencing is less commonly applied today because the raster software is more sophisticated. Still, care should be taken to avoid such obvious mistakes as multiplying grid cell numbers representing nominal categories by those representing ordinal, interval, or ratio categories. Such multiplication will always result in a 0. This property is often used deliberately to mask portions of a map that we don't wish to evaluate further. More detail concerning this issue can be found in DeMers (2002).

In vector, the process of reclassifying areas requires us to change both entities and attributes. The first task is to remove any lines that separate two classes that are going to be combined. This is called a **line dissolve** operation because it selects a particular line entity and dissolves or eliminates it. Then the attributes associated with the two polygons are rewritten for the new coverage as a single new attribute for both. Let's take a very simple example much like the one before but featuring only two polygons, one for wheat and the other for corn (Figure 9.1). Our purpose is to produce a single category called grain crops by "dissolving" (removing) the line separating the two original cate-

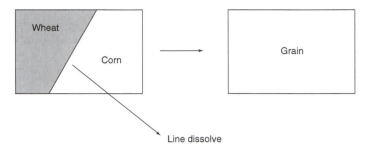

Figure 9.1   Reclassification and line dissolve. Reclassification through data aggregation: the regrouping of corn and wheat into a larger class called grains.

gories. We place the newly created category called grain crops in our attribute table and assign it explicitly to the new, larger polygon. We now have a new coverage with only a single categorical value. In most real applications, there will be far more line dissolves and far more attribute changes, but the process will be identical to the one we have just performed.

In both the raster and the vector reclassification of our polygons, an interesting pattern emerges. For both, we ended with fewer categories than we had originally. This result, called data **aggregation**, is a most useful and common simple reclassification. Imagine how you would go about trying to separate a large polygon into one portion with wheat and a second with corn. Unless you could compare this information to another map, it would be impossible to find the line separating them. And, if you knew where it was, you wouldn't need to perform an analysis at all; you would simply use the other coverage. There are useful methods for separating out more detail from coarse polygonal information. These methods require the use of two or more map layers compared with each other through a process called **overlay**, which we will cover in Chapter 12.

The method of reclassification we have used thus far is based on nominal data only. It is not difficult to envision how we could perform the same process of reclassification on ordinal, interval, or ratio data. In cartographic methods, this is done by creating range categories of data, often called **range-graded** classifications. Such a scenario merely requires us to recode these data on the basis of where they fit within the range classes. Just as we did with nominal data, we simply recode our grid cells or perform a line dissolve and attribute change in vector. We can also perform such operations as ordering the grid cell or polygon values, inverting them, or even using a process called **constrained math**, which permits the values of the target polygons to be operated on (multiplied, divided, etc.) by another number. But the process is fundamentally the same as what we found with nominal data.

All the methods thus far have one thing in common. The reclassification process is directed toward renaming the objects on the basis of the attribute values at their own location. It's a sort of local view (or what Dana Tomlin [1990] liked to call a worm's-eye view), in which each collection of grid cells or each polygon is seen as a distinct individual entity and the classification is constrained to the target area both in the use of initial values and in the reclassification itself. Reclassification on the basis of **locational information** is only one of four basic methods; the others are based on **position, size,** and **shape** information. But these concepts do not separate out as individual techniques. Instead, they are often combined with each other to provide a wide array of re-classification techniques. We will look at these techniques, starting with a general description of **neighborhood functions** (based on the idea of a bird's-eye view), and then proceed to examine several types including filters, two- and three-dimensional neighborhoods, and buffers. As you read along, notice the frequent similarities among the techniques as they are applied under different circumstances.

NEIGHBORHOOD FUNCTIONS

Reclassification based on locational attributes is very useful, but it limits us to the attributes within each object or feature. It would be nice if we could classify features based on a bird's-eye view. In other words, if we could fly over an area, we would be able to find out not only that a particular feature exists but where it is positioned within its surroundings. Such reclassification procedures are called neighborhood functions because the idea is to characterize each object as part of a larger neighborhood of objects based on some shared attribute. If you think about this for a moment, it is easy to recall growing up in a neighborhood. You might have classified that neighborhood based on size or because you happened to interact with the people in it. After all, we tend to interact more frequently with people who live closer to us than those who live far away. In some cases, neighborhoods are formally defined by political or economic criteria.

Neighborhoods may also be defined in terms of a unifying attribute for an entire area (such a classification is called a **total analysis of neighborhood**), or the focus may be on smaller portions of the total area (a **targeted analysis**) (De-Mers, 2002; Tomlin, 1990). A targeted analysis, also called **immediate neighborhoods**, includes only locations that are in immediate proximity (i.e., adjacent to) the target area or location. The total neighborhood analysis, also called **extended neighborhoods**, includes locations that are in immediate contact with the target area or location plus areas some distance beyond. Although this separation of neighborhood functions is both intellectually stimulating and quite useful to the advanced GIS analyst, at the introductory level it can be somewhat confusing. We will look at neighborhood functions operating on both two- and three-dimensional objects. We will also separate both of these types of neighborhood function into either **static neighborhood functions**, where the analysis takes place all at once for the selected target area, or **roving window neighborhood functions** (in raster), where the analysis takes place within the framework of a window that moves across the coverage. We saw an example of a windowing function in Chapter 8, when we characterized the edginess of the boundary for an area.

ROVING WINDOWS: FILTERS

As we saw in Chapter 8, some functions use a specified window to reclassify cells for a calculation of edginess. Such windowing functions are also called **filters**, especially when the window itself contains numbers against which the grid cells are to be compared. These can be used to reclassify whole raster grids based in a wide range of alternative filter values and combinations of values. This technique is used quite frequently in image processing (Lillesand and Kiefer, 1995), but it has equal utility in raster GIS operations. In particular, the filter is used, much as in remote sensing, to isolate edges or linear features (an approach called **high-pass filter**), to emphasize trends by eliminating small pockets of unusual values (often called **low-pass filter**), or even to give a measure of orientation **(directional filter)**. Although these filters can be statically placed over a target neighborhood to provide a neighborhood characterizing result (see the next section), most often they are applied as roving windows to characterize an entire coverage (Figure 9.2).

The high-pass filter (Figure 9.3) is designed to separate out detail in a raster coverage that may be obscured by nearby grid cells that contain relatively similar attribute values. In remote sensing, these values most likely measure electromagnetic reflectance. However, we can use almost any surface-related data. Let's say that we are interested in finding small topographic ridges in our raster database. With each grid cell containing a value for elevation, we want to highlight the differences between the slightly higher values of a ridge and the slightly lower values surrounding this feature. The standard method for performing a high-pass filter is to create a 3×3 filter with the weighting value of 9 at the center grid cell, in this case at location (2,2), and a weight of 21 for all the remaining grid cells. This filter is placed over each 3×3 matrix of grid cells in your coverage, and the members of each corresponding grid cell pair are multiplied. That is, the elevation value of the center grid cell in your topographic coverage is multiplied by the weighting factor of 9, and all the remaining topographic values are multiplied by 21. Next, these nine newly created values are summed to obtain the final, high-pass value, for the center grid cell. In other words, this single operation of nine grid cells operated on by the filter of nine grid cells produces a single value that is placed at the center of the new coverage.

The next step is to move the filter one grid cell to the right, so that the central cell in the topographic coverage is now (3,2). The calculations are repeated as before, resulting in a new high-pass value for column 3, row 2. The procedure is repeated for the entire coverage, with the result that all lower values are somewhat suppressed, whereas all higher values are enhanced or made larger.

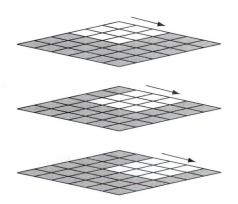

Figure 9.2 Roving window. A window of cells is placed at a starting point where the calculations are performed and the output cell value is assigned. Then the filter is moved one cell at a time across the entire raster coverage in serpentine fashion until it is finished.

High-pass filter moving window

	-1	-1	-1
	-1	9	-1
	-1	-1	-1

Original brightness values

	1	2	3	4	5	6	7	8	9	10	11	12	13
1	41	45	45	44	45	45	39	38	42	40	44	57	57
2	40	45	43	41	43	42	36	38	41	42	47	49	39
3	39	44	44	42	40	40	43	47	46	47	46	40	37
4	41	43	44	39	39	43	42	41	46	49	47	47	41
5	38	43	41	41	43	43	44	42	42	44	44	43	38
6	35	40	39	37	43	40	36	32	34	37	31	27	36
7	38	38	36	34	35	35	32	35	36	34	35	36	39
8	38	39	35	36	39	39	36	35	33	37	40	41	42
9	37	38	39	39	39	42	40	36	37	41	41	44	45
10	39	38	40	40	38	40	43	42	41	41	45	47	45
11	33	37	38	43	43	44	44	39	38	40	41	42	38
12	38	42	38	42	46	46	40	37	39	42	44	44	43
13	42	37	36	41	39	38	38	36	35	41	42	44	48

High-pass filtered brightness values

	1	2	3	4	5	6	7	8	9	10	11	12	13
1	31	60	53	45	56	71	30	27	59	18	20	11	145
2	26	64	37	23	48	47	-8	10	29	25	58	74	9
3	18	57	55	45	31	32	58	90	63	59	46	7	21
4	44	53	59	17	20	53	35	17	56	79	63	87	57
5	26	66	43	44	62	57	77	61	53	66	71	76	32
6	7	52	41	21	79	49	21	-13	4	33	-21	-59	15
7	41	42	26	6	11	15	0	41	47	23	32	33	40
8	39	52	16	28	52	53	30	30	6	36	51	47	48
9	27	37	46	45	38	64	47	17	27	54	33	50	60
10	59	41	48	41	12	27	60	60	55	45	68	82	72
11	-7	27	22	62	48	56	65	27	21	29	24	31	5
12	38	79	26	54	78	82	38	24	43	58	60	54	50
13	72	26	10	52	24	19	28	19	7	56	44	42	65

Note how some of the values are now much higher than before and others are much lower.

Figure 9.3 **High-pass filter.** Operation of a high-pass filter showing the use of a 3×3 matrix of numbers (the filter) designed to enhance the higher values and suppress the lower values. *Source:* A.H. Robinson et al., *Elements of Cartography,* 6th ed., John Wiley & Sons, New York, © 1995, modified from Figure 12.16, page 214. Used with permission.

What results is a coverage that shows the topographic ridge quite prominently. It can then be reclassified as you see fit, thus identifying as a specific topographic feature the region you have classified as a ridge.

Because edges and linear objects tend to come in different orientations (e.g., you may find a number of ridges related to some geological cycle of folding and erosion), it is sometimes desirable to "direct" your filter toward a particular orientation. If, for example, you are trying to highlight ridges that have an east-west orientation, you will want to enhance these features, leaving the others as low-contrast objects in your topographic coverage. To do this, you create a filter (we will stick to the 3×3 size) whose central row of numbers is positive and higher, whereas the numbers in the remainder are 1/9 (Figure 9.4a). For a northwest-southeast bias, you will have the higher numbers going from upper left to lower right on your spatial filter (Figure 9.4b). The same holds true for north-south and northeast-southwest bias, where the high numbers are oriented in the direction you are trying to enhance (Figure 9.4c, d). In this way, the higher numbers will be

Figure 9.4 **Filter configurations.** A variety of filter configurations can be used to identify linear objects in raster images. The orientations are used to isolate linear objects that are oriented in a particular way.

multiplied by the higher positive values and therefore enhanced or made larger, and the remainder will be suppressed by the operation with 21.

If you wish to suppress the higher values to remove meaningless topographic peaks, thus obtaining coverage that more closely resembles the overall trend in topography, you can use a filter in which there is less difference between values. The most common filter used for low-pass operations is a 3 × 3 filter whose numbers are all the same, commonly 1/9 because there are nine grid cells in the filter (Figure 9.5). Because the idea now is to make all the numbers closer together in value, you move the filter as before, one grid cell at a time through the entire image, but instead of adding the values to obtain the new number for the central grid cell, you average them. Essentially, then, with each pass you are obtaining an average number for the grid cells under the filter.

Although the normal method of filtering in raster databases employs the 3 × 3 filter and although set number combinations are programmed into standard software applications, there is no need to stick to either convention. Most software that contains the filtering capability will allow some flexibility in the size of filter and the numbers used. Smaller filters are more often used for edge enhancement, whereas larger ones are more common for low-pass filtering or averaging. The larger the filter size, the more averaged the numbers will become because of the larger set of numbers used for averaging. It often takes experimentation to decide on the utility of abandoning the default filter size and number combinations.

Figure 9.5 **Low-pass filter.** Multiplying each cell in the filter (a) by each cell in the original data (b) and totaling assigns the resultant cell value (c) to the central cell in the output matrix (i.e., 1/9 · 42 + 1/9 · 38 + 1/9 · 71 + 1/9 · 68 + 1/9 · 89 + 1/9 · 72 + 1/9 · 70 + 1/9 · 59 + 1/9 · 85 = 594/9 = 66).

Immediate Neighborhoods

We know that a GIS should be capable of measuring the size of a polygon or a fragmented region composed of numerous polygons. Suppose, however, that we are interested only in the identity of a region's polygons within a certain neighborhood or distance. For example, we are studying the spread of innovative farm practices to see whether a pattern of imitation emerges in which traditional farmers adopt new methods practiced by their neighbors. Let's say we are interested in neighborhoods of no-till agriculture. First we retrieve a coverage showing only locations in which no-till agriculture is being performed. Then we select a distance value as an estimate of where imitation is most likely to be observed. We might assume that no-till agriculture will be practiced only by connected neighbors, or we might assume that the idea will take root elsewhere as well. In either case, the GIS takes this value and spreads outward until it reaches the end of our selected search radius, all the while adding up the amount of no-till agriculture (essentially measuring the area of groups of polygons or polygons). The result will be small groups, medium groups, and large groups of fields that seem to indicate the sharing of this new practice.

The GIS has treated the original region of no-till agriculture so that all the polygons or grid cells that fall within the specified distance of one another are given the same attribute number. The groups are numbered on the basis of their rank compared with all other groups the software finds. Each group can later be reclassified by measuring its size. We might conclude that the farmers in the larger groups of no-till neighborhoods are more apt to talk to one another, or that the farms themselves are larger, or perhaps that these farmers knew some people at the nearby agricultural college. In any event, the values suggest that there is either a different mechanism for larger as opposed to smaller neighborhoods sharing these ideas, or some other functional reason for the difference in sizes of these clumps. The interpretation of the causes of these results will generally have to be further verified, but the neighborhood function of the GIS allowed us to recognize that differences exist.

In the preceding characterization of neighborhoods, we were looking at a single attribute to make our groups or clumps based on a specified distance. Frequently, however, we are less interested in groups of identical polygons or grid cells than in defining the similarities and differences within a specified neighborhood distance. As an example, let's say that we want to determine the average age of people in a specified region on the basis of census block data. By selecting a search radius, just as we did before, the software looks at the attributes for all the different census block polygons or grid cells and then performs a simple numerical average of these values. When it is done, it creates new coverage for average age on the basis of the values it calculated.

But average is not the only characteristic we can use to define new neighborhoods. Perhaps we are looking at a particular species of animal known to be particularly attracted to highly varied land cover types. The more diverse the cover types, the more these animals like the neighborhood, which provides a wide range of possible places to nest, to forage, and to take shelter from the sun and from predators. We may also know that this species needs a particular amount of land area in which to operate. We convert this known amount of area to a search radius and begin as before. In this case, the software looks at all the

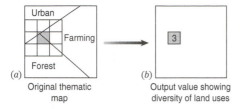

(a) Original thematic map   (b) Output value showing diversity of land uses

Figure 9.6   Focal function. By examining the surrounding grid cells *(a)*, this focal function counts the number of land uses in the neighborhood (3 × 3 cell matrix) and returns a single cell value of 3 *(b)*.

different land cover patterns within the search radius and counts them, returning the number of cover types as a measure of spatial or landscape diversity (Figure 9.6). Areas with the highest diversity would be the most likely place for the target species to nest. If we were trying to introduce this species into a region, we would most likely choose the areas of highest diversity.

If we can perform an average on the basis of our surrounding neighborhood polygons or grid cells, we can also perform a wide range of other calculations. We might want to know some maximum value for a neighborhood—for example, the maximum number of crimes for the neighborhood for a given year (Figure 9.7). Or we might want to know the minimum value, such as the lowest price for homes in the neighborhood to determine whether we could afford to live there. Other measures include total counts of all types in the neighborhood; the median, highest, and lowest frequencies; deviations from a central point to the mean of the surrounding values; and even the proportion of the neighborhood that shares attributes common to the central cell.

The measures we have identified in the last two paragraphs can operate in many different ways. If, for example, they return the value from analysis to the central point or target cell, that value (average, diversity, median, etc.) of the neighborhood is assigned to the central location in the neighborhood for a new coverage. We also do not necessarily need to assign a target distance at all, using instead a roving window approach to characterize the entire coverage on the basis of the same measures as before (average, diversity, median, etc.). In this case, the output values will not be returned to a central location but will be assigned throughout the new coverage, giving us an idea of the trends from one part of the coverage to another.

Grid coverage showing individual cell values before block function.

Grid coverage showing block values after a block Maximum function.

Figure 9.7   Block function. In some cases, we want to convert whole groups of cells on the basis of the immediate neighborhood. This particular block function uses a maximum value for the surrounding neighborhood cells, thus defining the 3 × 3 cell matrix on the basis of the highest value in the neighborhood. Notice that the blocks do not overlap. Block functions would be considered extended neighborhoods in Tomlin (1990), but they are called block functions in ARC/GRID (a raster-based GIS from Environmental Systems Research Institute [ESRI]).

In some cases we can characterize a neighborhood on the basis of the values contained within a single map, or we can use a separate layer for target cells and another for the neighborhood measure. For example, if we want to know the neighborhood diversity for known bird nest locations, we would choose the bird locations as target cells and another map of land cover as our measure of neighborhood. In addition, measures of neighborhoods such as average, minimum, and maximum are comparisons of the surrounding polygons with each other, whereas other cases (e.g., measures of deviation from the average, the proportion of a neighborhood that contains the same value as the target location) require the target location to be compared with the surroundings.

But the ability of the GIS to measure size and shape offers other ways of characterizing neighborhoods. A measure of size, for example, is often combined with a measure of the clumps of a single polygonal attribute value. Most often this information is used to rank or order the results of analytical techniques that group data together into localized or regionalized clumps. The sizes of clumps may be very important in our analysis. Ecologists, for example, are aware of the sizes of range required by wolves and other large predators (Forman and Godron, 1987). They are also familiar with the minimum requirements of individual forest stands to support a diversity of animal species or even to continue to exist as a forest patch. In some cases, there may be a need to mix a neighborhood measure of diversity (where higher diversity indicates better habitat) with a measure of size (where larger areas are preferred). Many other combinations can also be employed. Let's take a look at an example. While you read about this one, think of some others that might be of interest to you.

Jaguars are large predatory cats of a species that requires different types of habitat, especially jungle vegetation, for cover, and running water, because most of their diet consists of fish. For the sake of simplicity, let us say that jaguars require approximately 200 square miles of territory in which to live (a relatively accurate measure) and that they really need only two types of habitat—jungle forest and stream corridor. Let us further assume that researchers have indicated a requirement for territory that is at least 1 part stream corridor to 10 parts forest. We can identify the ratio between target points along the stream from a stream's coverage and the amount of surrounding (neighborhood) forest in the vegetation coverage that also includes stream corridor as a category. We can reclassify the vegetation coverage into stream corridor and forest vegetation to simplify our calculations. Then by using the neighborhood function that computes the ratio of the amount of neighboring area having the same category as the target location, we can identify the neighborhoods that have at least one-tenth stream corridor. Finally, now that we have created neighborhoods that suit the habitat-type needs of the jaguar, we can consider size. Neighborhoods that have at least 200 square miles of area with a stream corridor/forest ratio of 1:10 will be ideal sites for jaguar habitat.

Suppose, however, that the habitat neighborhood revealed by our analysis looks somewhat like an hourglass, having a narrow middle with the stream running through it. The areas outside our forest and stream corridor habitat are inhabited by humans, a species jaguars generally avoid. There may not be enough stream corridor in this bottleneck to support the animals' activities there. That is, as far as the jaguars are concerned, the two ends of the hourglass may be functionally separate habitat areas, neither of which is large enough for survival purposes. Using the ability of the GIS to measure distances across polygonal objects, we can perform an analysis that locates and classifies regions based on some value of narrowness, usually the narrowest portion of the polygon's shape. Thus we have employed another measure of neighborhood, this one based on shape.

Extended Neighborhoods

Immediate neighborhoods can be loosely defined as synonymous with contiguous regions. As we have already seen, we can reclassify neighborhoods in many ways using values available from within contiguous regions. However, we can extend the concept to include attributes associated with fragmented or discontinuous regions as well. For a clear understanding of a fragmented region, one might easily use the islands of Japan or the Hawaiian islands as examples of noncontiguous areas that share common characteristics. Our GIS functions applied to these types of regions is most often called **zonal** as opposed to the **focal** and **block functions** we discussed before.

The idea behind zonal operations is very similar to that of the focal operations. The available mathematical operators are essentially the same, and the results are typically stored as a single-cell (point) output. We begin as before by selecting a target location (most of a grid cell) upon which we will operate. Instead of selecting an adjacent neighborhood, however, we begin by classifying a region based on some measurable characteristic. We might characterize our fragmented region as a collection of forested patches that might be found appealing to some bird species if other conditions are satisfactory, so the bird will evaluate the forest patches to determine whether an appropriate nesting site is available at our target cell.

We examine the characteristics of all portions of the fragmented region to determine, for example, that there is a selected average tree density. This is the same thing we did before with immediate neighborhoods using the focal function, except now we "average" the values for the extended neighborhood (i.e., the fragments) and then pass the value on to the output target cell (Figure 9.7). Though similar in approach, the extended neighborhood allows for characterization of single-point locations based on often widely separated fragments of a region that share common attributes.

BUFFERS

Another very common method of reclassification is a process called **buffering**. A buffer is a polygon created through reclassification at a specified distance from a point, line, or area. Because it is based on the location, shape, and orientation of an existing object, we could easily have classified buffering as a method of reclassification based on position. However, a buffer can be more than just a measured distance from any other two-dimensional object; it can also be closely related and even controlled to some degree by the presence of friction surfaces, topography, barriers, and so on. That is, although buffering is based on position, it has other substantial components as well. An area sur-

Stream   Stream plus buffer

Figure 9.8   Line buffers. Stream and its buffer based solely on a distance measure selected by the user.

rounding a stream that indicates something about the stream corridor offers an example of a buffer. In Figure 9.8 the buffer was produced by reclassifying the area on either side to differentiate it from the amorphous background. Although the figure shows the stream plus the buffer, normally the buffer is generated as a separate feature and is often stored as a separate coverage as well. To produce this buffer merely requires that a specified distance be measured in all directions from the target object. We have seen how a GIS performs a distance measure in both raster and vector; creating the buffer is merely an extension of that procedure. But because of the utility of the technique and the frequency with which it is applied, most GIS software includes separate routines that allow the buffers to be produced.

Buffering is a matter of measuring a distance from another object, whether a point, a line, or another polygon. In the case of point features we measure out a uniform distance (in raster or vector) in all directions from that point (Figure 9.9). In some cases, we may want to produce a buffer around a linear feature, as in Figure 9.8. Or, returning to Figure 9.9, we could begin with an area and measure a set distance from its outer perimeter. There may even be a requirement for a buffer around a second buffer around still another buffer to produce a

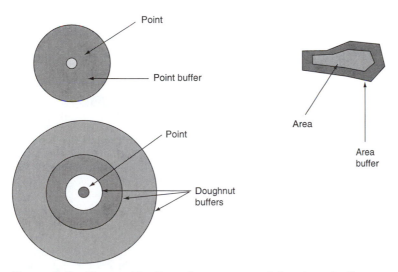

Figure 9.9   Types of buffers. Point, area, and doughnut buffers.

doughnut buffer, so called because of its resemblance to the pastry (also shown in Figure 9.9). This procedure is relatively simple in raster because each buffer is simply a number of grid cell distances beyond the existing feature or the previously created buffer.

In vector data structures, however, we must explicitly encode the topological information for each polygon we produce. In particular, we are required to provide topological information about the connections between polygons by explicitly placing nodes at the beginning and end of each bordering line segment. A doughnut buffering procedure attempts to create an island polygon that is not connected explicitly to the neighboring polygon. The difficulty of producing a doughnut buffer in vector is largely a function of the data model used, but by carefully following the software instructions, you can generally obtain this buffer. You may want to experiment on a test database before you try to produce a doughnut buffer on a real database.

Although we have seen that buffers can be extended outward from a point, line, or polygon, they are not limited to that direction. Buffers can be directed to extend both outward and inward at the same time (Figure 9.10*a*). In addition, where real-world buffers such as setbacks are used—for example, the setback from the center of a suburban street to some portion of a homeowner's property that is available to the city for lighting and utility work—buffers can easily be designed using a GIS (Figure 9.10*b*).

Now we have seen that buffers are essentially distances measured in any or all directions from point, line, and area objects. But we have not mentioned one of the most pressing problems related to creating buffers—that of distance. How big should the buffer be? This question often comes up in software training seminars when students are asked to produce a buffer around an object in a coverage provided with the tutorial. Unfortunately, the most common response to this question in such a setting is, "It doesn't matter; this is just to show you how to do a buffer." But, as we have seen, being able to build a buffer is not worth much if you have no idea what its purpose is or how large it should be. These basic conceptual questions are not really separate because the distance of our buffer is often dictated, or at least modified, by the purpose of the buffer. We'll look at these matters by focusing on the question of distance as the unifying factor.

How large should a buffer be? Some buffers are designed to indicate that around a given entity, for an unknown and perhaps unknowable distance, lies a region that needs to be protected, studied, guarded, or otherwise afforded special treatment. This scenario is not as uncommon as you might think. Many real-world buffers are just as arbitrary as the digital buffers we produce in our

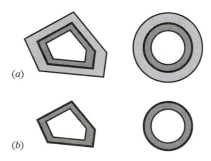

(a)

(b)

Figure 9.10 Bidirectional buffers and setbacks. Although buffers can be extended on either direction of the object *(a)*, they can also be directed toward the center of polygonal features to model setbacks *(b)*.

GIS. Yet in both cases, there is usually an underlying gut feeling behind our choice of a particular arbitrary distance. Contractors frequently build buffers around construction sites to protect passersby from falling debris and heavy machinery. The boundaries of areas believed to be contaminated areas by poison gas, nuclear disaster, and hazardous materials spills are generally imposed by governmental or law enforcement agencies. But like the areas cordoned off around spaceships in the 1950s science fiction movies, quite often these danger zones are only guesses—that is, **arbitrary buffers**. Most often, the guesses are based on gut feelings or anecdotal information of unknown source. They do share one characteristic, however, and it illustrates a time-honored approach—when in doubt, err on the conservative side. In other words, all these arbitrary buffers seem to be larger than they actually need to be. You probably notice this when you are forced to walk or drive around a construction buffer, or when a buffer surrounding a relatively small anomaly forces you to find another place to park. The extra distance that inconveniences the public is the added margin of safety for an arbitrary buffer.

Buffer distances can also be based on nearly any of the measures and reclassification procedures we have seen so far, whether two-dimensional or three-dimensional. For example, we could create a second type of buffer based on a functional rather than Euclidean distance from the target object. This would be a **causative buffer**, one based on a priori knowledge about the area within which the buffer is produced. Let's say, for example, that we are putting a buffer along a stream to indicate the potential for contaminating soil on either side of the stream. We know, however, that the soils on one side of the stream are heavily clay rich, whereas the soils on the other side are richer in sand. Because contaminants will pass through sand faster than through clay, the buffer must be based on the frictional or impedance quality of the clay soil. The result will be a buffer smaller in Euclidean diameter on the clay side and larger on the sand side, reflecting the difference in permeability of the soils. The use of friction surfaces and barriers is common when producing buffers, because they offer some justification for the buffer width we choose. However, as you remember from our discussion of friction surfaces, they tend to be poorly understood. Thus a buffer assigned on the basis of a friction or barrier impedance value may be of little more use than an arbitrary buffer of simple design.

A buffer can also be based on visibility measures. In that case, the buffer is selected not on the basis of an arbitrary value or on a poorly defined friction value but rather on a definable, measurable value—a **measurable buffer**. This is the third basic way a buffer distance can be determined. The measurements are not arbitrary but are quite accurate, being based solely on measurable phenomena. A common approach is to combine the second and third methods of buffering, relying on a measurable phenomenon whose interaction with the buffer area is poorly understood. For example, we know that trees work quite well as filters of contaminating materials along stream corridors. We know also that the more trees, the more effective the filter will be. So we could measure the density of vegetation along a stream corridor and then reclassify it to produce a number of neighborhoods, each representing a different class of tree density. Next, we could use these different tree densities as surrogates for friction surfaces, which we would then employ to modify the amount of buffering. However, we are not absolutely sure how directly the tree density is related to the amount of movement of contamination through the vegetation and into the streams. Thus we have some measurable values, and they can be logically applied, but we don't know the exact relationships among them. This is often the case when buffers are employed.

When the fourth and final method of buffering is employed, however, legal or otherwise mandated measures exist and must be adhered to—a **mandated buffer**. For example, if you build a home within the 100-year flood zone in some communities, you are not likely to be allowed to purchase flood insurance. Although the 100-year flood zone is a measurable value, the insurance companies could just as easily have selected the 75-year flood zone or the 150-year flood zone. In other words, the value itself is measurable; its selection over other values is arbitrary. To develop a buffer of this kind usually requires a database of the terrain and an ability to calculate the volume of water that would fill the stream floodplain if a flood of a size not likely to occur more than once in a century were actually to occur.

But other mandated buffers can be employed. We are told how close to a fire hydrant we can park and how much of our frontage property actually belongs to the local community. Building codes specify distances around utility areas and between and among buildings; conservation organizations obtain easements on properties; legal ownership buffers are established around rail lines and power corridors, and so on. You might be able to think of a number of others. In each case, there is a legal reason for placing the buffer a set distance from other objects.

No matter which of the four basic buffer methods (arbitrary, causative, measurable, and mandated) is used, the buffer may not be the same along the en-

Figure 9.11 Variable buffer. Here we see that three different buffer sizes are used to indicate the difference between trunk streams and tributary streams. Each of these different stream types has a different code, allowing different-sized buffers to be created.

tire length of linear objects or on all sides of a polygon. Such differences, exemplified by our different buffer widths along the stream based on soils, create a general class of buffers called **variable buffers** (Figure 9.11). The variable buffer can be based on friction, barriers, or any of our other neighborhood functions. It can be selected arbitrarily or on the basis of a measurable component of the landscape, or it can be mandated by law. In each case, special procedures must be invoked in the creation of the buffers. For vector, the nodes between line segments can most often be used to dictate buffer differences along the line segments. In raster, the grid cells need to be coded selectively so that a buffer for each group of grid cells can be established; these buffers most likely will be combined later into a larger buffer map layer.

Buffers are useful methods of reclassifying the landscape and are a common feature in many GIS analyses. The fundamental problem with buffers is that they frequently require us to know more about the interactions of our landscape's elements than we do. You should always attempt to overcome this obstacle by seeking out the best available knowledge about each situation before proceeding. The more you know, the more confident you will be about the distance you have buffered. If you have a legally mandated buffer distance, you need only be sure that your buffer corresponds to that mandate. But if your knowledge about the interactions of your landscape elements is weak, or if you have no knowledge on which to base selection of the buffer distance, your best bet is to make the buffer larger rather than smaller.

Terms

aggregation
arbitrary buffers
block function
buffering
causative buffer
constrained math
directional filter
doughnut buffer
extended neighborhoods
filters
focal function

high-pass filter
immediate neighborhoods
line dissolve
locational information
low-pass filter
mandated buffer
measurable buffer
neighborhood functions
overlay
position measures
range-graded

roving window neighborhood functions
shape measures
size measures
static neighborhood functions
targeted analysis
total analysis of neighborhood
variable buffers
zonal function

Review Questions

1. What two technological advances have heightened the importance of classifying spatial data? Why is it important that we view classification with a measured amount of cynicism?

2. Why are classification systems for earth-related data so varied? After stating some of the reasons, identify the common theme that runs through them all.

3. Give an example of a simple reclassification procedure that aggregates two or more classes together into a single class. How is this done in raster? in vector?

4. When one is reclassifying areas using a simple raster system, why is it important to keep track of the order in which the grid cells are reclassified? Give an illustrative example.

5. What is a line dissolve operation? How is it used for reclassifying areas? How does it work?

6. What is the relationship between reclassifying areas on the basis of nominal data types versus ordinal, interval, and ratio data types? What are range-graded data? How are they used in the reclassification process?

7. What are filters? How are they used for reclassification processes? What's the difference between static and roving window filters?

8. What is the difference between a high-pass and a low-pass filter? Describe how both work and give their respective purposes. What is a directionally biased filter? How does it work to reclassify features?

9. What are neighborhoods? Give some real-world neighborhoods with which you are familiar. What is the difference between immediate and extended neighborhoods? What do neighborhoods have to do with reclassification? What is the difference between total and targeted neighborhood analysis?

10. Describe a process of defining neighborhoods on the basis of a single attribute. What are some possible ways this could be used?

11. How could we use such mathematical procedures as average, maximum, and diversity to define a neighborhood? Create a table that shows a number of possible math-based measures of neighborhoods and an example of how each could be used. Describe how any one of these (e.g., diversity) actually works. Diagram the process.

12. Give examples of neighborhoods defined by a combination of possible methods (size, shape, diversity, etc.). Explain how this will be done: what steps are taken, how they are to be performed, and in what order. If you have access to a GIS, you may want to create a simple database to try this out.

13. What are buffers? How are simple buffers based only on distance calculated? What is a doughnut buffer? Why is it so difficult to execute in vector GIS?

14. Describe how buffers can be produced on the basis of functional distance, neighborhood classifications, and so on. What is a variable buffer? How is this done in vector? in raster? When would it be used?

15. What are the four basic methods of determining buffer distance? Describe them.

References

Anderson, J.R., E.E. Hardy, J.T. Roach, and R.E. Witmer, 1976. A Land Use and Land Cover Classification for Use with Remote Sensor Data. Geological Survey Professional Paper 964, U.S. Geological Survey. Washington, DC: U.S. Government Printing Office.

DeMers, M.N., 2002. *GIS Modeling in Raster.* New York: John Wiley & Sons.

Forman, R.T.T., and M. Godron, 1987. *Landscape Ecology.* New York: John Wiley & Sons.

Klemas, V.V., J.E. Dobson, R.L. Ferguson, and K.D. Haddad, 1993. "A Coastal Land Cover Classification System for the NOAA Coastwatch Change Analysis Project." *Journal of Coastal Research,* 9(3):862–872.

Küchler, A.W., 1956. "Classification and Purpose in Vegetation Maps." *Geographical Review,* 46:155–167.

Lillesand, T.M., and R.W. Kiefer, 1995. *Remote Sensing and Image Interpretation.* New York: John Wiley & Sons.

Sauer, C.O., 1921. "The Problem of Land Classification." *Annals, Association of American Geographers,* 11:3–16.

Soil Survey Staff, 1975. *Soil Taxonomy: A Basic System of Soil Classification for Making and Interpreting Soil Surveys.* U.S. Department of Agriculture, Soil Conservation Service, Agriculture Handbook No. 436. Washington, DC: U.S. Government Printing Office.

Tomlin, C.D., 1990. *Geographic Information Systems and Cartographic Modeling.* Englewood Cliffs, NJ: Prentice Hall.

United Nations Educational, Scientific, and Cultural Organization, 1973. *International Classification and Mapping of Vegetation.* Paris: UNESCO.

U.S. Department of Agriculture, Forest Service, 1993. *National Hierarchical Framework of Ecological Units.* Washington, DC: U.S. Government Printing Office.

U.S. Geological Survey, National Mapping Program, 1992. "Standards for Digital Line Graphs for Land Use and Land Cover, Technical Instructions, Referral ST0–1–2." Washington, DC: U.S. Government Printing Office.

Walter, H., and E. Box, 1976. "Global Classification of Natural Terrestrial Ecosystems." *Vegetatio,* 32:75–81.

Wilen, B.O., 1990. "The U.S. Fish and Wildlife Service's National Wetlands Inventory." *U.S. Fish and Wildlife Service, Biological Report,* 90(18):9–19.

CHAPTER 10

Statistical Surfaces

Thus far, we have concentrated on point, line, and area features, but we have made frequent reference to surface features as well. We have seen how the surface features can be modeled in the computer environment, and we have just found that they can be used to modify our ideas of what a neighborhood is and how it operates. Surfaces are frequently used to model the friction or impedance that we encounter as we travel. They can be grouped into **slopes** and **aspects**, as well as viewsheds and specific features such as valleys, hills, and watersheds. But the surfaces we will encounter will not always be topographic surfaces. As we will see, our geographic filter can distinguish surfaces that are **continuous** or **discrete**, **smooth** or rugged, physical or cultural. In other words, our definition of surfaces will expand to include data of any type that either exist or can be assumed to exist as changing values throughout an area.

The topographic surface is the most readily understandable, and throughout this chapter we see frequent reference to topography as a classic type of surface. Topographic features are developed because at locations throughout the areas they occupy there are measurable objects, in this case elevation values. Because topographic elevation values occur everywhere, we say that the surface is a continuous surface. However, as we try to record this information with a view to producing reasonable and quantifiable descriptions, we find a major dilemma, not unlike that associated with trying to describe trees in a forest or blades of grass in a grassland. So much data exists that we simply cannot record the complete set. And so, as before, we must produce a meaningful sample of the elevation values from which we can reconstruct the essence of the topography. A strong similarity can be seen between the properties we sample for topographic surfaces, such as barometric pressure, temperature, and humidity, and other continuous surfaces. These variables also occur everywhere, but we cannot record data for them at every location. Instead we select locations to represent the distribution. We will learn how this is done, to ensure that our travels can be recorded properly.

Although continuous surfaces present a particular set of problems for the GIS professional, many other objects do not occur continuously everywhere; rather, they are found as discrete objects at specific locations. Because they occur with a very high frequency, or because we want to record them for very large regions, however, these discrete data must also be sampled. In some cases, as with continuous data, we will take samples at specific point locations, whereas in other cases, we must collect data for whole areas at a time. For example, as we travel across the United States we will surely find differences in human population numbers for each county. Rather than taking the time to tour each county, locating individual people at every stop, we simply add up the number of people for each of the 3000 or so counties and use them as point samples. In both cases, we can produce maps that resemble the contours of a topographic map by assuming that they occur everywhere. In this way, we can produce either a contourlike map or a fishnet map that models the trend or form of the distribution. Although we know that the objects we measure are not continuous, it is useful to employ these techniques for the sake of ease in communicating the shapes and patterns of distribution. We will also see how these techniques can be used to predict distributions in places we have not sampled.

Because of the increased importance and availability of surface data, we need to study this topic separately, despite its inclusion in earlier chapters. We will learn about the kinds of data that can be used in a surface modeling framework, the different surface models that can be used to display them, how to make predictions about missing data in a surface, and how such extrapolations might be employed in decision making. We will also look at some commonly available surface data types and learn how they were derived, how we can use them in our work, and how they can be converted to data models of different types. And we will see how surface data can be created from point observations of a wide variety of natural and human phenomena to enhance their utility for modeling.

LEARNING OBJECTIVES

When you are finished with this chapter you should be able to:

1. Define and describe a statistical surface and explain some nontopographic forms.

2. Describe how statistical surfaces can be represented using isarithms, and tell how the method of sampling data changes how statistical surfaces are represented.

3. Explain how statistical surfaces can be represented by regular and irregular lattices and how these can be converted to other forms, such as discrete altitude matrices, digital elevation models, and TIN models.

4. Explain how a continuous surface is represented in raster format.

5. Understand and describe the different methods of interpolation, their uses, and their advantages and disadvantages.

6. Explain the four factors to be considered in any form of interpolation and understand how problems of each type can be solved.

7. Understand how visibility and intervisibility analysis work to produce viewsheds.

8. Describe Z-value slicing and explain how it is used.

9. Explain the use of the method of ordinates for determining volumes.

WHAT ARE SURFACES?

Surfaces are features that contain height Z values distributed throughout an area defined by sets of X and Y coordinate pairs. The Z values need not be restricted to elevation values. Instead, any measurable values (i.e., in ordinal, interval, and ratio data scales) that can be thought of as occurring throughout a definable area can be regarded as comprising a surface; the term generally used is **statistical surface** because the Z values constitute a statistical representation of the magnitude of the features under consideration (Robinson et al., 1995). Perhaps because the statistical surface extends our geographic filter to include such values as population, salary, animal densities, and barometric pressure, it is often considered to be among the most important cartographic concepts.

Behind the statistical surface is the idea that all the measured values (Z values) either occur continuously across the area of interest or can be assumed for the sake of mapping and modeling to occur continuously across it. Statistical surfaces that employ data that occur to some degree at every location inside the study area, such as elevation data, are called continuous (see Figure 2.3). Those that occur only as individuals but with some difference in numbers per unit area, such as the number of houses per square mile in each neighborhood, are called discrete (see Figure 2.3). Because conceptualizing the statistical surface for these two data types can be tricky, we will examine them separately.

"Continuous" data are said to occur at every possible location in the area. That is, it is possible to obtain a measurement, no matter how small, for this variable anywhere within the area in question. We know, for example, that air temperature occurs everywhere. At every location, we could obtain a measure of the temperature with a thermometer. If we did so, we would see that the values change gradually from place to place in a continuous stream. In some cases, the temperature changes slowly, perhaps less than a tenth of a degree per 100 meters. We say that these continuous data demonstrate a smooth surface, with little change in statistical information per unit distance. At times, however—when we cross under the space between two totally different air masses along a frontal line, for example—the temperature values change very abruptly. We call the surface produced by rapidly changing values a **rough surface** because a major change in the statistical data occurs with small changes in distance.

The obvious advantage of continuous data is the certainty that there are plenty of data to work with. In defining our statistical surface as continuous, we assume that there are an infinite number of data points from which we can make our measurements. This poses two closely related problems. First, it is physically impossible, even with the most advanced scientific instruments, to measure an infinite number of locations. Second, even if we could measure everywhere, no computer is capable of handling an infinite amount of data. So defining a continuous surface on the basis of an infinite number of data points must be replaced by a method of modeling that allows us to use samples taken at critical locations. These samples will serve to represent the most important changes in surface as a simplified representation.

SURFACE MAPPING

The statistical surface is most often displayed in analog form as an **isarithmic map**. This approach is the most common one for continuous types of surface data. We can envision isarithmic mapping as a series of lines surrounding our statistical surface. Each line, commonly called a **contour line** when applied to topography, represents all points that occur at the same elevation. The general term for lines connecting points of equal statistical value is **isarithm**. Figure 10.1 shows a portion of a digital elevation model of the Colorado Mountains illustrating the concept of isarithms, especially prominent in the upper left-hand corner of the model.

Viewed from the top, isarithms appear as a series of semiparallel lines. If they surround a topographic form, such as a hill, they will be closed and will encircle the feature; if not, they continue until they hit the edges of the map. They allow us to observe specific patterns and shapes based on the type of topography encountered. We see, for example, that isarithms that cross stream valleys tend to look like the letter *V* pointing uphill or upstream. We see that steep slopes show more isarithms spaced closer together horizontally than gently sloping surfaces. You can now see how the traditional contour lines can be used to represent neighborhoods of slope as well: steep slopes with closely spaced contour lines; gentle slopes with widely spaced contour lines (Figure 10.2).

Figure 10.1   Digital elevation model (DEM) of a portion of the Colorado Rocky Mountains. Note how the isarithms appearing in the upper left-hand portion of the map are unevenly spaced horizontally to adjust for their vertical sampling.

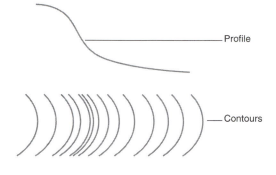

Figure 10.2 Contour spacing and surface configuration. Steep slopes are shown by closely spaced isarithms, and gentle slopes are shown by widely spaced isarithms. The distance between lines (the contour interval or class interval) is based on elevation differences, not horizontal differences.

We can also observe another important aspect of these isarithms by looking at Figure 10.1 again. It is no accident that the vertical distance between isarithms is the same for any two pairs of lines. The vertical distance between lines was preselected to allow the reader to more easily understand what the lines represent. Each contour map has its own vertical distance between isarithms, depending on whether there is a rapid or a slow change in elevation. This preselected value, called the **contour interval**, is used to divide or quantize the change in Z value by choosing a uniform amount of elevational change that will continue to be used at every step. Thus the person viewing the map knows that each contour line indicates a set difference in elevation rather than one that changes all the time. Imagine the isarithms that would result if we used a different, randomly selected distance value for each contour line drawn. If we had a steep mountain in our topographic surface, the lines would not necessarily be close together; instead, the distance could change arbitrarily, thus preventing us from interpreting closely spaced lines as steep slopes and widely spaced lines as gradual surfaces.

SAMPLING THE STATISTICAL SURFACE

Because continuous surfaces have an infinite number of possible Z values, it is essential to sample them for input to the GIS. Samples of Z surface data can be obtained in two major ways. The first method employs selected point values, in which case the isarithmic map produced is said to be **isometric**. This method is most commonly applied to elevational, barometric pressure, temperature, and other natural surface data. However, we may also wish to work from data aggregated by small areas such as population by county. Although we know that these data are discrete, we can treat them as if they were continuous. By assuming that each of these areas is a point, we can produce the same kind of isarithmic map that would result if the data had, in fact, been gathered at point locations. This type of isarithmic map, called an **isoplethic map**, requires us to determine where we are to place the points. We have already studied how centroids or centers of gravity can be calculated and used in determining such placement.

In the case of isoplethic maps, we choose the data collection point from within each area on the basis of its central point or on a central point weighted by the clustering of the data. But in isoplethic mapping we have to decide where we are going to select the data sampling points in the first place. There are two general approaches. The first employs a **regular** sampling procedure or grid, often called a **regular lattice** sampling. The points are later connected to form a triangular lattice **structure** much like trellises used in gardens to support climbing plants (Figure 10.3a). The regular lattice sampling approach has the advantage of simplicity in that there is no need to spend a large amount of time deciding where the points are to be placed. A distance for both X and Y directions is selected, and the points are measured where the X and Y lines cross. A major problem arises, however, when it is assumed that by sampling regularly throughout a surface, we can represent both smooth and rough surface features. This assumption is often invalid. A rough surface implies a rapid change in Z value with distance. We say that there is more information per unit area because every change in elevation is seen as a potential piece of information regarding the surface form. When the amount of surface information increases with distance, an identical need arises to sample that area more than an area exhibiting little change in Z value. For this reason, we will obtain a better understanding of surface shape if we use an alternative method of sampling called the **irregular lattice** (Figure 10.3b).

In an irregular lattice, we can determine the amount of sampling to be done in any part of the surface on the basis of prior knowledge of how rugged or smooth it is. We are not restricted in either the X or the Y direction. Neither are we restricted in the number of point samples we can make within any given area. It might be assumed that in the absence of restrictions, we are likely to make more samples using this method, but this is not necessarily the case. Most often we can obtain a good model of the surface with fewer data samples because we can reduce the number of samples for relatively smooth surfaces well below the number that would be needed for a regular lattice. In other words, whereas we increase sample density for areas with increased surface information, we can decrease the sample density for areas with decreased surface information. Increased sample density for smooth surfaces gives no relevant information; it just adds unnecessary time and effort as well as increased data storage later on.

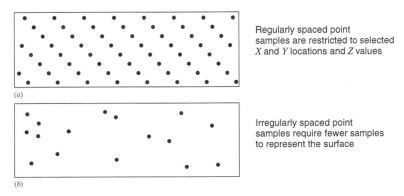

Regularly spaced point samples are restricted to selected X and Y locations and Z values

(a)

Irregularly spaced point samples require fewer samples to represent the surface

(b)

Figure 10.3   The lattice. (a) Regular and (b) irregular lattices of points. The points can be sample locations themselves or points as they are represented in the GIS database.

THE DEM

Now that we have established a procedure for selecting sample data points for creating a map of a statistical surface, we need to understand how a surface could be represented inside the computer, both in raster and in vector. We have introduced the TIN model, the basic vector data structure for representing surfaces in the computer. However, the TIN model is one of a number of methods of storing *Z*-value information, creating a group of products collectively called **DEMs**. Such methods are based either on mathematical models or on image models (including TIN) designed to more closely approximate how they are normally sampled in the field or represented on the paper map (Mark, 1978). Although mathematical models are quite useful, the currently available DEMs are most often based on image models. We will therefore focus on image models and leave the more advanced student to pursue the mathematical model representations. Image models come in two general types: those based on points and those based on lines.

Image models of *Z* surfaces based on lines are nearly the graphical equivalent of the traditional method of isarithmic mapping. In many cases, such models are produced by scanning or digitizing existing contour lines or other isarithms. The purpose is to extract the form of the surface from the lines that most commonly depict or describe that form. Once input, the data are stored either as line entities (each line with a common elevation attribute and each line segment connected by points) or as polygons of a particular elevational value. Because it is not particularly efficient to calculate slopes and aspects and to produce shaded relief output from such data models, it is more common to convert them to point model form, treating each point connecting each line segment as a sample location with an individual elevation value. This produces a **discrete altitude matrix**.

The discrete altitude matrix is a point image method that represents the surface by a number of points, each containing a single elevational value. Much like the methods of sampling topography in the field, the altitude matrix uses a number of data points sampled from stereo pairs of aerial photographs, usually employing an analytical stereoplotter that locates and measures elevational data based on the displacement of the two photographs (Kelly et al., 1977). Again paralleling the field sampling procedures, the analyst can use a regular lattice or an irregular lattice. Because the regular lattice oversamples where the topographic information is minimal and undersamples where it is abundant, the irregular lattice is preferred. Progressive sampling using automatic scanning stereoplotters that increase the number of samples in highly changing topographic regions offers an easy route to execution (Markarovic, 1973).

Irregular lattices can be converted to TIN modes under two different approaches. The first is to use the data points in the irregular lattice itself as the basis for the triangular facets in the TIN. This has the merits of requiring little additional input, instead allowing the GIS to create the triangular facets. Alternatively, some form of interpolation among data points can be used. The product of such an interpolation approach is a predictive set of data points that can be used to create the TIN facets. Although interpolation allows the addition of data points to the altitude matrix and, ultimately, to the TIN, predicted values are not as accurate as measured data points. Thus, any model produced from such a technique has added error. Even the order in which DEM data points are selected for insertion into the TIN can have an impact on the overall quality of the spatial model (Little and Shi, 2003).

DEMs are readily available for many parts of the world as altitude matrices with grids of 63.5 meters obtained from 1:250,000 scale topographic maps; they are becoming available in larger scale formats such as can be obtained from 1:25,000 maps and aerial photography (Appendix A). Among the advantages of using DEMs derived from interpolation techniques that create a regular matrix is ease of input to raster GIS. DEMs based entirely on an irregular lattice of point values and input to a raster GIS will have to be interpolated using raster techniques. We will look next at the representation of continuous surface data in a raster GIS, and then we will more closely examine how the interpolation process works.

RASTER SURFACES

Because raster data models divide the surface of the earth into distinct units, each unit or grid cell can contain only a single absolute elevational value. This, in effect, takes a continuous data variable and converts it to a discrete representation (Figure 10.4). Although each grid cell contains a single absolute elevational value, it also occupies space. To more accurately represent elevation in raster, you may have to select a relatively small grid cell size and increase the accuracy with which the point elevational values for each grid cell represent an appropriate elevation throughout the entire space occupied by the grid cell.

Although grid cell size is important in coding continuously varying data into a raster GIS, it is equally important to decide where within the grid cell area you want the actual elevation point to be located. You have options for indicating the exact location of the elevation point: at the center of the cell or in one of the four corners (Figure 10.4). For analyses that require the use of topography,

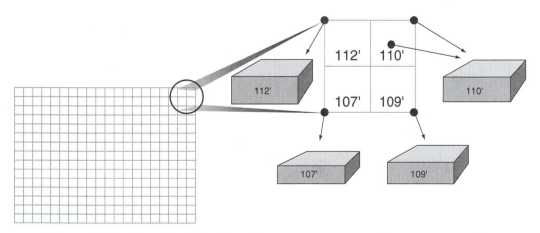

Figure 10.4 Quantized Z surface. Raster representation of a continuous surface. Note that for each grid cell, an absolute elevation is assigned. To ensure accurate analysis later on, it is important to decide specifically where the point elevation is located within the grid cell space.

however, the location will have an impact on the results of the calculation. For example, when we measured least-cost paths in Chapter 8, we began in the center of each grid cell because we were assuming that the elevational value was stored in that location. If your grid cells are coded so that the elevational value is at one of the four corners, your calculations of distance will be off by at least half a grid cell distance. Thus, in deciding where to place the elevational value within each grid cell, you should first examine how the raster GIS you are using actually calculates such measures as functional distance, slope, and aspect.

In many instances, data will be available for only a sample of grid locations, whether you are using raster or vector. In raster, you are most likely to obtain your elevational data as an altitude matrix in one of the two forms we discussed—regular or irregular lattice. If the regular lattice is small enough to fit your grid cell size, you can easily convert directly from the elevation points at each lattice vertex to a grid value for each of these points (again deciding beforehand where you want the elevational data to be located). When the data are provided in an irregular lattice form, you are going to have to estimate or predict all the missing values. This process, called interpolation, is needed because all the grid cells in your coverage must have elevational values. But as we will see next, interpolation is a useful analytical tool for modeling in its own right, as well as a good technique to combine with other analysis methods for more complex models.

INTERPOLATION

Because Z values for continuous surfaces are samples, whether they are topographic, economic, demographic, or climatic, we need to be able to create models that depict with relative accuracy the features we observe. In traditional cartography, for example, we need to be able to convert from point samples of elevation or other statistical surfaces to a form of display that uses isarithmic lines to show the form of the surface. We also need to be able to produce visual representations of other types, such as fishnet maps and shaded relief maps. And we need to be able to calculate slopes, aspects, and cross sections and to predict unknown elevations for objects that occur at places for which we do not have elevational data. Interpolation provides much of what is needed to perform these operations.

The process of interpolation can be conceptually very simple, but it requires an a priori assumption. Before we begin working with surface data, let's look at a mathematical examination of interpolation based on the idea of a mathematical progression. Progressions are numbers that occur in some identifiable sequence. Thus the mathematical sequence

$$1 \quad 2 \quad 3 \quad 4 \quad 5 \quad 6 \quad 7 \quad 8 \quad 9 \quad 10$$

is based on starting at 1 and adding 1 at each successive step. It is called a linear or arithmetic sequence because it increases by the same amount each time, and each increment is developed by simple arithmetic. Other arithmetic series could be

$$10 \quad 20 \quad 30 \quad 40 \quad 50 \quad 60 \quad 70 \quad 80 \quad 90 \quad 100$$

which uses 10 as the addition factor, or

$$1000 \quad 900 \quad 800 \quad 700 \quad 600 \quad 500 \quad 400 \quad 300 \quad 200 \quad 100$$

in which 100 is subtracted at each step. Still other arithmetic series may be created using multiplication or division.

Linear Interpolation

Within these simple series we can easily identify the a priori assumption mentioned earlier: namely, that the numbers at each successive step can be determined by a simple mathematical procedure. If we can discern the mathematical procedure, we can also insert any missing values. So, for example, if we are told that the following simple series is missing two values:

$$30 \quad 40 \quad 60 \quad 70 \quad 80 \quad 100$$

we can infer that the series is based on addition and that the missing values lie between 40 and 60 and between 80 and 100. Because this is a linear series, we can see that the missing numbers are 50 (40 + 10) and 90 (80 + 10). This is nearly identical to **linear interpolation**, the method of assigning values between points of known elevation spread over an area.

Take the simple example illustrated in Figure 10.5. We are looking at a single line transect of data points that range between 100 feet in elevation and 150 feet in elevation. If we assume that the surface changes in a linear fashion, just as in a simple series, and we have a linear progression, it is obvious that four numbers, spaced equal distances apart, can be interpolated between 100 and 150 feet. By segmenting the distances between these two points into five equal units, we can treat the distances as surrogates for changes in elevation. Therefore, at each unnumbered segment we need only insert a 10-foot elevation to obtain the missing values. If we do this for an entire area rather than just for a single line transect, we can obtain values for all the 10-foot contour interval locations. By drawing smooth lines to connect these segments, we can create contours of 100, 110, 120, 130, 140, and 150 feet. In other words, we can create an isarithmic map (contour map in this case) that allows us to visualize the elevational features.

Thus far we have worked with linear progressions, assuming that the surface changes in this linear fashion. Seldom do surface values conform to such a lin-

Figure 10.5   Linear interpolation. Linear interpolation to find missing values between 100 and 150 feet.

ear relationship. In some cases, the series is more logarithmic; in others, it is predictable only for small portions of the surface. Under such circumstances, the results of a linear interpolation method will not accurately depict the surface. And there are other needs for surface information that may require us to determine the overall trend in surface, rather than trying for the most accurate depiction. Some of these techniques can get complicated mathematically, and it is not our purpose to examine them in painful detail. Instead, we will stick to the conceptual level in examining a few of the nonlinear methods of interpolation so that we can get an idea of how they might best be used in GIS.

Methods of Nonlinear Interpolation

Nonlinear interpolation techniques are designed to eliminate the assumption of linearity called for in linear methods. We will look at three basic types of nonlinear interpolation method: weighting, trend surfaces, and **Kriging**. A number of advanced texts cover the many additional approaches in great depth (see e.g., Burrough, 1983; Davis, 1986). Here we look only at the general types.

Weighting methods assume that the closer together slope sample values are, the more likely they are to be affected by one another. For example, as we go up a hill, we note a much greater similarity in the general trend in elevation values close to you than there would be if we were to try to compare your local elevation to points far away. As we go downhill, there will be a similar change in elevation values for neighboring points. Nearing the bottom of the hill, however, we quickly notice that the elevation values change rather quickly at the base of the hill, whereas the plain beyond the hill once again takes on a certain similarity in elevational change. To more accurately depict the topography, we need to select points within a neighborhood that demonstrate this surface similarity. This is done by a number of search techniques, including defining neighborhood by a predefined distance or radius from each point, predetermining the number of sample data points, or selecting a certain number of points in quadrants or even octants (i.e., one point for each quadrant is used for the interpolation) (Clarke, 1990).

Whatever method is used, the computer will have to measure the distance between each pair of points and from every kernel or starting point. The elevation value at each point is then weighted by the square of the distance so that closer values will lend more weight to the calculation of the new elevation than closer distances (Figure 10.6). Many modifications of this approach are available. Some reduce the amount of distance calculations by employing a "learned search" approach (Hodgson, 1989); others modify the distance by weighting factors other than the square (e.g., cubed or higher powers); and still others include barriers that simulate coastlines, cliffs, or other impassable features likely to affect the output of an interpolation (Shepard, 1968). The barrier method is especially useful in developing surface models that can account for these local objects. And as in the use of barriers for other modeling tasks, the interpolation cannot pass through the barrier in its search for neighboring weights and distances.

Under some circumstances, we are more interested in general trends in Z surface rather than in the exact modeling of individual undulations and minor

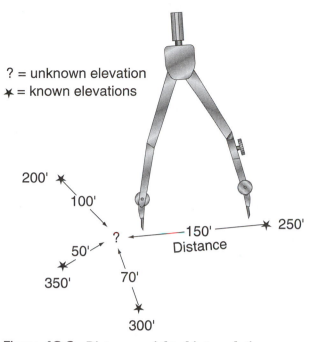

? = unknown elevation
✳ = known elevations

200' ✳
100'
50' ✳
350' ✳
70'
300' ✳
150'
250' ✳
Distance
?

Figure 10.6   Distance-weighted interpolation method. Note how the closer Z values have more weight than those far away. For example, the missing value is more likely to be closer to 350 feet in elevation because of its proximity to that elevation.

surface changes. For example, we might want to know the general trend in population across a country, to support demographic research, or whether a buried seam of coal trends toward the surface, to indicate how much overlying material needs to be removed for surface mining operations. The most common approach to this type of surface characterization is called **trend surface**. As in the weighting method, in trend surfaces we use sets of points identified within a specified region. The region is based on any of the methods already discussed for weighted methods. Within each region, a surface of best fit is applied on the basis of mathematical equations such as **polynomials** or **splines**. These equations are best described as nonlinear progressions that approximate curves or other forms of numerical series. To develop the trend, we examine each of the values in the region and fit it to the mathematical equation. From the equation used to estimate the surface of best fit, a single value is estimated and assigned to the kernel. As we have seen, the process continues for other target points or kernels, and the trend surface can then be extended for the entire coverage.

The number assigned to the target cell or kernel may be a simple average of the overall surface values within the region, or it may be weighted on the basis of the particular direction in which the trend is moving. Trend surfaces can be relatively flat, showing an overall trend for the entire coverage, or they can be relatively complex. The type of equation used will determine the amount of undulation in the surface. The simpler the trend surface looks, the lower the degree it is said to have. For example, a first-degree trend surface will show a single plane that slopes across the coverage—that is, it trends in a single direction. If it trends in two directions, it is said to be a second-degree trend surface (Figure 10.7).

The final method of interpolation, known as Kriging, optimizes the interpolation procedure on the basis of the statistical nature of the surface (Oliver and Oliver, 1990). Kriging uses the idea of the regionalized variable (Blais and Carlier, 1967; Matheron, 1967), which varies from place to place with some apparent continuity, but cannot be modeled with a single smooth mathematical equation. As it turns out, many topographic surfaces fit this description, as do changes in the grade of ores, variations in soil qualities, and even a number of vegetative variables. Kriging treats each of these surfaces as if it were composed of three separate values. The first, called the **drift** or **structure** of the surface, treats the surface as a general trend in a particular direction. Next, Kriging assumes that there will be small variations from this general trend, such as small peaks and depressions in the overall surface that are random but still related to one another spatially. (We say they are spatially autocorrelated.) Finally, we have **random noise** that is neither associated with the overall trend nor spatially autocorrelated. Clarke (1990) aptly illustrated this set of values by means of an analogy: if we are hiking up a mountain, the topography changes in an upward direction between the starting point and the summit; this is the drift. But along the way, we find local drops denting the surface and accompanied by random but correlated elevations. Also along the way, we find boulders that must be stepped over, which can be thought of as elevation noise because they are not directly related to the underlying surface structure causing the elevational change in the first place (Figure 10.8).

First-degree trend surface

Second-degree trend surface

Third-degree trend surface

Figure 10.7   Degrees of trend surface. First-, second-, and third-degree trend surfaces based on the complexity of the polynomial equation used to represent the surface.

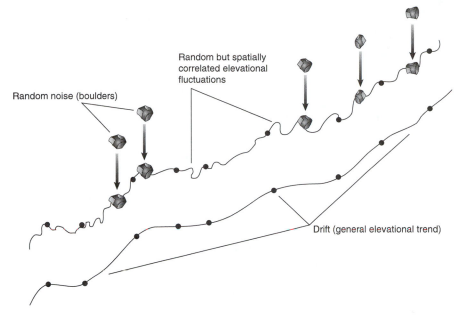

Random but spatially
correlated elevational
fluctuations

Random noise (boulders)

Drift (general elevational trend)

Figure 10.8   Elements of Kriging. Drift (the general trend), random but spatially correlated elevational fluctuations (the small deviations from the trend), and random noise (boulders) exemplified by a hiking trail up a mountainside.

Now that we have three different variables to work with, each must be operated on individually. The drift is estimated using a mathematical equation that most closely resembles the overall change in the surface, much like a trend surface. An expected, elevational distance is measured with the use of a statistical graphing technique called the **semivariogram** (Figure 10.9), which plots the distance between samples, called the **lag**, on the horizontal axis; the vertical axis gives the **semivariance**, which is defined as half the variance (square of the standard deviation) between each elevational value and each of its neighbors. Thus the semivariance is a measure of the interdependency of the elevational values based on how close together they are. A curve of best fit is then placed through the data points to approximate their locations (giving us a measure of the spatially correlated random component). If you look closely at the semivariogram, you will notice that when the distance between samples is small, the semivariance is also small. This means that the elevational values are very similar and are therefore highly dependent on one another because of their close spatial proximity. As the distance (lag) between points increases, there is a rapid increase in the semivariance, meaning that the spatial dependency of values drops rapidly. Eventually, a critical value of lag known as the **range** occurs, at which point the variance levels off and stays essentially flat. Within the range (i.e., from 0 lag to the point at which the curve levels off), the closer together the samples are, the more similar they probably will be. Beyond the range, the distance between sites makes no difference; the sites are totally unrelated at any of these larger distances. This information gives us a measure of what neighborhood needs to be applied (e.g., with weighted interpolation techniques) to encompass all points whose elevational values are going to be related.

Figure 10.9 Example of a semivariogram. Semivariogram showing the relationship between sample points (dots) and the fitted line. Note how within the lag the *Z* values are related to one another and beyond the lag there is no relationship at all because the dots are too far apart. *Source:* Adapted from P.A. Burrough, *Principles of Geographical Information Systems for Land Resources Assessment,* Oxford University Press, Oxford Science Publications, Monographs on Soil and Resources Survey, No. 12, © 1986; Figure 8.11.

Another important feature in the semivariogram is that the fitted curve does not run directly through the origin. Both mathematically and conceptually, if you have no distance between samples, there should be no variance because the items would essentially be the same point sample. But remember: the curve is an estimate of the locations. The difference between an expected variance value of 0 at 0 lag and the predicted positive value is an estimate of the residual, spatially uncorrelated "noise" variance and is called the **nugget variance.** As Burrough (1986:157) put it, this nugget variance "combines the residual variations of measurement errors together with spatial variations that occur over distances much shorter than the sample spacing, and that consequently cannot be resolved."

Now that we have defined the three important components of the regionalized variable with the semivariogram, we can determine the weights needed to perform interpolation for local regions. Unlike the weighting methods employed before, however, the weights for interpolation within neighborhoods are chosen to minimize the estimation variance for any linear combination of elevation samples. This variance can be obtained directly from the model that created the semivariogram in the first place.

Kriging comes in two general forms. The general form, *universal Kriging,* is most often used when the surface is estimated from irregularly distributed samples where trends exist (a condition called nonstationarity). *Punctate Kriging* is the elementary form and assumes that the data exhibit stationarity (lack of a trend), are isotropic, and are collected at equally spaced point locations (Davis, 1986). Most often, punctate Kriging is used specifically for finding point estimates from other point estimates, rather than for defining surfaces.

Because Kriging is an exact method of interpolation, it frequently gives a relatively accurate measure of the elevations of missing values. This exactness often comes at a cost in time and computing resources, however. Even so, Kriging has another advantage over other interpolation methods in that it provides not only interpolated values but also an estimate of the potential amount of error for the output. One might conclude that given its ability to supply an error measure and the exactness of the computed surface, one might be well advised to use the method all the time. This is not the case, however. When there is a great deal of local noise due to measurement error or large variations in elevation between sample points, the technique cannot easily produce a semivariogram curve. Under such circumstances, the results of Kriging are not likely to be substantially better than those of any other method.

No matter which interpolation technique you use, most GIS software will provide at least a minimum number of approaches for your use. In vector data models, most often TIN, the process of interpolation is most easily performed by isolating individual points and their associated elevational values and converting them to an altitude point matrix. From that point matrix, the interpolation procedures can easily operate on the selected algorithms, just as described. Actually, the TIN model itself can also perform interpolation (McCullagh and Ross, 1980), but the method is somewhat more involved, and we leave such techniques for an advanced GIS course. In raster grids with missing values, the grid cells that contain elevational values are treated like points located inside each grid cell area (e.g., in the center). Then the distance between grid cells can be measured as described earlier, and the elevational values are again interpolated on the basis of the techniques described. In this case, the grid cells receiving new elevational values are assumed to be placed in the same locations as the original target grid cells. When your GIS does not provide the desired interpolation technique, there is usually a method for converting the point matrix to a form compatible with specialty software designed expressly for working with surface data. They can be converted back to your original software for later processing. For some basic overviews of interpolation methods, you might consult Lam (1983) and Flowerdew and Green (1992).

Problems of Interpolation

We have now seen a number of methods of interpolation, some of which are more exact than others. When performing any of them, however, four factors need to be considered:

1. The number of control points
2. The location of control points
3. The problem of saddle points
4. The area containing data points

Generally, the more sample points we have, the more accurate the interpolation will be, and the more our model will be like the surface. However, only a limited number of samples can be made for any surface. Eventually, one reaches a point of diminishing returns: having more points doesn't substantially improve the quality of the output, and indeed it quite often increases the computation time and the data volume. In some cases, too much data will tend to produce unusual results because clusters of points in areas where the data are easy to collect are likely to yield a surface representation that is unevenly generalized and therefore unevenly accurate. In other words, having more data points does not always improve accuracy; Figure 10.10 shows that at a certain number of points, the accuracy actually falls.

The number of control or target points is frequently a function of the nature of the surface. We have seen that the more complex the surface, the more data points you need. But for important features of particular interest, such as depressions and stream valleys, we should also place more data points to be sure of capturing the necessary detail. In addition, although the location of sample points relative to one another has an impact on the accuracy of the interpolation routine, the relationship is not perfectly linear (Figure 10.11).

The problem of sample placement is even more severe when we consider interpolation from data collected by area to produce an isoplethic map. We have seen that our GIS offers the capability of determining the centroid of each polygon or the center of distribution of each polygon. When the data points are relatively evenly distributed, it is easiest to use the centroid-of-cell method. And the center-of-gravity method is most useful when the sample points are clustered or unevenly distributed. With both methods, however, there is always the chance that the center will occur outside the sample polygon, especially if the polygons are of unusual shape. When this occurs, the easiest solution usually is to pull the centroid or center of gravity just inside the polygon at its nearest possible location. This will probably have to be done interactively within the GIS environment.

The **saddle-point problem**, sometimes called the alternative choice problem, arises when both members of one pair of diagonally opposite Z values

Figure 10.10 Accuracy of isarithmic map versus number of data points. Characteristic curve of a hypothetical relationship between the number of points and the accuracy of an isarithmic map. *Source:* A.H. Robinson et al., *Elements of Cartography,* 6th ed., John Wiley & Sons, New York, © 1995. Adapted from Figure 26.29, page 513. Used with permission.

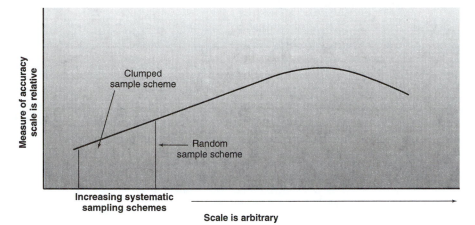

Figure 10.11   Sample point scatter and isarithm accuracy. Characteristic curve of a hypothetical relationship between the distance between data points and the accuracy of an isarithmic map. *Source:* A.H. Robinson et al., *Elements of Cartography,* 6th ed., John Wiley & Sons, New York, © 1995. Adapted from Figure 26.30, page 513. Used with permission.

forming the corners of a rectangle are located below and both members of the second pair lie above the value the interpolation algorithm is attempting to solve (Figure 10.12). This generally occurs only in linear interpolation because any distance weighting will likely solve the problem. When it does occur, however, the computer software is presented with two probable solutions to the same question of where to assign the contour line (Figure 10.12). A simple way to handle this problem is to average the interpolated values produced from the diagonally placed control points and then place this average value at the center of the diagonal. The interpolation can therefore proceed with a single solution to the problem because it now has additional information with which to calculate the interpolated values (Figure 10.13).

The final problem that must be considered in interpolation is a common one in GIS operations involving the area within which the data points are collected.

Figure 10.12   Saddle point problem. The saddle-point problem, where *Z* values are arranged rectangularly. Note the two possible solutions to the same interpolation problem. *Source:* A.H. Robinson et al., *Elements of Cartography,* 6th ed., John Wiley & Sons, New York, © 1995. Adapted from Figure 26.27, page 512. Used with permission.

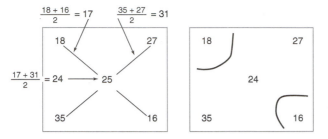

Figure 10.13   Solution to saddle-point problem.
The solution uses an average value placed at the exact
center.   *Source:* A.H. Robinson et al., *Elements of Cartography,* 6th ed., John Wiley & Sons, New York, © 1995. Adapted
from Figure 26.28, page 512. Used with permission.

More specifically, for the interpolation to work properly, the data points that
are to be estimated through a process of interpolation must have control points
on all sides. If, as is often the case, we select a study area for analysis and use
that same study area to perform interpolation, we will soon need to interpolate
points near the margins of the study area. When we approach the map margin,
we see that the interpolation routine is faced with control points on two or
three sides of our unknown elevation points because the map border precludes
any data points beyond the margin. As we have seen, best interpolation results
are obtained when we can search a neighborhood in all directions for selection
of control points and determination of weights. In the absence of these surrounding data points, the algorithm will use whatever is available, biasing the
results away from the border.

As an example, let's try to perform interpolations along the left margin of a
map. We will assume that this margin cuts directly through a hill that rises from
the center of the map toward the left margin. The adjacent map sheet indicates
that the hill continues to climb for some distance beyond the left margin of our
study area. But because there are no data points to the left of the map border,
the algorithm will search for its neighbors to the right, above, and below the
point to be estimated, and the interpolated values will be influenced in these
three directions. That is, the matrix generated by the algorithm probably will
use elevational values that are lower than those that, in fact, occur on the adjacent map sheet. Therefore, when the interpolation is performed, it will tend to
underestimate the missing elevations, producing a surface map that is of little
use at the margins. Any slope calculations, azimuthal calculations, line-of-sight,
least-cost path, or surface calculations will be in error near the margins as well.

Despite the obvious problems encountered when performing surface calculations under such circumstances, even experienced GIS users occasionally fall
into the trap of selecting surface data points from within the original study area
only. Sometimes this procedural error occurs because surface data were not
part of the original design, sometimes because the study area was selected on
the basis of the confines of a single map (even, perhaps, a topographic map),
and sometimes because of time limitations. No doubt there are other reasons,
but the result is the same—a compromised database and poor analytical results.

The solution to this problem is as obvious as the problem itself. Simply extend
the borders of the elevation coverage beyond the boundaries of the initial study

area (Figure 10.14). It is not necessary to expand the entire study area to accommodate the need for increased surface control points. Instead, we create a surface coverage whose outside margins are larger than the original study area, and then we perform the interpolation to produce the necessary surface configuration. Once we are done, the outside margins of the original study area can be used as a sort of cookie cutter, to shave off the edges (of the elevation model, slope model, viewshed, etc.). The idea is to operate on the larger elevational coverage to obtain whatever coverage will be used in conjunction with the other, smaller coverages. Determining the additional size of the surface coverage needed to produce reasonable interpolation results is difficult. In theory, it would be useful to perform some type of neighborhood analysis, perhaps like that used in the Kriging operation, to determine how far apart data points must be to ensure that they do not influence their neighboring elevational values. However, in practice, this is time consuming and, ultimately, impractical. The best advice is to err on the conservative side by expanding the area sufficiently to ensure good calculations. Most often, extending the margins 10 percent on all sides will guarantee good results, but the complexity of the surface must be factored into this decision. The more rapidly elevation changes along the margins, the wider the margins should be, to compensate for the weights the neighboring control points will put on the calculation of missing values when weighting or Kriging techniques are used. No standard guidelines exist here. You are left to use your own experience and firsthand knowledge of your data. Any additional border will increase the quality of your analysis and is always preferred to no additional borders at all.

Clip after contouring

Select larger study area

——————	Actual interpolated contour lines
- - - - - - -	Locations where contour lines will be created during interpolation
•	Hypothetical point elevation values used for interpolation

Figure 10.14   Avoiding interpolation error at map margins. Solution to the problem of missing data for interpolation at the margins of a study area: we extend the study area and perform interpolation or other surface procedures; then the completed coverage can be clipped and subjected to operations by additional coverages, with accurate results.

TERRAIN RECLASSIFICATION

Four basic characteristics of three-dimensional surfaces are commonly used for characterizing neighborhoods: steepness of slope, azimuth or orientation, shape or form, and intervisibility. To one degree or another, all can be performed with both raster and vector GIS, again depending on the sophistication of the software. In many cases, as with reclassifying neighborhoods with two-dimensional coverage data, these characteristics may very well be used in combination. We will look at each one individually and finish with a discussion of how they might be combined for more complex analyses.

Steepness of Slope

If you are planning to build a cabin on a mountain, you will likely want to know where the flattest parts lie, to be sure that your new home doesn't decide to relocate to the foot of the mountain. Or if you intend to cut some timber along a mountainside, you may want to use the slopes to roll the freshly cut pieces downhill (not a generally accepted technique, by the way). Or maybe you are planning a ski resort and want to put up three different levels of slope: Snowplow Only, Extra Padding Needed, and Call Out the Rescue Squad. In all these scenarios, you need to know something about slope; however, you are not interested in the shortest route down but rather in a general overview of the locations of steep, moderately steep, and relatively flat areas. The process is relatively simple conceptually: you need to know the relationship between the horizontal distance (measured in vector or raster) and the vertical change in elevation from bottom to top. A common way of expressing slope is rise over reach (rise/reach), where rise is the change in elevation and reach is the horizontal distance. To do this in vector, you will need a data model similar to the TIN model we discussed earlier. Raster can readily manage this, although compensation must be made for some minor errors due to grid cell quantization of space.

The general method of calculating slope is to compute a surface of best fit through neighboring points and measure the change in elevation per unit distance (Clarke, 1990). Specifically, the GIS will calculate the rise/reach value throughout the entire coverage, creating a set of categories of slope amount, much as we would do when defining class limits. If we wish fewer slope categories than are actually developed, we can reclassify the set produced by the GIS. Although techniques designed to characterize different neighborhoods by the amount of slope on a topographic surface are in common use, the surface need not be a topographic one. Our idea of a surface can be generalized to apply to any type of surface data that are measurable at the ordinal, interval, or ratio levels, called a statistical surface, which is a surface representation of these spatially distributed statistical data. Thus, we could analyze the steepness of slope of population change, or precipitation, or barometric pressure—any value that either is or can be assumed to be continuous across our coverage. By definition, however, it is clear that nominal data are not included. Imagine a raster database with attribute numbers representing land use types, where 1 = agriculture, 2 = urban, 3 = manufacturing, and so on. Any surface produced from this set of attributes is meaningless because the numbers, although they themselves may seem to be ordinal, interval, or ratio, do not truly represent these levels of data. Instead, they have been wrongly applied to noncomparable nominal categories. This point is made strongly to emphasize that lack of understanding of basic levels of geographic data may lead to incredibly worthless results.

A couple of quick examples of the proper use of slope as a definition of neighborhood might prove instructive. Let's say we are trying to find slopes that are less than 25 percent (e.g., 25 meters' change in elevation over 100 meters in horizontal distance) in order to define areas possibly suitable as sites for a cabin. In vector, using a TIN-like data model, we can perform any number of interpolation algorithms to provide what essentially amounts to a contour map. To determine slope, the software simply compares the horizontal distance between the vertices of each of the TIN facets with the respective horizontal coordinates. Since the TIN model stores these calculated values in its attribute tables, no calculations need be performed once the TIN is in place. Each facet value can be retrieved, and the slopes can be grouped or classed as those with less than 25 percent slope and those with 25 percent or more slope. We can then rename these as "badslope" (≥ 25-percent slope = too steep for building), and "goodslope" (≤ 25-percent slope = acceptable for building). In this way, we have reclassified neighborhoods on the basis of slope, to facilitate decision making in the single area of suitability for building. There are numerous ways of manipulating the slope information, in particular those using nonlinear interpolation methods (more later) and those that generalize topography to indicate the trends in surface, called trend surface analysis.

Both simple and complex methods of reclassifying neighborhoods based solely on slope can also be performed in raster GIS. The simplest method is to use a search of the eight immediate neighbor cells of each target cell. This is most often done by looking at all grid cells in the database and examining their neighbor cells, so that the slope values for the entire coverage can be performed. The software fits a plane through the eight immediate neighbor cells by finding either the greatest slope value for the neighborhood of grid cells or an average slope (normally the option is left to the user). For each group of cells, the software uses the grid cell resolution as the measure of distance and then compares the attribute values (elevations) from the central cell to all the surrounding cells (Figure 10.15). If, for example, we are building a ski resort and we want slopes no greater than 15 percent for the use of inexperienced or novice skiers, between 16 and 25 percent for talented amateurs, and between 26 and 45 percent for professionals, we can select these three classes through a simple reclassification process. We would identify all neighborhoods whose maximum (or average) slopes are between 0 and 15 percent, 16 and 25 percent, and 26 and 45 percent. All slope values greater than 45 percent would be called "badslopes" because they are too steep for skiing.

As with vector, raster GIS software may also allow for nonlinear interpolation methods such as Kriging and surface-fitting processes. Many of these algorithms are based on different search methods employed while performing the analysis (Hodgson, 1989). If your software does not allow these methods as separate algorithms, there may be a way to produce a close approximation of some of these techniques by using multiple applications of the original technique.

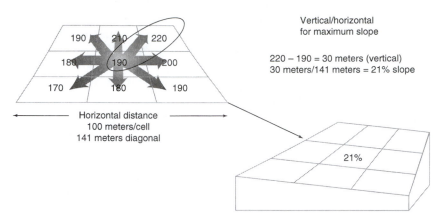

Figure 10.15   Trend surface in raster. Raster method of slope determination based on fitting a surface to the eight neighbors of a central target cell. This example shows how a raster-based GIS will indicate the maximum slope on the basis of a comparison of a central target cell with its neighbors.

For example, it is possible to "smooth out" the slopes to obtain a trend surface of sorts by using whatever algorithm your software uses first, and then repeating the process on the product of the first application. In other words, you perform a number of iterations of the slope calculation. The more iterations, the less rugged your surface will appear. If you keep it up, the final surface will appear as a flat plane. The technical jargon indicates that you will be moving toward a first-degree trend surface, from some higher order of surface.

Azimuth or Orientation (Aspect)

Because surfaces exhibit slopes, these features are, by definition, oriented in a particular direction, called the **aspect**. The two concepts of slope and aspect are inseparable from a physical as well as an analytical perspective. Without a slope, there is no topographic aspect. We have seen that azimuth or orientation can be a useful technique for classifying neighborhoods on the basis of two-dimensional features. The same can be said for terrain features. There are numerous applications of this technique. For example, biogeographers and ecologists are aware of a noticeable difference between the vegetation on slopes that face north and slopes that face south (Brown and Gibson, 1983). The primary reasons for this differential entail the availability of sunlight to green plants, but our interest in the phenomenon is that GIS will allow us to separate out north- versus south-facing slopes for comparison to related coverages such as soil and vegetation. Another example of the utility of knowing slope orientation is found when we try to build wind generators. We want to locate these cheap alternative producers of electricity high on the slopes of hills so that they have ready access to the wind, but they need to be on slopes that face into the prevailing winds rather than be sheltered from them. Geologists frequently want to know the prevailing slopes of fault blocks, or exposed folds, as a path to understanding the underlying subsurface processes. Or a grower may want to place an orchard on the sunny side of a hill to be able to take advantage of the maximum amount of sunshine. All these determinations and many more can be performed through the use of neighborhood functions that classify sloping surfaces based on their aspect.

In a vector GIS that uses a TIN-like data model, working with aspects is relatively simple. Each of the facets of the TIN model has a specific slope and aspect. The aspect is defined as the compass direction associated with each sloping surface (Evans, 1980). When the measurements we have seen are used for determining aspect, the TIN aspect values can be retrieved without additional steps. And, as before, we can group these aspects into classes. For example, if we are biogeographers interested in only north- and south-facing slopes, we can create three separate classes based on criteria selected earlier. Thus, north-facing slopes could be classed within a single neighborhood if they ranged in azimuth from 345 degrees to 15 degrees, and slopes facing south could be classed into a single neighborhood if the range were between 165 and 195 degrees. This means that the classes have a 30-degree range, or 15 degrees on either side of the cardinal direction. All the remaining aspect values can then be classed as "badaspect" because the biogeographer does not need them for analysis. You may want to select your own class limits for these neighborhoods on the basis of individual requirements.

In raster, the most common method is to again perform an analysis, throughout the entire coverage, that repeatedly compares a central cell to its neighbors. In this case, the surface fitted through the nine-cell matrix evaluates either an average or a maximum direction by looking at the high- and low-elevation values within the matrix. If, for example, the highest value is located at the top center of the matrix and the lowest at the bottom center, the solution for this portion of the matrix is that the general aspect is south (assuming that your coverage is oriented north and south). Or you may find that the highest and lowest grid cell elevation values are located in the upper left and lower right, respectively. This will give a southeast aspect to your sloping surface. The results of raster analysis of aspect can come either in degrees (0–360) or in a simpler set of vector values resembling the Freeman chain codes, where, for example, north, south, east, and west would be 0, 2, 4, and 6, respectively, and northeast, southeast, southwest, and northwest would be 1, 3, 5, and 7, respectively. The actual numbers and methods will depend on the software.

Shape or Form

Another useful method of reclassifying statistical surfaces is to provide a measure of their form. The simplest method of visualizing surface form is to produce a **cross-sectional profile** of the surface. This is a common practice in many courses in map reading, geography, and geology, where students are asked to render the profile of a topographic surface along a line drawn between two points. This is done by transferring each elevational value to a sheet of graph paper where the horizontal is exactly the same width as the line between the points and the vertical axis is scaled to some **vertical exaggeration** (usually a whole-number power) of the original surface elevation values. The process is easy to perform in a vector system using a TIN model, where the line is drawn along some portion of the coverage (it need not be a straight line). The software then generates a profile identical to that produced with analog maps in the manual method. Keep in mind, however, that this visual is not a coverage, and the results are used merely as a means of interpreting what is there. We know, for example, that *V* shapes in profiles are most often associated with stream valleys and that *U* shapes indicate glacial troughs. The more you know about the surface processes you are working with, the better your interpretation of the output is likely to be.

Your raster GIS may not have a method of producing a cross-sectional profile. Many, however, use an alternative method that creates a raster coverage by comparing a central target cell to two of its immediate neighbor cells. You can select the orientation of this search to be able to characterize a set of profiles for grid cells by looking at horizontal rows, vertical columns, or even diagonal sets of grid cells. The software will take each threesome and compare the central cell to its two immediate neighbors. Most often, the software will describe each cell either on the basis of its skyward angle (measured in degrees between each pair of surface elevational values) or by characterizing it as one of a number of form types. Form types are flat, rising, falling, monoclinal rising (flat then rising), monoclinal falling (flat then falling), rising then falling (convex as in a hill), and falling then rising (concave, as in a valley) and are characterized by code numbers placed in the new coverage (Figure 10.16). The code numbers must then be interpreted manually.

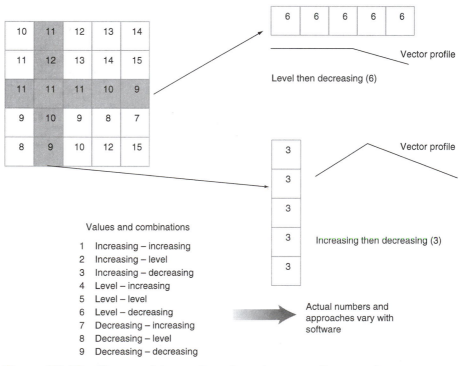

Figure 10.16   Characterizing surface shape in raster. Raster and vector output for characterizing a neighborhood based on form. In vector, the output is a cross-sectional profile, whereas in raster, the result is a coverage that shows the relationships between the target cell and its two neighbors. The two neighbors are selected as vertical, horizontal, or diagonal neighbor cells.

Both surface form techniques, whether raster or vector, are designed to produce neighborhoods based on changes in surface value that the user can interpret to represent specific features. Thus ridges, channels, peaks, watersheds, and so on may need to be identified as specific topographic features for later analysis. Watersheds, for example, are defined as all areas that drain into a stream network. More than simply draining into the stream channel, watersheds tend to function ecologically as single, uniform regions. Ecologists, hydrologists, engineers, pollution and flood control experts, and many other specialists need to be able to define these areas precisely. We have only seen the simplest method, but other more sophisticated methods have been applied to just this type of problem. As your skills improve, you may want to read up on these (Band, 1986, 1989; Douglas, 1986; White et al., 1992).

Visibility and Intervisibility

As we have seen from Chapter 9, many neighborhood measures characterize all or part of a map itself. Another type of neighborhood measure uses the terrain in a like manner. The process called **visibility** and **intervisibility analysis** recognizes that if you are located at a particular point on a topographic surface, there are portions of the terrain you can see (viewshed) and others you cannot see. The generalized term for the process is **viewshed analysis**, whereby one defines the regions that are visible from a particular point in the terrain. The many uses for this method include siting television, radio, and cellular telephone transmitters and receiving stations, locating towers for observing forest fires, routing highways that are not visible to nearby residents, and planning for artillery emplacements (Clarke, 1990). Any planning objective that requires anthropogenic features to be either visible or concealed will find utility in intervisibility analysis.

In vector, the simplest method is to connect a viewing location (observer location) to each possible target in the coverage. Next you perform **ray tracing**; that is, you follow the line (ray) from each target point back to the starting point, looking for elevations that are higher (Clarke, 1990). Higher points would obscure the observer's view of what is behind it (Figure 10.17*a*). There are many possible ways to determine intervisibility in vector; the large number of calculations involved makes the ray tracing method a simpler and more useful technique, though perhaps less accurate. Intervisibility performed in vector requires the use of a TIN data model in which the surface is defined by the triangular vertices. A number of algorithms for this process have been provided for vector data structures, including TIN (DeFloriani et al., 1986; Sutherland et al., 1974). It is not our purpose here to go into these algorithms but rather to show the concept, so let's take a quick look at a possible application of intervisibility in vector to see what it does and how it might be applied.

Suppose you are planning to build a home in the foothills of a mountain range and you want to be able to see as much terrain as possible from your front porch. You have limited your candidate locations to three. For each location, identified in your GIS terrain coverage (TIN model), the software looks out in all directions at the vertices of the model and identifies them. It then retrieves the elevation values for all these points. Next it compares these eleva-

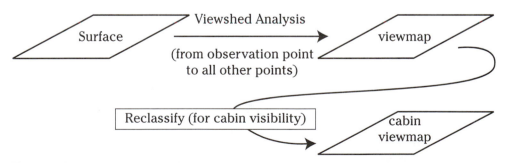

Figure 10.17 Visibility analysis. Viewsheds are maps of visible areas on a map produced by a process called ray tracing. The general method (a) can also be modified to include the addition of obstructions like trees (b) or can include the possibility of an elevated observer (c) such as observing out terrain from a lighthouse.

tions to the elevation of the potential building site. All areas that are higher in elevation (and all those beyond those locations) are not visible to you and must be reclassified as invisible; all the remaining areas are visible. The resulting polygon map shows you how much area is visible (Figure 10.18) for each coverage tested.

Raster methods of viewshed analysis operate in much the same way, but they are less elegant and more computationally expensive. The process begins

Figure 10.18 Viewshed flowchart. An example viewshed analysis flowchart for a mountain cabin example. Notice that the flowchart requires at least two operations (the original viewshed operation) plus a reclassification process specific to the needs of the mountain cabin.

by defining a viewer cell as a separate coverage against which the elevation coverage will be tested. Starting at the location of the viewer cell, the software evaluates the elevation (in the second coverage) that corresponds to that location. Then it moves out in all directions, one grid cell at a time, comparing the elevation values of each new grid cell it encounters with the elevation value of the viewer grid cell. Each time it encounters a grid cell with an elevation value higher than the viewer grid cell elevation, it reclassifies it as not visible and codes it to indicate a nonvisible grid cell. Each time it encounters a grid cell with an elevation value lower than the viewer elevation, it places a different code, indicating that the area represented by the cell is visible from the viewer's location. This is the simplest way to perform this method; there are others, and each yields different, but computationally valid, results (Anderson, 1982; Dozier et al., 1981). You might want to check your software's documentation to evaluate the utility in your work of one type over another (Fisher, 1993).

Most applications of viewshed analysis are based solely on topographic surfaces, but in some cases the topographic surface will have forest cover with known individual heights or grouped heights associated with the trees. To perform intervisibility where the heights of these or other obstructing objects are known, the elevation coverage values must include the obstruction heights (Figure 10.17*b*). These can be added in both vector and raster, usually by means of a mathematically based (addition) combination of the two themes (see Chapter 12). Where topography itself is not important, the heights of the obstructions can be used alone to determine the degree of visibility of all points compared to those of the viewer. The applications of viewshed analysis are endless, and the technique is quite common in surface neighborhood analysis.

SLICING THE STATISTICAL SURFACE

As we learned earlier in the chapter, a common method of displaying surface information makes use of isarithms. We also saw that the isarithms are commonly drawn at set intervals (a contour interval in topographic maps). The interval chosen allows us to depict the shape of the surface, but we always assume that the elevational values between the contour lines exhibit some continuously changing value because we assume that the surface itself is continuous. We also assume that the interval was selected to best display the surface features. Most GIS software, whether raster or vector, offers the ability to change the isopleth interval or even to convert the area under each individual isopleth interval to a flat surface, eliminating the need to assume a continuous surface. For simplicity we will call this general group of techniques **slicing**, and we will envision their execution as applying a sharp knife horizontally through the surface.

Slicing can be simply a matter of selecting a different isoplethic class interval set to permit us to look at surface features differently (Figure 10.19). We might, for example, want to make the vertical distance between isopleths larger to bring out overall shapes of features without great amounts of confusing detail. This useful visualization technique gives an impression of trends without necessitating the actual computation of a trend surface for the area. Alternatively, to see more of the detail in the surface, we can decrease the vertical distance

Figure 10.19 **Slicing the *Z* surface.** Slicing through a *Z* surface to produce different contour intervals.

between the isopleths. This latter approach assumes that the sample data are sufficiently detailed in the first place.

A more common application of slicing could more appropriately be called a neighborhood function, as discussed in Chapter 9. We look at it here because in its performance it relates more closely to the interpolation procedure. This approach assumes that by placing contour lines at individual intervals, we have effectively reduced a continuous surface to a discrete surface more closely resembling a stair-step feature. Why would we want to degrade our data from continuous to discrete? Perhaps the following example problem requiring neighborhoods will help explain.

Let's say you are working for a consulting firm under contract to a foreign government. You have been asked to define appropriate land uses for a large portion of land on the basis of a combination of soil capabilities and elevation classes. Elevation classes have been chosen rather than slope classes partly because most soil surveys incorporate slope into the soil capability classes and partly because the client's agricultural advisers know that some plants grow better at certain elevations than at others. On the basis of each crop's elevational requirements (let's say there are five crops), you can divide the surface into five elevational regions by slicing through it in four places. Once you have performed the slice, the groups can be reclassified by their influence on the five crops. So you now have five groups of areas, based on elevational classes as they affect the five crops, and called "crop1elevations," "crop2elevations," and so on. Thus, you have converted the ratio data of the surface to nominal data on the basis of the interactions between crops and elevations.

These, as we will see in Chapter 12, can now be combined with other coverages to make decisions about where the crops should be planted.

CUT AND FILL

Although interpolation is useful for predicting surfaces, and even for selecting appropriate elevational ranges for crops, it can also potentially be applied to another set of problems related to the volume under surface features, or the volume of material related to cutting and filling (as in removing ore and backfilling to produce a flat surface during reclamation activities). Both types of procedure require knowledge of two separate surfaces, a top and a bottom. By producing difference values for both surfaces and applying contour slicing or integral calculus techniques, we can determine the volumes. Let's look at how this is done.

Suppose our backyard pool has vertical sides and covers an area of 150 square meters. The pool is 2 meters deep across the entire surface area. By multiplying the depth by the area, we obtain the overall volume of the pool: $2 \times 150 = 300$ square meters.

But many things that are measured across a large area are not uniformly distributed. For example, consider precipitation. Let's say that we have 15 rain gauges located throughout a county of approximately 100 square kilometers (1 million square meters), and after collecting rainfall data we want to find the total amount of precipitation at every location for the entire year. Because of the variability of precipitation, we are likely to find amounts ranging in value from perhaps 50 centimeters (0.5 meter) to as much as 150 centimeters (1.5 meters). To simplify our calculations of the total volume of rainfall throughout the area we can use a technique called the **method of ordinates** that totals all these values (Muehrcke and Muehrcke, 1992). By summing all the values, we create a flat surface similar to a backyard pool. Suppose, then, that our average precipitation throughout the county is 100 centimeters (1 meter). We can then multiply 1 meter by 1 million square meters to obtain a value of 1 million cubic meters of water.

But suppose we want a more exact calculation of the countywide volume of precipitation. Figure 10.20 outlines the steps. We first record the coordinates for each of our rain gauges and produce an altitude matrix, just as we would do for elevation data. Then we use an appropriate method of interpolation to generate a set of contours, which we slice into equal parts, guided by the contour intervals. If we then assume that all slices are of equal elevation (discrete), we can perform the same simple volume calculation for each of the separate contour slices. By simply adding these together, we obtain a somewhat more accurate measure of the overall volume than if we assumed the entire county received an average amount of precipitation. Because both of these techniques conceive of our volumes as having vertical surfaces, the calculation of volume is very simple in either vector or raster. For each slice of the volume, we simply multiply the area attribute value by the depth value.

Calculations of volume are essentially the same for a hill or a lake in that one of the surfaces is essentially flat. This, of course, simplifies the problem somewhat. But if your purpose is to determine how much ore there is in a subterranean de-

Separate the region into portions

Calculate the area
of each portion

Multiply that by the
amount of precipitation

Add these products to
find the total volume
of precipitation

150 cm

100 cm

50 cm

Figure 10.20   Method of ordinates. The method of ordinates used to determine the volume of water in an area.

posit, both the upper and lower surfaces will be irregular and in all likelihood will have to be interpolated as well. Still, you can use the same approaches as before, except that you can no longer assume that one of the surfaces is flat. By simply subtracting the upper elevational values from the lower, you can obtain a variable height value. Then, by noting the area and allowing the integration equations inside the software to do most of the work, you can get the needed result.

Although the conceptual model of volume determination is a simple one, you should be aware that it is not always implemented into commercial GIS. Some commercial products specialize in this sort of software, usually packages related to geological sciences and mining. If you either don't have these techniques available or don't want to piece them together with a number of different steps, you most often will find that the available volumetric software data can be converted to the GIS and back again at will. This practice is becoming more prevalent.

Terms

aspect	isoplethic map	saddle-point problem
continuous surface	Kriging	semivariance
contour interval	lag	semivariogram
contour line	linear interpolation	slicing
cross-sectional profile	method of ordinates	slope
DEM	nonlinear interpolation	smooth surface
discrete surface	nugget variance	splines
discrete altitude matrix	polynomials	statistical surface
drift	random noise	structure
intervisibility analysis	range	trend surface
irregular lattice	ray tracing	vertical exaggeration
isarithm	regular grid	viewshed analysis
isarithmic map	regular lattice	visibility analysis
isometric	rough surface	weighting methods

Review Questions

1. Define a statistical surface. Describe a nontopographic statistical surface. What is the difference between a continuous and a discrete statistical surface?

2. What is the difference between a smooth and a rough statistical surface? What is the importance of each in terms of point sampling to represent the surface configuration?

3. What is an isarithm? How do the inferences we can make from closely spaced isarithms differ from what widely spaced isarithms tell us? What is the specific term assigned to isarithms that measure topographic elevation?

4. What is a contour interval? Why must a contour interval remain the same throughout a map?

5. What is the difference between an isarithmic map and an isoplethic map? What assumptions must be made before we can use isolines for isoplethic maps?

6. What is the difference between a regular and an irregular lattice in terms of sampling for surface data? What are the advantages and disadvantages of each?

7. What are DEMs? What is the relationship between a discrete altitude matrix and a TIN model?

8. Describe and diagram the method of representing continuous surface data with a raster data model. Why is it important to know where in each grid cell the actual point locations for topographic data are intended to lie?

9. Describe and diagram linear interpolation. What are the major drawbacks to linear interpolation?

10. Why might we be more inclined to use nonlinear interpolation instead of linear interpolation? What are the three basic types of nonlinear interpolation routine?

11. Describe and diagram the use of weighted methods of nonlinear interpolation. Why might we want to include barriers in weighted interpolation methods?

12. What are trend surfaces? When might we want to use this interpolation approach rather than weighted methods? Give an example showing its utility.

13. What is the difference between Kriging and weighted interpolation methods? What is a semivariogram, and what does it tell us about the drift, random spatially correlated variations in the surface, and random noise?

14. Provide a diagram of a semivariogram based on hypothetical data. Explain the relationship between lag and semivariance. What can we say about data within and beyond the range? What does this tell us about selecting the neighborhood for interpolation?

15. What is the difference between universal and punctate Kriging? When is Kriging not likely to give better results than any other form of interpolation?

16. Give some concrete examples of the use of interpolation beyond simply creating an isarithmic map.

17. What four things do you need to look out for in any interpolation routine? Describe the problem of sample placement in isoplethic mapping. What is the saddle-point problem? How can it be rectified?

18. Describe the problem of no data points at the map margin as it applies to interpolation results. How can such inadequate results be avoided?

19. What is slope? How is it implemented in a vector GIS? in raster? How can slope be used to define neighborhoods?

20. What is aspect? How is it used to create neighborhoods? Give an example of using aspect to define a neighborhood, describing both how it is done and what the output will tell you about your data.

21. How can the shape or form of surface data be displayed in vector? in raster? What is the general term for this output operation? What possible uses can be made from these techniques as they are applied to reclassifying areas?

22. What is a viewshed? How does one calculate a viewshed in vector? in raster? Describe some uses of intervisibility analysis.

23. What is *Z*-value slicing? Other than changing the class interval (e.g., the contour interval), how can slicing be used for creating neighborhoods? Give an example.

24. Describe the method of ordinates for determining the volume of a *Z* surface feature. Explain how this could be done in a raster and a vector GIS.

References

Anderson, D.P., 1982. "Hidden Line Elimination in Projected Grid Surfaces." *ACM Transactions, Graphics,* 1(4):274–291.

Band, L.E., 1986. "Topographic Partitioning of Watersheds with Digital Elevation Models." *Water Resources Research,* 22(1):15–24.

Band, L.E., 1989. "Spatial Aggregation of Complex Terrain." *Geographical Analysis,* 21(4):279–293.

Blais, R.A., and P.A. Carlier, 1967. "Applications of Geostatistics in Ore Evaluation." *Ore Reserve Estimation and Grade Control,* 9:41–68.

Brown, J.H., and A.G. Gibson, 1983. *Biogeography.* St. Louis, MO: C.V. Mosby.

Burrough, P.A., 1983. *Geographical Information Systems for Natural Resources Assessment.* New York: Oxford University Press.

Burrough, P.A., 1986. *Principles of Geographical Information Systems for Land Resources Assessment.* Oxford Science Publications, Monographs on Soil and Resources Survey. New York: Oxford University Press.

Clarke, K.C., 1990. *Analytical and Computer Cartography.* Englewood Cliffs, NJ: Prentice Hall.

Davis, J.C., 1986. *Statistics and Data Analysis in Geology,* 2nd ed. New York: John Wiley & Sons.

DeFloriani, L., B. Falcidieno, and C. Pienovi, 1986. "A Visibility-Based Model for Terrain Features." In *Proceedings of the Second International Symposium on Spatial Data Handling.* International Geographical Union Commission on Geographical Data Sensing and Processing the International Cartographic Association, Seattle, WA, July 5–10, pp. 235–250.

Douglas, D.H., 1986. "Experiments to Locate Ridges and Channels to Create a New Type of Digital Elevation Model." *Cartographica,* 23(4):29–61.

Dozier, J., J. Bruno, and P. Downey, 1981. "A Faster Solution to the Horizon Problem." *Computers and Geosciences,* 7(2):145–151.

Evans, I.S., 1980. "An Integrated System of Terrain Analysis and Slope Mapping." *Zeitschrift fur geomorphologie* (supplement), 36:274–295.

Fisher, P.F., 1993. "Algorithm and Implementation Uncertainty in Viewshed Analysis." *International Journal of Geographical Information Systems,* 7(4):331–347.

Flowerdew, R., and M. Green, 1992. "Developments in Areal Interpolation Methods and GIS." *Annals of Regional Science,* 26:67–78.

Hodgson, M.E., 1989. "Searching Methods for Rapid Grid Interpolation." *Professional Geographer,* 41(1):51–61.

Kelly, R.E., P.R.H. McConnell, and S.J. Midenberger, 1977. "The Gestalt Photomapping System." *Photogrammetric Engineering and Remote Sensing,* 43(11):1407–1417.

Lam, N.S., 1983. "Spatial Interpolation Methods: A Review." *American Cartographer,* 10:129–149.

Little, J.J., and P. Shi, 2003. "Ordering Points for Incremental TIN Construction from DEMs," *GeoInformatica* 7(1):33–53.

Mark, D.M., 1978. "Concepts of Data Structure for Digital Terrain Models." In *Proceedings of the DTM Symposium,* American Society of Photogrammetry, American Congress on Survey and Mapping, St. Louis, MO, pp. 24–31.

Markarovic, B., 1973. "Progressive Sampling for Digital Terrain Models." *ITCJ* 1873–3:397–416.

Matheron, G., 1967. "Principles of Geostatistics." *Economic Geology,* 58:1246–1266.

McCullagh, M.J., and C.G. Ross, 1980. "Delaunay Triangulation of a Random Data Set for Isarithmic Mapping." *Cartographic Journal,* 17(2):93–99.

Muehrcke, P.C., and J.O. Muehrcke, 1992. *Map Use: Reading, Analysis, Interpretation.* Madison, WI: J.P. Publications.

Oliver, M.A., and R.W. Oliver, 1990. "Kriging: A Method of Interpolation for Geographic Information Systems." *International Journal of Geographical Information Systems,* 4(3):313–332.

Shepard, D., 1968. "A Two-Dimensional Interpolation Function for Irregularly Spaced Data." In *Proceedings of the Twenty-Third National Conference of the Association for Computing Machinery,* pp. 517–524.

Sutherland, I.E., R.F. Sproull, and R.A. Schumacker, 1974. "A Characterization of Ten Hidden-Surface Algorithms." *Computing Surveys,* 6(1):1–55.

White, D.A., R.A. Smith, C.V. Price, R.B. Alexander, and K.W. Robinson, 1992. "A Spatial Model to Aggregate Point-Source and Nonpoint-Source Water-Quality Data for Large Areas." *Computers and Geosciences,* 18(8):1055–1073.

CHAPTER 11

Spatial Arrangement

Thus far we have focused on characterizing the objects we have seen. But to evaluate our environment properly we also need to know the relationships among the individual items we see and their intervening spaces. Rather than looking at the amount of space occupied by a single object, or how the object's space is configured (i.e., its shape), we will now concern ourselves with the overall arrangement of objects in space. Such arrangements can be characterized by how many objects there are in a particular area or whether they are distributed evenly or in groups. We will look at the distance relationships between objects themselves and how they are related to the overall size of the area they occupy.

Such distributional patterns can be seen in many situations. We know, for example, that some human population distributions are widely dispersed, as in farm households in a rural area. Other population distributions are more concentrated in larger collections we call cities. Plants and animals may occur individually in evenly distributed arrangements, or they too may form more concentrated groups. Even physical features such as sediment types and landscape features—streams and hills, mountains and valleys—may occur as widely spaced individuals or as larger groups. Anthropogenic objects like roads, fences, and houses can occur in patterns as well. As our geographic filter increases, we will see still more. Observing that there are differences in **spatial arrangements** of objects allows us to ask questions about what these patterns are, how they can be classified, and how they might provide insights into the processes that made them.

If we revisit the areas where we first observed arrangements of objects, either in the real world or, as we would do in GIS, through multitemporal coverages of the same objects, we see that the observed patterns change. Regions that once appeared to show diffused and scattered patterns may now show evidence of coalescence. Objects that were once randomly organized in space may now begin to occur in **regular**, repeating patterns. Some distributional patterns may show growth and expansion, whereas others may be shrinking or vanishing. Patches or areas may merge; single linear objects may begin to connect to form networks; stabilized dunes may begin moving and diffusing. In all these cases time is an important component in our understanding of spatial arrangement. And we soon wonder what the processes are that cause the change from one arrangement to another. We can ask many questions: what the directions of change might be, whether there are driving forces that we can understand, what the upper or lower limits of these driving forces might be, and so on.

In this chapter we will be looking primarily at the arrangements of objects in a single map. If you are patient, you will soon see how the arrangements of individual objects can best be compared to one another. At that point you will start thinking about how the GIS will allow you to compare objects in other maps.

LEARNING OBJECTIVES

When you are finished with this chapter you should be able to:

1. Clarify what we mean when we say we are interested in arrangements of objects. Why is this knowledge useful to us?

2. Define a uniform distribution. Tell what makes a uniform distribution regular versus random. Define a clustered distribution.

3. State the significance of regular, random, and clustered distributions in terms of the potential processes that caused them. Give some examples of each distributional pattern and give a hypothesis that might describe the causes.

4. Tell the purpose of quadrat analysis. Describe how it is done and name the statistical test we use to confirm or reject our results.

5. Tell the purpose of the variance–mean ratio in point distributions and state how it differs from traditional quadrat analysis.

6. State what nearest neighbor analysis tells us that quadrat analysis does not and describe how the two techniques are similar.

7. Give the purpose of Thiessen polygons. Diagram and describe how they are created.

8. Give three concrete, real-world examples of the use of distance or adjacency as a method of retrieval of cartographic features.

9. Describe the process of performing a nearest neighbor statistic on line patterns and state what this quantity tells us about the distribution of our lines. Describe some situations in which the nearest neighbor statistic for lines might give misleading results.

10. Describe the use of line intersect methods for analyzing line pattern distributions. Define a random walk.

11. Define a vector resultant and state what it tells us about the patterns of linear objects. Define resultant length and discuss how it compares to resultant force problems in physics.

12. Define the mean resultant length, explain how it differs from the raw resultant length, and tell when we would use it. Answer questions such as the following: What would a large mean resultant length tell us? What is the difference between circular variance and mean resultant length?

13. Describe how we can adjust our measurements for mean direction, mean resultant length, and circular variance to account for orientations that can be measured in either of two different directions.

14. Describe how the gamma index is performed and spell out what it tells us about a given network. What does a gamma value of 0.48 tell us in terms of the amount of connectedness in our network?

15. Tell what the alpha index is, how it is different from the gamma index, and how it is the same. Give some non-road-network examples of how both these indices might be useful.

16. Describe, in general terms, the gravity model. Explain how distance affects the interactions of point objects and how the magnitude of the nodes changes the interactions between them. Give an example of how the gravity model might be applied.

17. Describe the types of attributes that might need to be assigned to the nodes and the arcs between nodes in work with routing and allocation problems.

POINT, AREA, AND LINE ARRANGEMENTS

Spatial arrangement is the placement, ordering, concentration, connectedness, or **dispersion** of multiple objects within a confined geographic space. Until now, most of the analytical techniques we have examined have dealt either with individual objects or with collections of objects when they might be defined as regions or neighborhoods or conceived of as statistical surfaces. We have even touched very briefly on the possible **interactions** of objects, regions, surfaces, and neighborhoods with those of other map layers. However, most of the objects we encounter in a single coverage also have a definable spatial patterning that might indicate possible mechanisms for producing them.

In a purist approach, the term **spatial arrangement** generally refers to the simple cartographic display of spatially distributed objects. In such a "map as communication" mode, we would say that the map shows us where the objects are located and illustrates the shape of this distribution. That is enough, we might say. But we intuitively recognize that more can be known. There is more that can be used to describe the interactions of each individual object to its neighbors and the relationship of all these objects to the whole space in which they reside. And if we can find ways to measure these relationships, we can discover how to isolate and understand the possible mechanisms that produce such patterns.

POINT PATTERNS

Perhaps the most common techniques for analyzing spatial distributions are applied to point patterns. Point objects can be individual trees, houses, animals, street lights, and even cities, depending on the scale (Figure 11.1*a*). As we will see later, point objects can be represented in linear, or area form, as well (Figure 11.1*b*).

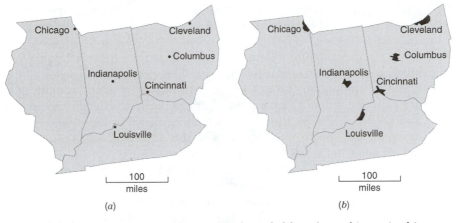

Figure 11.1   Point and area representation of cities. Area objects, in this case cities, can be viewed *(a)* as points or *(b)* as lines, depending on the scale at which they are presented. This shows the close association between the analytical techniques that can be applied.

The simplest measure of point patterns is a matter of determining the **density** of the distribution. This is done by dividing the number of points by the total area in which the points exist. This may yield a useful descriptor of the numbers within the distribution, but it does little to characterize itself. Population density, housing density, tree density, and so on are commonly used to provide a measure of the compactness of points. With that information, we can compare the densities to those of comparable point objects in other areas to make comparisons of the different mechanisms that might be operating. Or we might compare the points in the same location but at different times to give us an idea of change in density through time. We might find, for example, that population density is increasing in urban areas through time, or that housing density is increasing, or that tree density decreases as the trees mature and compete for space and sunlight. Even this simple statistic, easily calculated in either vector or raster, can give us many useful insights into our data.

No matter what the overall density of the distribution, we may be interested in the form of the distribution in addition to the numbers per unit area. Point patterns are said to occur in one of four possible arrangements, each with a specific set of criteria. A pattern is **uniform** if the number of points per unit area in one small subarea is the same as the number per unit area in each other subarea. If the points occur on a grid, separated by exactly the same distance throughout the entire area, the uniform pattern is said to be regular, as we saw with a regular grid in sampling surface data points. In other cases, uniformly distributed points may occur in **random** locations scattered throughout the study area. Or, in other cases, the points are grouped in tight arrangements in a **clustered** pattern (see Figure 2.8).

Quadrat Analysis

Uniform point patterns are defined based on the relationships among uniform subareas called **quadrats** of the larger area. This is a very common method of analyzing discretely occurring biological data. Each quadrat contains a number of points defined by, for example, individual herbaceous plants, ant mounds, or any other type of point phenomenon. If each uniform quadrat contains the same number of point objects, we say the distribution is uniform for the overall study area. Uniform distribution is not likely to be found among biological phenomena, since living organisms tend to migrate toward specific locations that are rich in nutrients, have good drainage, contain certain types of soil texture, and so on. If distribution is in fact uniform, we could suppose that no major mechanisms control the locations of the objects. Thus we need to be able to test to see whether distribution is uniform.

In the standard method of distribution testing, called **quadrat analysis**, we assume that roughly the same number of objects will be found in each subarea. We determine how many objects are likely to be in each subarea by dividing the total number of data points by the number of subareas. The number we get will be our "expected distribution" if it can be characterized as uniformly distributed. In a fairly simple statistical test that can be applied to this set of data, called the chi-square (χ^2) test, we subtract the expected number of points for each quadrat from the observed numbers $O,$ square the result, divide by the expected number, and take the summation. This translates into a simple formula:

$$\chi^2 = \Sigma\ \frac{(Q - E)^2}{E}$$

The results of this mathematical manipulation give us a value that we can test against predefined critical values in a standard table. If the number we get is not substantially different from what we would expect, the distribution is uniform; a notable difference, however, demonstrates some nonuniform pattern that might suggest an underlying process. Although this technique might be considered to be purely statistical, it can be performed in some GIS software, especially when the data model is raster. Many specialty software packages can perform this analysis as well. If you are unfamiliar with the use of statistics, especially as they apply to spatial data, you might want to consult a simple text on the use of statistics in spatial data analysis. However, for simplicity's sake, it is easiest to simply remember that the larger the chi-square value, the less likely the data occur in a uniform pattern.

Although most of you have probably assumed that the output of GIS analysis would be another map, in this case the output is simply a single value that must be evaluated to determine whether a uniform distribution can be seen. At this point, you might be asking yourself this question: If my distribution is not uniform, what mechanism might contribute to the anomalies? Most often the point distributions we see relate to other mapped variables (map layers) for the same study area. In other words, additional relationships may exist among map layers. These potentially related map layers can be additional points, but often they are maps of areal distributions of phenomena. In our biological examples, these might be soil variables. We will then need to compare the points in the single coverage to the polygons in a second coverage. This useful technique is covered in the next chapter, but you might want to think about it now.

Although quadrat analysis can help us decide whether a uniform distribution exists in a given situation, it can be used to provide us with more information as well. A special technique, based on quadrats, is called the **variance–mean**

ratio (VMR). VMR is an index that shows the relationships between the frequency of subarea variability and the average number of points in each subarea (Lavin, 1979; Shea, 1983). It is calculated by dividing the variance of the subarea point frequencies (defined by the weighted standard deviation) by the mean for each subarea (VMR = var/mean). As before, the test statistic is chi-square and is calculated by multiplying the VMR by one less than the total number of subareas. High chi-square values indicate that our distribution is clustered; that is, the variability between the number of points in each subarea and the average for the entire area is very high. Alternatively, small chi-square values mean that the distribution is more evenly dispersed or uniform. Intermediate values indicate a distribution more closely associated with a random process, where some quadrats seem to have slightly more and some slightly less than an average value.

As before, the results of the analysis suggest that if the distribution is not statistically random (i.e., if it is either uniform or clustered), you may be able to determine a possible causative mechanism by wisely selecting a set of variables to compare to your point coverage. Uniform distributions, for example, tell us that point objects are sprinkled more or less evenly within each subarea; but uniform distributions can occur in regular patterns as might typify trees in an orchard, or randomly, as is more characteristic of trees in a forest. In randomly distributed points, we will not find the same number of points in each subarea, whereas in regularly distributed points, we generally will.

Nearest Neighbor Analysis

Thus far, we have confined our characterization of point distribution patterns to the numbers of points occurring within subareas. In other words, we are studying the distribution of points by comparing the areas they occupy. However, it is also instructive to examine the locational relationships from one point to another. To do this, we most often rely on an alternate method of point pattern analysis, **nearest neighbor analysis**, a common procedure for determining the distance of each point to its nearest neighbor and comparing that value to an average between-neighbor distance. Calculating this statistic involves simply determining the average or mean nearest neighbor distance between each possible pair of near neighbor points. (Near neighbors are defined as the closest point to each test point you select.) The mean nearest neighbor distance provides a measure or index of the spacing between points in the distribution. This in itself is useful because obviously point objects will conflict if they are too close together. For example, we know that many animals have requirements for space expressed as "range." When one animal's range overlaps that of another of the same species, conflict is likely. You have probably seen this many times in your backyard, where members of the same species of bird routinely attack one another.

As before, however, it would be useful to be able to extract additional information from the statistic. As it turns out, just as in quadrat analysis, we can compare the nearest neighbor spacing index to three possible patterns that might occur: regular, random, and clustered. This technique can be generally explained for each of these cases by creating an index against which to compare your results as follows. For an index of random distribution, simply divide 1 by twice the square root of the density of points (number of points/area). If you want a test of maximum dispersion (regular distribution), divide 1.07453 by the square root of the point density. Finally, for a test of maximum clustering, where the points would normally be on top of one another, we can simply assume the value is of the divisor 0. Any value you obtain for your index that is not 0 will be a positive number. A simple comparison of your average nearest neighbor distance to that of these three index values will give you an idea where along the continuum they fall.

Let's take a quick look at how this works with the set of data in Table 11.1 and the associated plot (Figure 11.2). We have six data points, each with a single X, Y coordinate pair, all falling within an area of 25 square units. The average nearest neighborhood value from these data turns out to be approximately 1.4. For randomly distributed points, the index is found from 1 divided by 2 times the square root of the density (in this case, 6 data points per 25 square units, or 0.24). Since the square root of 0.24 is 0.4895, the final result for our random nearest neighbor distance is 1.02. Our average nearest neighbor of 1.4 is somewhat higher than the random value, 1.02. Our test index for perfectly dispersed points is 1.07453 divided by 0.489 (the square root of the density), to give a rounded value of 2.19. This would be the value if our point arrangement was perfectly uniform. Our average nearest neighbor is much smaller than that. And, of course, a perfectly clustered distribution is 0, and our average nearest neighbor is much larger than that. So what we find is that our average nearest neighbor is slightly more dispersed than random, or somewhat between truly uniform and random. In other words, it is starting to take on a more regular configuration but as yet is still rather randomly distributed.

Again, these values give only a measure of the patterning of these point features. Most often we will need to decide how we might later compare these values to other map layers or simply use the statistic as a descriptor for further statistical analysis. The nearest neighbor statistic is an absolute measure; hence, it is not immediately amenable to comparison with nearest neighbor statistics for other point distributional patterns. The nearest neighbor index can be standardized to permit such comparisons (McGrew and Monroe, 1993), but

TABLE 11.1 Calculating Nearest Neighbor Distance[a]

Point	Coordinates		Nearest Neighbor	Nearest Neighbor Distance (NND)
	X	Y		
A	0.7	1.0	B	1.6
B	1.25	3.0	C	1.4
C	2.5	3.7	D	1.3
D	3.3	2.75	C	1.3
E	4.0	4.0	C	1.34
F	3.8	1.0	D	1.5
Average NND				8.44
Random average NND				1.4
				1.02

[a]Nearest neighbor distance calculated by using the Pythagorean theorem.

Figure 11.2 Nearest neighbor point coordinates. Locations of data points for determining nearest neighbor distance (see Table 11.1). Each point (e.g., point *A*) has an associated nearest neighbor (in this case, point *B*). The distances are measured using the Pythagorean theorem.

the techniques are beyond normal use in a GIS setting. In addition, other methods can be used to determine the degree of clustering on the basis of other statistics (Davis, 1986; Griffiths, 1962, 1966; Ripley, 1981), but again, these are more detailed than we need for an introduction to the use of nearest neighbor analysis in GIS.

THIESSEN POLYGONS

Points can be organized into a more regional context without relying on the use of other map layers for comparison by means of **Thiessen polygons** (also referred to as **Dirichlet diagrams** and **Voronoi diagrams**). Rather than simply looking at the arrangement of points, we can grow polygons around each point to illustrate the "region of influence" it might have on other points in the coverage. For example, as we will see later when working with the gravity model, we assume that the distances between points impose an attraction on the neighbors. In addition, the magnitude of the node—for example, the size of a city—is often directly related to the degree of influence. We will restrict ourselves to the case of all points considered to be of equal magnitude. This simplifies the description, allowing us to understand it in this simple case. When your skills increase, you can carry the technique to its logical conclusion by weighting the points.

Creating Thiessen polygons is relatively simple conceptually, but the process can become quite involved as the number of points increases. For insight as to how to proceed to develop the polygons, let's first take a look at what these shapes are to represent. If we have a number of point objects, such as towns (again, assuming for simplicity that all are the same size), we can imagine that each point is surrounded by a single irregular polygon. But the polygon has an important property—every location within it is closer to the point encircled than to any other point in the coverage. By contrast, then, every location outside each individual polygon is closer to some other point than to the enclosed point. In other words, the boundaries of each polygon give to the point it surrounds the most compact possible area of influence. In most cases, each point in the coverage will have its own Thiessen polygon, sometimes called a **proximal region**, to show its exclusive area of influence (Clarke, 1990). Now, think about how we might be able to do this.

As illustration, let's take a small sample of points (Figure 11.3). With five points as a starting place, we might conceptualize how the Thiessen polygons are derived by envisioning a bubble growing outward from each point, much like the bubbles formed from bubble pipes. If you look carefully, you will see that each of the interfaces between bubbles forms a straight line (if viewed orthogonally). This straight line is oriented perpendicular to a line that connects each pair of neighboring points. If you measure the distance between the two end points of the perpendicular line, you will see that the distances are identical on either side of the line forming the interface. In other words, the edges of the polygons are formed by an equidistant (bisector) line perpendicular to the line connecting each pair of points. Algorithms for producing Thiessen polygons have been implemented in both CAC and GIS systems for decades, both in vector systems (Brassel and Reif, 1979) and even in quadtree data structures (Mark, 1987). Those inclined to examine details of algorithms might consult those references.

For the rest of us, perhaps the most important question is this: What do we do with Thiessen polygons? We can gain some initial ideas by looking at the reason for their development in the first place. The Thiessen polygon was named after the climatologist A.H. Thiessen, who sought to interpolate climate data from highly uneven distributions of weather station data. In other words, he was trying to describe and analyze point data with area-based symbols and analytical techniques. Thus, if we have only a few scattered points of data but we want to characterize regions on the basis of these points, we use Thiessen poly-

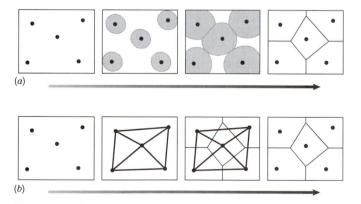

Figure 11.3 Creating Thiessen polygons. Construction of a Thiessen polygon from five points: *(a)* arrangement of the data points, *(b)* construction of the related Thiessen polygons. Note the similarity between the form of the Thiessen polygons and the pattern we might get from a bubble pipe. *Source:* Figure derived from Environmental Systems Research Institute, Inc. (ESRI) drawings.

gons to divide our space into a polygonal form. Because we assume that in each polygon, the influence of the enclosed point is absolute, we can treat the data as a polygonal map.

Most uses for Thiessen polygons involve determining the influence of point data representing shopping centers, industries, or other economically based activities. If we modify the position of our perpendicular line on the basis of the size or other magnitude of each point, our Thiessen polygons become even more representative of the actual influence of industries or shopping on the surrounding space. With such information, the economic placement specialist can determine, for example, how much of a city's population (on the basis of proximity) will likely frequent a planned shopping center. Thiessen polygons are not restricted to economic geography tasks, however: Hutchings and Discombe (1986) found that this device offered a practical approach to the detection of spatial patterns in vegetation. The use of this technique is likely to increase as the functional capabilities of the GIS become more well known to the users.

AREA PATTERNS

Until now, we have looked only at the distribution of points as they are arranged in a defined space. But we know that points are not the only objects that can be distributed in a spatial context. If we change our scale somewhat, our points become areas, and their patterns can then be analyzed by similar methods. We can begin to analyze area patterns much as we did with points— by determining the density of polygons per unit area of our study area. In performing a density measure of polygons, however, we must first measure the area of each class of polygons in which we are interested. We can do this in raster or vector, just as we learned in Chapter 4. Then, knowing the area of each category of polygons, we divide the area of each polygon type (i.e., each region) by the overall area of the coverage. This gives a percentage rather than a number per unit area as in point density. It is possible, of course, to calculate the number of polygons (or clumps of grid cells) per unit area, but because of the potential for widely varying polygon sizes, this approach is not likely to be of much use.

Again, with areas, we might be interested in more than just a measure of the density of the polygons. We may very well be interested in the arrangement and pattern shapes created by groups of polygons. Groups of polygons arranged differently in space may suggest causes for such arrangements, much as they do for point patterns. Examples of potentially interacting polygons could be new plowing innovations at certain farms, cities, and towns and the movement of goods and services among and within them, or even water sources distributed within an area that might provide good spots for overwintering birds. Before we can examine the interactions of polygonal objects with one another, however, we must know something about how they might be arranged. As with point patterns, areas can be clustered, dispersed (regular), and randomly spaced relative to one another (see Figure 2.8). In addition, area patterns can be connected to one another or separated by some definable distance. We will look at some methods of analyzing these pattern metrics.

DISTANCE AND ADJACENCY

Arrangement is inherently based on distance. Among the techniques commonly available for evaluating arrangement is the distance-based retrieval method. For example, one could readily query the software to select and highlight all points, lines, or polygons within a particular distance of another feature. We would do so to learn, for example, if there were any gas stations within 1000 meters of a highway (a point-to-line comparison). Or we might want to know whether there were substantial numbers of people at risk within 1000 meters of hazardous cargo routes on a highway, or we might want to search for polygons representing rest homes or subsidized housing (a polygon-to-line comparison). Another scenario might include a request for all high-income neighborhoods within a certain distance of the local mall so that the shop owners could perform a market analysis (a polygon-to-polygon comparison). All of these scenarios and many more require only a selection based on distance. Points can be retrieved based on distance to other points, lines to other lines, and so on.

Adjacency is a special case of this arrangement measure and the distance-based retrieval method that implements it. The most common form of this measure is to begin with a polygon or a number of polygons and to determine which of these are adjacent to other polygons. Let us say, for example, that you want to know which land parcels are adjacent to a lake (essentially a measure of lake-front property); we would select the lake first and query all polygons that are immediately connected (adjacent) to the lake (Figure 11.4). As in previous chapters, a simple flowchart is included to show how this might be illustrated (Figure 11.5).

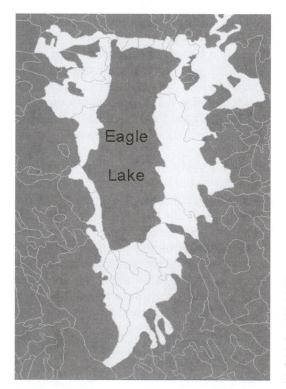

Figure 11.4 Adjacency. Notice how only the polygons that are touching Eagle Lake are highlighted. Unlike buffering, this technique only locates and selects polygons that are within a particular distance. In this case, the distance was set to zero so adjacent polygons would be selected.

Figure 11.5   Adjacency flowchart. The query measures distance from the lake, but the distance is set to 0. Notice how the flowchart itself simplifies the notation.

We will see in Chapter 12 that the arrangements or associations of points or lines to polygons are also possible. For example, intersection can also be seen as a special case of arrangement. In most cases, however, such measures are performed by comparing multiple themes; we will therefore reserve our discussion of these types of techniques until the next chapter.

OTHER POLYGONAL ARRANGEMENT MEASURES

Analyses of polygon arrangements can become quite sophisticated, and software links to a GIS have provided opportunities to perform them (Baker and Cai, 1992; McGarigal and Marks, 1994). The landscape ecology community is a frequent consumer of these techniques, usually treating polygons as patches, especially with respect to a larger, more uniform background, a so-called matrix. You might want to consult Forman and Godron (1986) for an overview of some of these methods as you acquire sophistication in GIS analysis. In general, you will find measures of polygonal **isolation**, measures of **accessibility** (connectedness of polygons to lines), polygon interactions, and dispersion. Since many of these measures are borrowed from the literature in geography, biogeography, ecology, forestry, and other disciplines, the examples will be varied enough to give you a feel for how these additional measures might be used.

LINEAR PATTERNS

Just as we have seen with points and areas, lines occupy space and occur in particular configurations and patterns on the land as well. We see patterns of lines all the time, but often we miss them. Streets and highways form lines with particular definable patterns that we recognize as being related to the networks produced by humans to move people and things between the spatially arranged points called towns. We see fencelines also occurring in particular configurations and numbers depending on field size, lot size, and shapes of the polygons they trace (Simpson et al., 1994). Striations on exposed bedrock show parallel lines indicating the dragging of rock below the glaciers as they scoured the land thousands of years ago. The mechanisms that caused each of these line patterns can best be understood if we first define the specific patterns and concentrations before us. Next, we will look at some of the more common measurements of line patterns available to the GIS professional.

Line Densities

Because of the added dimensionality of lines (as with polygons), especially lengths and orientations, linear analysis is somewhat more complicated than the study of points. Some researchers have tried to examine the distributions of line lengths (Aitchison and Brown, 1969), and others have looked at line spacing (Dacey, 1967; Miles, 1964), much akin to the analysis of nearest neighbors in point distributions (Davis, 1986). We will examine these as well as the larger body of knowledge of line orientation in the next sections. Meanwhile, we need to introduce the simplest measure of line patterning—line density.

Like point and polygon arrangements in an area, lines can occur frequently or infrequently with regard to the space they occupy. To measure this density value for points, which are considered to be dimensionless, we merely divided the number of points by the area. When we worked with polygons, however, we accounted for the area occupied by each polygon by adding all the areas and dividing this sum by the area of the entire map. In the same way, to measure the density of lines (with one dimension), we divide the sum of all the line lengths by the area of the map coverage. Our measure of line density will then be measured in units such as meters per hectare or even miles per square mile. Except for comparing with other values for different regions or for the same region at different time periods, there is not much we can do with this information. We need to know more about the distributional patterns of the lines, just as we did when we were working with points and polygons. Next, we will examine the nearest neighbor statistic as it is applied to lines rather than points. You will see that although the calculations are somewhat different because of the added dimensionality nature of lines, the results provide a useful characterization of line patterns.

Nearest Neighbors and Line Intercepts

We have seen how the nearest neighbor statistic can be used to characterize the spatial arrangements of points. The distribution of pairs of lines can be determined in much the same way, although our computations are somewhat more complex because unlike points, lines have dimension. It might seem simple to adjust for this added dimension by simply selecting the center of each line and performing a nearest neighbor on that point alone. Because the linear objects are often of different lengths, however, this procedure is not going to give us a true picture of the arrangement of the lines themselves. From a statistical standpoint it is often considered useful to sample in a random fashion. Following that approach, our first task in measuring nearest neighbors between line objects is to select a random point on each line on the map (or on each line segment if the lines are not straight). Next, the distance between each point and its nearest neighbor is drawn perpendicular to that line (Davis, 1986) (Figure 11.6). Then we measure these distances and calculate the mean nearest neighbor distance for all the distances. As with all nearest neighbor statistics, we need to be able to test this value against a random distribution. Dacey (1967) determined values for expected nearest neighbor distance, expected

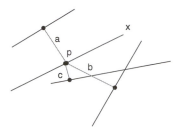

Figure 11.6 Nearest neighbor distances between lines. Finding the nearest neighbor between lines using a randomly selected point along each line as the sample location. *Source:* J.C. Davis, *Statistics and Data Analysis in Geology,* 2nd ed., John Wiley & Sons, New York, © 1986. Adapted from Figure 5.16, page 313. Used with permission.

variance, and standard error in a random distribution of lines. These values allow us to compare the expected and the observed and to generate a statistic against which to test our hypothesis of randomness (Davis, 1986). Those who find such a measure useful are referred to Davis (1986) for a description of the equations.

This test will work for most line patterns, whether the features are straight lines or curved lines, but it has some limitations. If lines in your coverage are very sinuous, and especially if they change directions frequently, this approach is less than successful. In addition, for the test to be useful, the lines should be at least 1.5 times the length of the mean distance between the lines. If the number of lines on the coverage is small, the estimate of density used in nearest neighbor analysis should be adjusted by a weighting factor of $(n = 1)/n,$ where n is the number of lines in the pattern. So rather than simply dividing the sum of the lengths by the area, we use the formula

$$\frac{(n - 1)L}{nA}$$

where

L is the sum of the lengths

A is the area

This adjusted line density value will improve the quality of the nearest neighbor statistic.

Line intersect methods are alternative methods for analyzing distributions of lines. One simple approach is to convert the two-dimensional pattern into a one-dimensional sequence by drawing a sample line across the map and noting where that line intersects the coverage line objects. At least two basic methods can be used to produce sample lines (Getis and Boots, 1978). The first is to randomly select a pair of coordinates and connect them with lines. A second method is to draw a radius at a randomly chosen angle and starting point, then measure a random distance from the center, finally constructing a perpendicular to that radial line (Davis, 1986). Once the random test line has been drawn over the coverage lines, the distribution of intersection intervals can be tested through the use of simple statistics like the **runs test** or other tests of sequences of data. An alternative to a single line is to use a zigzag line that crosses the coverage two or three times. The zigzag path (often called a **random walk**) will again produce a series of line intersects, the distances of which can be tested with any simple statistical test for sequences of data (Figure 11.7).

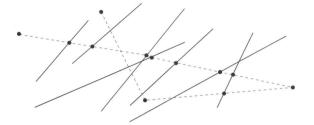

Figure 11.7 Random walk method to evaluate line patterns. Modification of the line intersect using a zigzag line called a random walk to identify sample points. *Source:* J.C. Davis, *Statistics and Data Analysis in Geology,* 2nd ed., John Wiley & Sons, New York, © 1986. Adapted from Figure 5.17, page 314. Used with permission.

If, after performing our nearest neighbor statistic or line intersect method, we find that our distribution is not a random one, there is reason to believe that a process other than random chance contributed to the observed distribution. Here again, it is useful to consider how the linear patterns might be compared to patterns of other objects in difference map layers (Chapter 12).

DIRECTIONALITY OF LINEAR AND AREAL OBJECTS

Linear objects can exhibit more than just distributional patterns on the landscape. Often we find that such features as sedimentary bedding planes, glacial striations, water-deposited pebbles, glacial boulder trains, fencerows, street grids, and forest blowdowns exhibit a preferred orientation. These features and many more are oriented in a particular direction, often strongly related to a functional force. We know, for example, that tree shelterbelts in agricultural regions are often placed in directions opposite the prevailing winds, that trees felled from hurricanes give evidence of the wind directions occurring within the storm, and that glacial striations are indicative of the movement of the glacier. But when we analyze orientation, we see that if lines are oriented in, say, a predominantly north-south direction, we have two directions to choose from. If these linear objects are one-way streets, however, the orientation doesn't tell us the single direction in which traffic must flow. So, besides orientation we also need to consider directionality. We can also think of the distributions of linear objects as either two-dimensional (i.e., distributed purely in azimuthal directions) or three-dimensional (i.e., distributed in an azimuthal direction combined with an angular direction in a sphere). Although both are important, we will concentrate on the simpler two-dimensional forms and leave the spherical directional and orientation measures for more advanced treatments (Davis, 1986).

In traditional statistical analysis of orientation, the data from maps of linear objects are transferred to a circular graph called a **rose diagram**, which plots all linear features starting at the center of a circle and draws each observation as a single line. In some rose diagrams, the length of the line also indicates the magnitude (such as wind speed) or length of the feature (as in the lengths of hedgerows or fences). But although rose diagrams are a useful framework for the visual inspection of directions, the measurements themselves obtained directly from our coverage are much more useful for numerical analysis. Our first analytical technique is to figure out what the **resultant vector** is. This is the procedure your high school text used to describe the usual example of three people pulling in different directions on a large object. In that instance, the vector gives us an idea of where the object is likely to move and with what speed.

Let's take an example of directional analysis in two dimensions by considering a large number of trees that were blown down by straight-line winds associated with frontal activity. Each blown-down tree might be plotted as a line feature in a map layer; the positions of the top and bottom of each tree would be recorded as well, giving us both the orientation and the direction of every downed tree (Figure 11.8). The meteorologists want to deduce the overall direction of the winds on the basis of the felled trees, but the trees are not all perfectly oriented in a single direction. Our first task is to analyze the vector resultant of all the blown-down trees.

The vector for each tree is defined by an angle θ (theta) measured from the base of a tree to its top. We multiply the X coordinate of each tree by the cosine of theta and each Y coordinate by the sine of theta. To find the resultant vector, we sum all these values for both coordinates, and resultant vector values X_r and Y_r show the dominant direction of the end points of all the trees in the blowdown. Figure 11.9 shows a resultant vector of **R** obtained from three values of theta, the vectors **A**, **B**, and **C**.

We may want to go further, however, and determine a mean direction θ based on the resultant vector. As an average of the directions for all the downed trees, it is analogous to the mean of any other set of data. The formula for mean is

$$\bar{\theta} = \tan^{-1}\left(\frac{Y_r}{X_r}\right)$$

(a)

(b)

Figure 11.8 **Line patterns of felled trees.** A portion of a photograph of a portion of Mount St. Helens after the eruption (a). The photograph could be digitized to show the linear objects and their general trend (b).

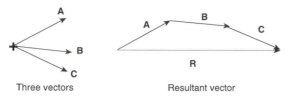

Three vectors   Resultant vector

Figure 11.9   Resultant vector. Resultant vector for three individual linear features.   *Source:* J.C. Davis, *Statistics and Data Analysis in Geology,* 2nd ed., John Wiley & Sons, New York, © 1986. Adapted from Figure 5.20, page 317. Used with permission.

Because the mean direction of our vectors depends not only on the dispersion of the trees but also on the number of observations, we can standardize these values by dividing the coordinates of each resultant vector by the number of line objects in the coverage. This will allow us to compare two different areas. For example, by comparing the average vectors for downed trees from two study areas, we could determine whether the winds were generally coming from the same direction.

As with any sample of points in which the mean serves as a measure of the central tendency of the data, or of the trend of the data to group around some central point, we can use the average to develop other statistics that define the amount of spread away from the mean. Figure 11.10 shows two cases of three vectors being used to determine the resultant vector **R**. When the vectors are fairly closely spaced, the resultant vector is very long, whereas the widely dispersed vectors give a resultant vector that is considerably shorter. Think about this in terms of three people pulling on ropes to move an object. The closer the people are and the more nearly they are pulling in the same direction, the more force results, to pull the object along. On the other hand, if the people are pulling in quite different directions, the object may move, but the amount of force applied is considerably less. In other words, the resultant vector is smaller.

We can determine the length of the resultant force by using the Pythagorean theorem on our resultants (X_r and Y_r). The formula is simply

$$R = \sqrt{X^2 + Y^2}$$

Closely spaced around a single direction

Widely dispersed vectors

Figure 11.10   Vector dispersal. Resultant vectors determined when individual lines are close together and widely spread.   *Source:* J.C. Davis, *Statistics and Data Analysis in Geology,* 2nd ed., John Wiley & Sons, New York, © 1986. Adapted from Figure 5.21, page 318. Used with permission.

where R is the **resultant length**. Thus, not only do we know the average direction in which the trees lie but we also have a measure of the compactness of the distribution. The more compact the distribution, the longer the line. Again, this is analogous to the three ropes pulling an object. The closer together the ropes (in other words, the more closely spaced the vectors), the stronger the force (or the greater the length of the resulting vector).

To compare the resultant length of our vector to that of another site, we again need to standardize the data. We find the **mean resultant length** $\bar{R}$ by dividing the resultant length R, found with the Pythagorean theorem, by the number of observations, n. This value will range from 0 to 1. The mean resultant length resembles variance in linear statistics inasmuch as it is a measure of spatial dispersion about a mean value. The value of mean resultant length, however, is expressed in the opposite direction; that is, large values indicate that the lines are closely spaced, and small values mean they are more dispersed in direction. Thus a large mean resultant value in our tree example would mean that the winds were nearly perfectly undirectional, whereas a small value would suggest eddies in the overall wind pattern. If the inverse nature of the mean resultant length seems counterintuitive, you can use its complement, defined by 1 minus the mean resultant length. This value, called the **circular variance**, will give increasing values as the spread of vectors increases. If you are statistically inclined, you might see some possibilities for directional analogs to standard deviation, mode, and median. Gaile and Burt (1980) provided equations for such measures.

One final problem remains with respect to orientation. As we saw earlier, the orientation of any linear object gives two possible (and opposite) directions. Unlike our trees, however, some features can be expressed in either of two directions. Let us say, for example, that two researchers are looking at different portions of a set of fencerows. One researcher states that most of the fences are oriented toward the north. The second, however, says they are oriented toward the south. Both are correct: that is, if we took one vector that was oriented directly north and the other directly south and used our equations for mean vector length, the vectors would tend to cancel each other.

Fortunately, Krumbein (1939) found a simple way to resolve this problem. He observed that when he doubled the angles, no matter which direction was originally used to record the data, the same angle was recorded. Suppose that a fencerow is oriented northwest to southeast (that is, 315 degrees to 135 degrees). If we multiply each of these values by 2, we obtain the same number: $315 \times 2 = 630$ degrees (2360 degrees $\times$ 270 degrees) and 135 degrees $\times 2 = 270$ degrees. The calculations for mean direction, mean resultant length, and circular variance will now be doubled. To get our true values for these measures, all we need to do is divide by 2.

These simple measures of directionality and dispersion can be tested for randomness (Batschelet, 1965; Gumbel et al., 1953) and for specific trends (Stephens, 1969) through standard processes of hypothesis testing. Other sets of directional data can be compared to those for our original coverage (Gaile and Burt, 1980; Mardia, 1972). These tests are beyond the scope of this book; the works just cited offer detailed discussion of these more complex techniques.

In the context of GIS, we are most likely to use the measures simply to help us characterize the distributions inside our coverage. Or, more likely, we will use them to discover useful comparisons to other map layers to help us understand the causes. Commercial GISs frequently provide tools that will allow us either to calculate these directional statistics directly or to pass preliminary data from the GIS to other software designed specifically for such analyses. You should be aware, however, that an increasing number of software systems, especially among the geosciences, are designed for handling directional statistics. Raster GIS systems are not particularly well suited to this type of analysis, but most topologically based vector GIS systems will allow you to at least compile some of the preliminary values (e.g., angles for line segments), which can be stored in the database system directly as attribute values that can be passed to other software systems.

CONNECTIVITY OF LINEAR OBJECTS

An important aspect of the spatial arrangement of lines is their ability to form networks. Networks occur in many different forms and are both natural and anthropogenic. Roads and rail lines are networks that serve to move and transport people and materials from place to place. Telephone lines are networks to allow the movement of information and communication. Rivers and streams are networks for aquatic creatures, and even hedgerows may act as networks to allow the movement and migration of small mammals through a landscape. The list of systems that may be defined as networks is quite large. Although the density of such features and their orientation may be of interest to us, we need to be able to analyze the actual connections made by these features and the amount of **connectivity** provided from place to place. Most of you have probably encountered the frustration of traveling in a large city or across country and finding that a direct path to a location you wish to visit does not exist. Even a quick look at a road map of North America shows a great many places that can be visited only through very long and indirect paths. That is, the areas are characterized by a relative lack of connectivity.

Connectivity is a measure of the complexity of a network. Several devices may be used to calculate this value (Haggett et al., 1977; Lowe and Moryadas, 1975; Sugihara, 1983; Taaffe and Gauthier, 1973). The two most common ones are the **gamma index** and the **alpha index**.

The gamma (γ) index compares the number of links, L, in a given network to the maximum possible number of links between nodes. To calculate this, we simply produce a ratio of the two. First we count the number of links (line segments between nodes) that actually exist in the coverage. As you might guess, this is very difficult in raster GIS but easily accomplished in vector. Once we have the number of actual links, we need to determine the number of possible links in the coverage. This is not as difficult as it might appear because there are limits to the number of links, determined entirely by the number of nodes. If, for example, only three nodes are present, only three links are possible (Figure 11.11), but if we add another node, we see that three additional links are possible, for a total of six (Forman and Godron, 1986). So if we assume that no new intersections (line crossings) are formed, the maximum number of possible links is increased by three each time (Figure 11.11). In other words, the maximum number of links, L_{max}, is always $3(V - 2)$, where V is the number of nodes.

(a)   (b)   (c)

-N-

500m
⊢―――┬―――┤
1000ft

Figure 11.11 Gamma index. Two different networks of hedgerows based on the same set of nodes. The network on the left *(a)* is minimally connected and has no circuits, whereas the network on the right *(b)* has higher connectivity and has circuits that provide alternate routes for travel. *(c)* Shows the study area for each set of networks. (From Forman and Godron, 1986.) *Source:* R.T.T. Forman and M. Godron, *Landscape Ecology,* John Wiley & Sons, New York, © 1986. Adapted from Figure 11.12, page 418. Used with permission.

To find the gamma index, we simply divide the number of links, *L,* by the maximum number of links, L_{max}.

$$\gamma = \frac{L}{L_{max}} = \frac{L}{3(V-2)}$$

The gamma index will range from a low of 0, where none of the nodes is connected, to 1.0, where all possible links between all nodes are present (Forman and Godron, 1986).

Figure 11.11, illustrating a network with 16 possible nodes, reveals the difference in connectivity for two different cases. The case in Figure 11.11*a* shows 15 links among the 16 nodes. This gives us a connectivity of

$$\gamma = \frac{L}{3(V-2)} = \frac{15}{3(16-2)} = \frac{15}{42} = 0.36$$

whereas the case in Figure 11.11*b* has 20 links for the 16 nodes, yielding a connectivity of

$$\gamma = \frac{L}{3(V-2)} = \frac{20}{3(16-2)} = \frac{20}{42} = 0.48$$

So the first network is about one-third connected (36 percent), whereas the second is almost half connected (48 percent).

In Figure 11.11, we were looking at hedgerows, which might be used as corridors for small mammals. The same arrangement of features could easily have been a road net developed for human transportation. The results would have been the same. In the first network, the route from the starting point at *R* to the ending point at *S* would be relatively long, compared to that for the second network with its increased connectivity. If we were transportation planners, this difference would suggest that increased connectivity also would improve the movement of goods and services between nodes.

Connectivity, however, is not enough to fully characterize networks. Recall our discussion of the formation within networks of circuits—loops providing alternate paths for travel from node to node. Perhaps the best example of a circuit would be a beltway, one of the circular highways surrounding urban areas in the United States. As we travel across the country, we must decide whether to travel through or around each city. If we choose to go around, thus avoiding the urban traffic, we have two possible routes: we can turn either left or right, to go around the city counterclockwise or clockwise. Thus the loop feature offers choices for travel, allowing us to select the shortest route to a destination on the other side of the city.

The index designed to measure **circuitry**—the degree to which nodes are connected by circuits of alternative routes—is called the alpha (α) index. Similar to the gamma index, this measure is a ratio of the existing number of circuits to the maximum possible number. We see that a network with no circuits present contains one link fewer than the number of nodes: ($L = V - 1$). If you look again at Figure 11.11, you see that this is true where the first network (α) has 15 links and 16 nodes. This figure is minimally connected in that it has the fewest number of links possible, given the number of nodes, but each is still connected to at least one link. By adding a link to this network, we create a circuit or loop. As such, when a circuit is present, $L > (V - 1)$. Therefore, the number of circuits present in the network (i.e., the number of links present minus the number of links in our minimally connected network) can be given by ($L - V$) + 1. Thus we know the actual number of the circuits in the network (Forman and Godron, 1986).

An alpha index for the amount of circuitry is given by the ratio of the actual number of circuits present to the maximum number possible in the network. Therefore, since the number of possible links is determined by $3(V - 2)$ and the minimum number of links in a minimally connected network is expressed as $V - 1$, we can determine the maximum possible number of circuits (by combining these two simple formulas). That is, we subtract the number of links in the minimally connected network, $V - 1$, from $3(V - 2)$, the maximum number of links, and from $3(V - 2) - (V - 1) = 2V - 5$, we obtain the final formula for the alpha index:

$$\alpha = \frac{\text{actual number of circuits}}{\text{maximum number of circuits}} = \frac{(L - V) + 1}{2V - 5}$$

As before, the values for the alpha index range between 0 (a network without any circuits) and 1.0 (a network with the maximum number of circuits present).

If we again use the network from Figure 11.11, the alpha value for Figure 11.11*a* is

$$\alpha = \frac{(L - V) + 1}{2V - 5} = \frac{(15 - 16) + 1}{(2 \times 16) - 5} = \frac{0}{27} = 0$$

showing us that no circuits are present. And the alpha value for Figure 11.11*b* is

$$\alpha = \frac{(L - V) + 1}{2V - 5} = \frac{(20 - 16) + 1}{(2 \times 16) - 5} = \frac{5}{27} = 0.19$$

showing that there is 19-percent circuitry out of a possible 100 percent. So Figure 11.11*a*, shows no options for travel; movement from point *R* to point *S* has only a single option. For Figure 11.11*b,* on the other hand, we have numerous alternative routes, some longer than others, to get from *R* to *S.*

Because of the addition of links needed to produce circuits, it is entirely possible to consider the alpha index as an alternative method of measuring connectedness. However, because the alpha and gamma indices give us a different look at the overall network patterning, they might more appropriately be combined to provide an overall measure of the amount of **network complexity**. To perform the alpha or the gamma index in a GIS requires that the GIS be a vector system. The relationship between these two indices and the vector GIS model is particularly strong because, like the vector GIS model itself, the indices are based on topological properties. In this case, the primary topological foundations are based on **graph theory**, which, as we saw when we looked at the DIME topological model, is concerned less with the lengths of lines or their shapes than with the degree of connection between them.

For modeling transportation, we need to know more than just how connected the network is—for example, the lengths of the lines between nodes, something about the direction in which travel can proceed along these lines, and impedance values for these lines. In addition, numerous other simple indices, derived mostly from transportation and communication theory, can be used for characterizing the connectivity of networks. We can, for example, calculate the linkage intensity per node or the number of alternate routes between nodes; we can determine which node has the most links **(central place)**, and we can ascertain delineations of regions based on connectivity and accessibility (Haggett et al., 1977; Lowe and Moryadas, 1975). Of course, all these could be combined with each other and with the measures of linear dispersion, arrangement, and orientation to give a more detailed view of networks. As you conduct more work with your GIS, you will see how these approaches might be applied; you can consult the references at the end of this chapter to learn more. In the meantime, we will look at one last measure that is in common use for modeling relationships among linear objects, based somewhat on the connectivity we have just examined.

GRAVITY MODEL

In our discussion of connectivity, we made the simple assumption that all the nodes were of equal size or importance. This allowed us to concentrate on the degree to which the links connected these nodes or the number of circuits and alternate paths available. However, think about the nature of towns and the nature of cities. Large cities are obviously more likely to provide more opportunities for shopping, more access to the arts, and more occasions to attend professional sporting events. That is why we tend to go to cities for shopping and entertainment. But towns and cities are not the only pairs of nodes that can exhibit different amounts of attraction. Take, for example, a small pond versus a lake. The lake is likely to attract more waterfowl than will be found near the pond.

Regarding the two examples of contrasting nodes, you should note that we have defined a magnitude for each that we can use to separate them out in a hierarchical fashion. Both large cities and lakes have a greater likelihood of drawing activity, whether it is performed by ducks or people, than do their smaller counterparts. The magnitude of the attraction can be thought of much as astronomers envision the gravitational attraction between two celestial bodies. The larger the body, the greater the gravitational attraction between itself and neighboring bodies. We know this is true because astronauts who landed on the lunar surface were experiencing a gravitational attraction only about one-sixth that of earth.

Converting the concept of gravitational attraction to use in a two-dimensional setting provides the measure of the interaction between two nodes in a GIS coverage that we call the **gravity model**. In its simplest form, the gravity model takes the form of the following formula:

$$L_{ij} = K\frac{P_i P_j}{d^2}$$

where

L_{ij} is the interaction between nodes i and j

P_i is the magnitude of node i

P_j is the magnitude at node j

d is the distance between the two nodes

the constant K relates the equation to the types of object being studied (population size, animals, etc.).

The values of P can be represented by a force of interaction between the nodes, such as the demand for products, the amount of retail sales for city shopping centers, or the amount of wetland habitat for waterfowl.

As with gravity, the larger the magnitude of the nodes, the greater the interaction between them is likely to be. Conversely, the greater the distance between the two nodes, the less attraction between them. You might again think about the draw to a large city for shopping as an example. The larger the city, the more shopping it is likely to offer, but the farther away you are from a given city, the less likely you are to visit it regardless of the bargains available.

Some GIS software systems, both raster and vector, have the capability for developing gravity models between points. Quite a number of variations can be found on the simple model you have seen here, but they are best studied in more advanced texts on spatial analysis. Although most gravity models are used for economic placement analysis, much like Thiessen polygons, quite a number of applications are possible. Investigators have used gravity models to describe the flow of passengers by air and land between cities, the volume of telephone calls, flows of birds, and the seeds the birds disperse among woodlots (Buell et al., 1971; Carkin et al., 1978; McDonnell, 1984; McQuilkin, 1940; Whitcomb, 1977). As long as your concern is the movement of objects between nodes of different magnitude, the gravity model may be of use to you.

ROUTING AND ALLOCATION

Among the most useful applications of networks in GIS are the related tasks of **routing** and **allocation**. Routing involves finding the shortest path between any two (in the simplest case) nodes in a network (Figure 11.12). Because the nodes can be assigned weighting factors as in the gravity model, a route might be between one point and the nearest point with the highest weight (as in the amount of demand for a product). (See Lupien et al., 1987, for an excellent discussion of routing and allocation.)

Each link in the net can also be assigned an impedance value, much like a friction surface but imposed only on the line itself. The impedance value might be related to a speed limit along a street. By using an accumulated distance (see Chapter 8 for a discussion), based on a combination of calculated distance and the impedance factor, the most efficient route can be found, rather than just the shortest. Nodes can also be coded with stops (indicative of traffic signals or stop signs), turning impedances based on the difficulty of turning left or right at an intersection, and even barriers preventing movement and forcing traffic along another path. As with calculating distances along surfaces, all these measures require prior knowledge of the nature of the streets, intersections, and other nodes. Frequently, the weights and impedances are somewhat arbitrary or are based on intuitive knowledge rather than on absolute certainty of how they will affect travel.

Although routing can be done in raster, it is much easier if performed in a vector system. Because of the close relationship of graph theory topology and the topological vector data model, the two perform well together. You should also be aware that given the many possible routes, especially where circuits are involved, there are likely to be a number of possible ways for finding your route. A discussion of these ways could fill a volume on its own.

Allocation is a process that can be used to define, for example, the location of a market, the areal extent of a water treatment center, or the boundaries of a series of fire station service areas. A network structure within a vector GIS is most often used. The idea here is that the capacity of a given service is distributed throughout the net. Each link in the network has a specified number of items that must be served. For example, each street segment contains a number of homes that would need to receive water service from the water treatment center. Or each house could be regarded as a fire-prone structure that

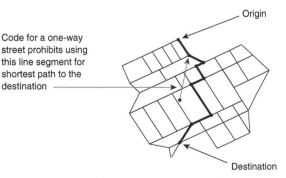

Origin

Code for a one-way
street prohibits using
this line segment for
shortest path to the
destination

Destination

Figure 11.12 Shortest path along a network.
A simple road network showing the solution for
the shortest path algorithm: notice the code entered along the diagonal road to indicate that this
one-way street is going in the wrong direction for
it to be usable for the shortest path solution.

may need the services of the nearby fire station (Lupien et al., 1987). In addition, each service or market center is reasonably capable of operating at a certain maximum capacity. Thus the maximum number of units that each service area can accommodate is compared to the number of service sites that occur along the network links. Say, for example, that a fire station can service only 100 homes with a reasonable level of safety. Beginning at the service location, the GIS will travel outward, following the nearest roads, counting the number of houses that occur along those links until it reaches its capacity of 100. If each street contains 10, the GIS will designate the 10 nearest streets (each with 10 homes) to ensure efficient allocation of the maximum capacity for the service center (Figure 11.13). The specified amount of route might include the number of people to service (i.e., number of residents per house), the road miles that can be traveled, and a travel time limit that must not be exceeded. If our road net were completely uniform (no impedances, no stops, no changes in speed limit, etc.), calculating this allocation would be a simple matter of deciding on the criteria for allocation and pushing our boundary outward until they are met. For example, if we wanted to allocate the routes for newspaper carriers so that each person's vehicle would have to travel only a certain number of road miles, the software would simply add up the road miles as each route spread outward from the starting point, until that value was reached, whereupon the links in the road that were allocated would be assigned attribute codes to correspond to the different carriers.

Most real allocation problems are quite complicated. In some cases, our primary interest might be the number of people we can service in a given amount of time. If you were planning to allocate newspaper routes, you would need, for example, to be able to distribute newspapers to the subscribers by a specified time of day, say 9 A.M. And, of course, no road network is totally uniform. Some roads have higher speed limits than others, others have stop signs, some include turns that are slower than others, and so on. And you do need to know that individual addresses are somehow connected to the streets you intend to service (the process of linking addresses to streets is called **address matching**). The importance of quality geocoding for effective address matching cannot be understated here. Most of the work of determining these allocations requires us to encode all the appropriate attributes, impedance values, turning impedances, and other variables. Once this has been done, the software is quite capable of providing a reasonable result. You must watch out for missing or incorrectly coded attributes along the lines, or incorrect address matches between individual homes and the streets. The time required to explain the

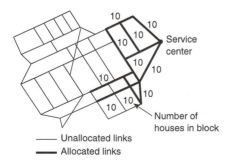

— Unallocated links
— Allocated links

Figure 11.13 Allocations of networks. A simple road network showing how links containing 10 homes each are allocated to a service center (say a fire station) that can service only 100 homes safely.

possible methods, applications, pitfalls, and remedies would probably fill a book the size of this one, so I leave you to experience these on your own when you begin working with linear objects in a transportational setting.

THE MISSING VARIABLE: USING OTHER MAP LAYERS

Only a single, simple reminder is necessary at this point. We have seen how points, areas, and lines can be examined for their arrangements, connectedness, and orientations. And we have seen how we might be able to characterize the neighborhoods of these collections of objects. Your ability to use the information you have teased out of these analyses is often only as good as the comparable data available in other map layers. This is especially true if you are trying to determine the relationships of objects to a causative mechanism. In the next chapter, we will learn how comparisons among different map layers can be made. There are far more comparisons than you might imagine, and their variety should provide ample opportunity to test your abilities to relate variables to one another.

Terms

accessibility	graph theory	regular
address matching	gravity model	resultant length
allocation	interactions	resultant vector
alpha index	isolation	rose diagram
central place	line intersect methods	routing
circuitry	mean resultant length	runs test
circular variance	nearest neighbor analysis	spatial arrangement
clustered	network complexity	Thiessen polygons
connectivity	proximal region	uniform
density	quadrat analysis	variance–mean ratio (VMR)
Dirichlet diagrams	quadrats	Voronoi diagrams
dispersion	random	
gamma index	random walk	

Review Questions

1. What do we mean when we speak of an interest in arrangements of objects? Why is this knowledge useful to us?

2. How do we define a uniform distribution? What makes a uniform distribution regular versus random? What is a clustered distribution?

3. What is the significance of regular, random, and clustered distributions in terms of the processes that could have caused them? Give some exam-

ples of each distributional pattern and give a hypothesis that might describe the causes.

4. What is the purpose of quadrat analysis? Describe how it is done. What statistical test do we use to confirm or reject our results?

5. What is the purpose of VMR in point distributions? How does VMR differ from traditional quadrat analysis?

6. What does nearest neighbor analysis tell us that quadrat analysis does not? How are the two techniques similar?

7. What is the purpose of Thiessen polygons? Diagram some samples and describe how they are created.

8. Explain, with examples, how the concept of adjacency and distance might be employed without highly sophisticated specialty software such as Fragstats.

9. Describe the process of extracting a nearest neighbor statistic from line patterns. What does this quantity tell us about the distribution of our lines? Describe some situations in which the nearest neighbor statistic for lines might give misleading results.

10. Describe the use of line intersect methods for analyzing line pattern distributions. What is a random walk?

11. What is a vector resultant? What does it tell us about the patterns of linear objects? What is the resultant length? How does it compare to resultant force problems in physics?

12. What is the mean resultant length? How does it differ from the resultant length? When would we use it? What would a large mean resultant length tell us? What is the difference between circular variance and mean resultant length?

13. How can we adjust our measurements for mean direction, mean resultant length, and circular variance to account for orientations that can be measured in either of two directions?

14. Describe how the gamma index is performed. What does it tell us about our network? What does a gamma value of 0.48 tell us in terms of the amount of connectedness in our network?

15. What is the alpha index? How is it different from the gamma index? How is it the same? Give some nonroad network examples of how both of these indices might be useful.

16. Describe, in general terms, what the gravity model is. How does distance affect the interactions of point objects? How does the magnitude of the nodes change the interactions between them? Give an example of how the gravity model might be applied.

17. Describe the types of attribute that might need to be assigned to the nodes and the arcs between nodes in routing and allocation problems.

References

Aitchison, J., and J.A.C. Brown, 1969. *The Lognormal Distribution, with Special Reference to Its Uses in Economics.* Cambridge: Cambridge University Press.

Baker, W.L., and Y. Cai, 1992. "The r. le Programs for Multiscale Analysis of Landscape Structure Using the GRASS Geographical Information System." *Landscape Ecology,* 7(4):291–301.

Batschelet, E., 1965. "Statistical Methods for the Analysis of Problems in Animal Orientation and Certain Biological Rhythms." *American Institute of Biological Sciences Monograph.* Washington, DC: AIBS.

Brassel, K.E., and D. Reif, 1979. "A Procedure to Generate Thiessen Polygons." *Geographical Analysis,* 11(3):289–303.

Buell, M.F., H.F. Buell, J.A. Small, and T.G. Siccama, 1971. "Invasion of Trees in Secondary Succession on the New Jersey Piedmont." *Bulletin of the Torrey Botanical Club,* 98:67–74.

Carkin, R.E., J.F. Franklin, J. Booth, and C.E. Smith, 1978. "Seeding Habits of Upper-Slope Tree Species. IV. Seed Flight of Noble Fir and Pacific Silver Fir." U.S. Forest Service Research Note PNW–312. Portland, OR: USFS.

Clarke, K.C., 1990. *Analytical and Computer Cartography,* Englewood Cliffs, NJ: Prentice Hall.

Cliff, A.D., 1970. "Computing the Spatial Correspondence Between Geographical Patterns." *Trans. Inst. British GEOG.,* 50: 143–154.

Dacey, M.F., 1967. "Description of Line Patterns." *Northwestern Studies in Geography,* 13:277–287.

Davis, J.C., 1986. *Statistics and Data Analysis in Geology,* 2nd ed. New York: John Wiley & Sons.

DeMers, M.N., J.W. Simpson, R.E.J. Boerner, A. Silva, L.A. Berns, and F.J. Artigas, 1996. "Fencerows, Edges, and Implications of Changing Connectivity: A Prototype on Two Contiguous Ohio Landscapes." *Conservation Biology,* 9(5):1159–1168.

Forman, R.T.T., and M. Godron, 1986. *Landscape Ecology.* New York: John Wiley & Sons.

Gaile, G.L., and J.E. Burt, 1980. "Directional Statistics: Concepts and Techniques in Modern Geography." *Geo Abstracts,* No. 25, University of East Anglia, Norwich, England.

Getis, A., and B. Boots, 1978. *Models of Spatial Processes: An Approach to the Study of Point, Line and Area Patterns.* Cambridge: Cambridge University Press.

Griffiths, J.C., 1962. "Frequency Distributions of Some Natural Resource Materials." *Pennsylvania State University, Mineral Industries Experiment Station Circular,* 63:174–198.

Griffiths, J.C., 1966. "Exploration for Natural Resources." *Journal of the Operations Research Society of America,* 14(2):189–209.

Gumbel, E.J., J.A. Greenwood, and D. Durand, 1953. "The Circular Normal Distribution: Tables and Theory." *Journal of the American Statistical Society,* 48:131–152.

Haggett, P., A.D. Cliff, and A. Frey, 1977. *Locational Analysis in Human Geography.* New York: John Wiley & Sons.

Hutchings, M.J., and R.J. Discombe, 1986. "The Detection of Spatial Pattern in Plant Populations," *Journal of Biogeography,* 13:225–236.

Krumbein, W.C., 1939. "Preferred Orientation of Pebbles in Sedimentary Deposits." *Journal of Geology,* 47:673–706.

Lavin, S., 1979. "Region Perception Variability on Choropleth Maps: Pattern Complexity Effects." Unpublished Ph.D. dissertation, Department of Geography, University of Kansas, Lawrence.

Lowe, J.C., and S. Moryadas, 1975. *The Geography of Movement.* Boston: Houghton Mifflin.

Lupien, A.E., W.H. Moreland, and J. Dangermond, 1987. "Network Analysis in Geographic Information Systems," *Photogrammetric Engineering and Remote Sensing,* 53(10):1417–1421.

Mardia, K.V., 1972. *Statistics of Directional Data.* London: Academic Press Ltd.

Mark, D., 1987. "Recursive Algorithm for Determination of Proximal (Thiessen) Polygons in Any Metric Space." *Geographical Analysis,* 19(3):264–272.

McDonnell, M.J., 1984. "Interactions Between Landscape Elements: Dispersal of Bird-Disseminated Plants in Post-agricultural Landscapes." In *Proceedings of the First International Seminar on Methodology in Landscape Ecological Research and Planning,* Vol. 2, J. Brandt and P. Agger, Eds. Roskilde, Denmark: Roskilde Universitetsforlag GeoRuc, pp. 47–58.

McGarigal, K., and B.J. Marks, 1994. FRAGSTATS, Spatial Pattern Analysis Program for Quantifying Landscape Structure, Version Two. Forest Science Department, Oregon State University, Corvallis.

McQuilkin, W.E., 1940. "The Natural Establishment of Pine in Abandoned Fields in the Piedmont Plateau Region." *Ecology,* 21:135–149.

Miles, R.E., 1964. "Random Polygons Determined by Lines in a Plane, I and II." *Proceedings of the National Academy of Sciences,* 52:901–907, 1157–1160.

Ripley, B.D., 1981. *Spatial Statistics.* New York: John Wiley & Sons.

Shea, K.S., 1983. "Region Perception Variability on Dot Maps: An Analysis of the Influence of Dot Size." Unpublished master's thesis, Department of Geography, University of Kansas, Lawrence.

Simpson, J.W., R.E.J. Boerner, M.N. DeMers, L.A. Berns, F.J. Artigas, and A. Silva, 1994. "48 Years of Landscape Change on Two Contiguous Ohio Landscapes." *Landscape Ecology,* 9(4):261–270.

Stephens, M.A., 1967. "Tests for the Dispersion and for the Model Vector of a Distribution on a Sphere." *Biometrika,* 54:211–223.

Stephens, M.A., 1969. "Tests for Randomness of Directions Against Two Circular Alternatives." *Journal of the American Statistical Association,* 64(325):280–289.

Sugihara, G., 1983. "Peeling Apart Nature." *Nature,* 304:94.

Taaffe, E.J., and H.J. Gauthier Jr., 1973. *Geography of Transportation.* Englewood Cliffs, NJ: Prentice Hall.

Unwin, D., 1981. *Introductory Spatial Analysis.* New York: Methuen & Company.

Whitcomb, R.F., 1977. "Island Biogeography and 'Habitat Islands' of Eastern Forest." *American Birds,* 31:3–5.

CHAPTER 12

Comparing Variables Among Maps

The process of **overlay** requires both graphic and attribute comparisons. Some of the details can become quite complicated, especially the computer graphics associated with vector overlay. The following descriptions of the overlay process should give you a feel for how the computer accomplishes this task. Rather than detailing what is essentially computer graphics rather than geography, the chapter imparts a general, conceptual understanding. You are encouraged to consult the numerous texts on computer graphics or to consider taking a course or two if your interests lie in this direction.

This chapter does not detail all possible methods of logically or mathematically combining themes. Instead, it focuses on a few that are readily understandable. Experiment! If you are using software for your course, try out different methods on the same set of data. Relate such trials to specific questions you are asking, to see which method gives you the best answer. There is no substitute for experience. And experience gained on even the simplest GIS packages is easily transferable to the larger, more powerful commercial software. You will find that the amount of time required to learn the largest system will be greatly reduced if you learn how overlay is done conceptually. After that, it is merely a matter of finding the documentation that says something like "OVERLAY" and determining which buttons will execute the correct approach for a particular type of comparison among cartographic variables.

LEARNING OBJECTIVES

When you are finished with this chapter you should be able to:

1. Understand the relationship between manual and automated overlay approaches.

2. Describe the overall advantages of overlay as a GIS analytical technique and describe the advantages and disadvantages of raster versus vector overlay approaches.

3. Describe examples of the use of point-in-polygon and line-in-polygon overlay operations.

4. Describe examples of the use of polygon overlay.

5. Provide detailed explanations of the different types of overlay in both raster and vector.

6. Understand the difference between CAD-type overlay operations and GIS-type overlay operations.

7. Describe how dasymetric mapping could be used in GIS.

8. Understand the problems associated with overlay operations and be able to discuss the types of error associated with them.

THE CARTOGRAPHIC OVERLAY

Among the most powerful and most frequently used features of the modern geographic information system is its ability to place the cartographic representation of thematic information of a selected theme over that of another. This process, commonly called **overlay**, is so intuitive that its application long preceded the advent of modern electronic geographic information systems. A very early application of the analog overlay technique dates back to the days of the American Revolution. During the battle of Yorktown in 1781, George Washington engaged the services of French cartographer Louis-Alexandre Berthier, who employed map overlay to examine the relationship between Washington's troops and those of British General Charles Cornwallis (Wikle, 1991). This was certainly not the first use of the manual overlay operation, but it shows that the idea has been around literally for centuries.

The conceptual framework behind spatial correspondence of different phenomena has been available for some categories of mapped data, but certainly not all. Sauer (1925), for example, established a model for the interactions of general categories of earth-related data in his work on the morphology of landscape. His research established the existence of strong correlations among human activity, landform, and other physical parameters. Although Sauer did not formalize these correlations for use with maps, it was clear that he saw connections among distributions of these phenomena on the surface of the earth. Later researchers formalized this design into a wide variety of approaches named, for example, sieve mapping (Tyrwhitt, 1950) and **biophysical mapping** (Hills et al., 1967).

Among the most influential practitioners of this early developing approach was Ian McHarg, whose work on environmental phenomena has spawned an entire school of thought among today's landscape architects, allowing considerably more work to be performed using the computer than was originally possible from field observation and single coverage cartography alone (Simpson, 1989). McHarg (1971) retained an analog method, applying clear acetate overlays of selected mapped environmental phenomena to allow him to evaluate the environmental sensitivity of an area. The darker the area of each coverage, the higher the sensitivity. As more layers were placed on top of one another, the areas of high environmental sensitivity became progressively darker. The area was then remapped showing a gradation from higher to lower sensitivity, and decision makers could use the map product in assessing available alternatives (Steinitz et al., 1976) (Figure 12.1).

Much remains to be learned about the causal relationships among spatial phenomena. This subject comprises a large part of what geographers do. Geographic information systems are now providing easily available map overlay procedures that may result in the development of new hypotheses, new theories, and perhaps even some new laws about these pattern similarities. Many specialists in other fields are also learning more about corresponding spatial patterns that, until recently, had not been observed, and perhaps were even un-

SLOPE
ZONE 1   Areas with slopes in excess of 10%.
ZONE 2   Areas with slopes less than 10%
         but in excess of 2¹/₂%.
ZONE 3   Areas with slopes less than 2¹/₂%.

SURFACE DRAINAGE
ZONE 1   Surface-water features—streams,
         lakes, and ponds.
ZONE 2   Natural drainage channels and
         areas of constricted drainage.
ZONE 3   Absence of surface water or
         pronounced drainage channels.

SOIL DRAINAGE
ZONE 1   Salt marshes, brackish marshes,
         swamps, and other low-lying
         areas with poor drainage.
ZONE 2   Areas with high-water table.
ZONE 3   Areas with good internal drainage.

BEDROCK FOUNDATION
ZONE 1   Areas identified as marshlands are
         the most obstructive to the high-
         way; they have an extremely low
         compressive strength.
ZONE 2   The Cretaceous sediments: sands,
         clays, gravels; and shale.
ZONE 3   The most suitable foundation con-
         ditions are available on crystalline
         rocks: serpentine and diabase.

SOIL FOUNDATION
ZONE 1   Silts and clays are a major obstruc-
         tion to the highway; they have
         poor stability and low compres-
         sive strngth.
ZONE 2   Sandy loams and gravelly sandy to
         fine sandy loams.
ZONE 3   Gravelly sand or silt loams and
         gravelly to stony sandy loams.

SUSCEPTIBILITY TO EROSION
ZONE 1   All slopes in excess of 10% and
         gravelly sandy to fine sandy loam soils.
ZONE 2   Gravelly sand or silt loam soils and
         areas with slopes in excess of 2¹/₂%
         on gravelly to stony sandy loams.
ZONE 3   Other soils with finer texture and
         flat topography.

COMPOSITE: PHYSIOGRAPHIC OBSTRUCTIONS

Figure 12.1   Acetate overlay to determine environmental sensitivity. Example of Ian McHarg's use of manual acetate overlay operations to illustrate increased sensitivity as a number of categories of physical parameters overlap. This map shows the overlapping of slope, surface drainage, bedrock foundation, soil foundation, and susceptibility to erosion for the Richmond Parkway study area in New Jersey.   *Source:* Ian McHarg, *Design with Nature,* John Wiley & Sons, New York, © 1995. Reprinted with permission.

observable without automation. As more disciplinary specialists learn about unique spatial patterns, we will have an increasingly detailed set of rules about which layers can legitimately be overlayed and which ones can reasonably permit statements about cause and effect. Until then, we would be wise to stay alert, especially because the modern GIS provides dozens of methods for quickly and easily overlaying maps. This power to overlay can easily be translated into a large number of opportunities to make mistakes and to overlay either nonassociated or autocorrelated layers, subject to false conclusions. It can also increase our ability to knowingly lie with maps well beyond what was possible before GIS could be used to compare spatial phenomena (Monmonier, 1991). In the following pages, we will look closely at our options for overlay operations, consider our limitations, and examine what we can reasonably overlay, when to overlay, and the potential problems of overlay.

POINT-IN-POLYGON AND LINE-IN-POLYGON OPERATIONS

Traditionally, we envision map overlay in one of two ways—analog or digital—and primarily as a method of comparing areas (polygons). This approach works well if all our data are polygonal. Often, however, point and line data must be compared either to each other or to the locations of polygons. Thus a means of comparing these disparate cartographic forms might prove useful to the GIS practitioner. A wildlife biologist, for example may wish to know whether a particular species of bird (point data) has a preference for a particular type, size, or configuration of vegetated patch (polygon). The necessary operation is to overlay the point data on the polygonal data—called a point-in-polygon operation.

A second example would be a crime analyst trying to determine whether certain neighborhoods are experiencing an increase in incidence of purse snatching. By plotting historical crime event records of purse snatching and overlaying them on the neighborhood polygons, our crime analyst notices that a particular neighborhood (we'll call it Pleasantfield) seems to have a concentration of this type of crime despite being a relatively average neighborhood. Later, by mapping these same data on a monthly basis, he notices a concentration of purse snatching in December. He also notices that this pattern seems to disappear starting in 1983—the same year that a new shopping mall was built in Pleasantfield. As he compares a map of point locations of purse snatching with a detailed map of the Pleasantfield neighborhood, he sees not only that a spatial relationship exists between these crime data and the Pleasantfield polygon but also that certain high traffic streets leading to and from the mall and its parking lot seem to experience the largest number of purse-snatching events in the neighborhood. This latter comparison shows the importance of certain lines (in this case, streets) occurring inside polygons (in this case neighborhoods).

Our detective has demonstrated the strong relationship between points and polygons, and we can see how useful comparing these entities can be. He has also demonstrated the close relationship between comparing variables within a single coverage, as we did in Chapter 11, with comparing those between layers. In examining his maps, he discovered not only that there were more points in a certain polygon but also that the points themselves were grouped in close proximity to themselves and to linear features in the area.

A third example will illustrate the possibility of correspondence between line phenomena and polygons, or a line-in-polygon procedure. Suppose you are an historical geographer in the Minneapolis–St. Paul metropolitan area. You are interested in influences on city growth, and you know that the Twin Cities area has experienced several periods of growth. You also know that the architecture during different time periods is easily identifiable and quite distinctive. Your first step is to map the regions identifiable by different architectural periods as separate polygons. The mapped pattern immediately shows three major zones, in which residential expansion seems to have proceeded amoeba-like, with long, fat tendrils spreading out from a more central core area. These tendrils seem to have taken unique, period-dependent directions. Because of the linear nature of the polygons, you begin to suspect an underlying linear feature as the cause for the patterns.

You compile a series of maps based on the most prominent linear patterns in any urban environment—the transportation nets including rail lines, streetcar rail lines, and major highways. Next you overlay the patterns to view possible correspondences. Results show that the polygons with the earliest growth spurt also have the major rail lines that existed in the area at that time. You next overlay the streetcar lines and immediately see a similar pattern for the second growth spurt. Finally, after overlaying the major highways, primarily the interstate system begun in the 1950s, you see once again the strong spatial relationships between transportation networks and urban expansion (Adams, 1970) (Figure 12.2).

Application of map overlay techniques to urban history scenarios, as above, allows the researcher to formulate hypotheses concerning the fundamental causes of urban expansion patterns. These techniques also facilitate in-depth analysis to prove that visual pattern similarities indicate true cause-and-effect relationships. We see then that the ability to perform these map overlay techniques with a computer has both practical and theoretical benefits.

Figure 12.2   Line-in-polygon operation. Schematic showing how a line-in-polygon overlay for the Minneapolis–St. Paul area might appear. The periods of urban expansion are illustrated by polygons, whereas the lines within the polygons are illustrative of the strong spatial relationship between growth periods and major period transportation nets.   *Source:* Adapted from John S. Adams, 1970. "Residential Structure of Midwestern Cities." *Annals, Association of American Geographers,* 60(1):37–62, Blackwell Publishers. Used with permission.

POLYGON OVERLAY

As we have seen, the traditional approach to overlay is one of comparing polygons on one map layer to polygons on another. The predominance of polygon overlay procedures among GIS systems today derives from the practical considerations for the early GIS work carried out under the auspices of the Canadian government during the development of CGIS. Canada's needs were primarily to examine spatial relationships of large regions on an area-by-area basis. This is still a major thrust among GIS practitioners, and as a result, vendors have developed systems whose strengths in overlay operations are still fundamentally tied to polygon overlays of these types. Thus there are many varied approaches to performing polygon overlay operations, each answering specific user needs. Before we explain the more technical aspects of overlay in general and polygon overlay specifically, let us continue to identify applications by looking at case studies.

A regional planner wishes to prepare a plan for controlled expansion of population in a rural farming county expected to experience massive urbanization over the next 20 years. Her mandate is to allow growth in areas that have soils capable of supporting houses with basements and yet to preserve the highest quality soils for agriculture whenever possible. She also must plan to prevent development of lands that are owned by the federal government, are already being used as farmland, contain known archaeological sites, or comprise the habitats of threatened or endangered species. The state constitution permits the county government to develop zoning regulations that prohibit urbanization in areas it wishes to preserve for other uses.

The first task of the regional planner is to determine what areas should be zoned against urban land uses. After researching the foregoing requirements, our planner creates maps detailing the quality of soils for building houses with basements, areas of soils with high agricultural potential, existing farmland areas, known archaeological sites, habitats of threatened or endangered species, and areas presently owned by the federal government. To produce a map showing where urbanization will be zoned against, she must block out all areas that cannot support houses because of the existing restrictions. She prepares a Mylar overlay map of areas having soils that cannot support houses with basements *or* high-quality soils capable of agriculture *or* existing farmland *or* archaeological sites *or* habitats of threatened *or* endangered species *or* federal ownership. Each of these factors could be shaded, giving a map with two primary categories: open, or suitable for expansion; and shaded, or under some restriction (Figure 12.3). The shaded areas, then, are land parcels that the county government will zone against urban land uses.

You will notice that the connectors between the restrictions above were all the same: the word *or,* in italic type. This convention indicates that a logical mathematical function is operating. In set theory, the word *or* corresponds to a "union" operation that includes all these factors. We will look at the *or* function in more detail when we consider how one would automate the procedure just performed.

There is an alternative approach to the overlay operation. Let us say that our planner decided to come at the problem from the opposite direction. That is, she chose to indicate areas that will allow urbanization. It is easy to see that the

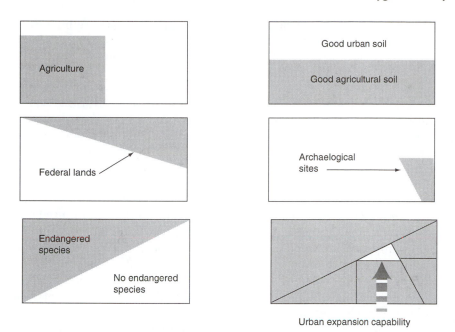

Figure 12.3 Logical "or" overlay. Shaded polygons showing relationship among polygons that are to be zoned against urban expansion. This illustrates the use of the logical "or" search in raster overlay operations.

open areas in Figure 12.3 would answer this question quite nicely, but for illustration we will have our planner create from scratch a map showing areas that will allow urbanization—that is, a map indicating all areas that do not carry restrictions. Stated differently, she must find all areas that have soils capable of supporting basements *and* have no soils capable of supporting farmland *and* have no existing farmland *and* have no archaeological sites *and* have no threatened or endangered species *and* are not owned by the federal government. By preparing Mylar overlays of her maps, where areas not having urbanization restrictions exist are shaded dark, she will be able to see areas that are capable of supporting urbanization as white (or open) areas, just as before. The operational difference is that this time the planner is looking only for areas that fit all the criteria. Such an *and* approach is a method of using Boolean logic for set theory associated with intersection. Union and intersection are frequently used operations in automated GIS software, and we will revisit them when we come to the automation of overlay operations.

   In the preceding *or* and *and* examples, all the variables in each map were weighted equally and could formally be called **exclusionary variables**; that is, they were designed to prevent the performance of a particular type of activity within zones that they occupy. Although this method of overlay comparison is very common, it is rather limiting to users because of the Boolean logic and the nominal level of the data applied to the variables. In many applications of overlay, we find that the binary responses generated by Boolean logic are unrealistically restrictive. Take the map used to indicate the quality of soils in which houses could be built in our zoning scenario. If you look closely at a normal soil survey report, the engineering properties of soils as they are related to building houses exist as a set of ranked classes ranging from severe limitations to moderate to slight and even no limitations. This ordinal ranking of areas gives planners flexibility by allowing them to decide whether houses should be built in areas with moderate soil limitations. After all, if modern technology can overcome the limitations defined in an earlier statute, planners certainly could consider for urbanization areas that do not conflict with other variables at all but have moderate soil limitations as their only limiting factor.

We could modify our analysis by assigning a higher importance to areas on the map that exhibit severe soils limitations for building houses with basements, a medium level of importance for those with only moderate limitations, and a low level of importance to areas with slight or no soils limitations. This approach, which we will call **a mathematically based overlay**, allows us to assign weights to individual variables to account for the differential impacts they might have on our decision. To do this manually, each factor would be assigned a shading pattern indicating its degree of importance as a factor in the model. For our zoning example, this would mean that soils poorly suited for building houses because of severe limitations might be rendered in black, whereas the other categories would have systematically lighter shades of gray, with white indicating no limitations.

As we look at our model, we see that other factors might also be classed as ordinal variables rather than Boolean, depending on how they are interpreted. Capability of soil for agriculture is another variable that obviously could be assigned ordinal value based on its range of limitations. But some of the others could also be raised to the ordinal measurement scale. Take the existence of archaeological sites. By creating a series of circular regions or buffers around an existing archaeological site as we did in Chapter 9, we could easily change an entire land parcel from one on which urban expansion is totally banned to one with a variety of levels of limitation. Thus we could create a first buffer of, say, 200 meters around the site, on which no housing at all would be permitted. The second buffer, of another 200 meters, might be classed as "possible use for limited housing." And still a third buffer of another 200 meters could indicate that the area could be used for most types of single-family dwellings, but not for high-density housing. The area outside the outermost ring in our doughnut buffer could then allow the building of homes of any kind, including apartment complexes.

Once again, this series of distinctions could be imposed manually by shading each category darker as the degree of restriction becomes more severe, much like the McHarg methodology. This would require our planner to measure outward from each archaeological site, draw in each successive polygon, and shade it appropriately. Although this procedure is possible, it is tedious and time consuming. The desired effect can be obtained easily in either a raster or vector GIS.

But if our planner is to give a realistic appraisal of areas that should not undergo urbanization, all the factors should be scrutinized and weighted to reflect what is really important to the planning mission. We have thus far assumed, incorrectly, that all the mapped layers are of equal importance to the problem of land planning. This is clearly not true. Take the factor concerning the present existence of agriculture. In the current version of the overlay operation, the existence of agriculture could, by itself, prohibit the zoning of land for nonagricultural use because the importance of the presence of crops or livestock is considered to be of equal weight with all the other factors. We could easily modify this perspective by assigning weights to each theme on the basis of its importance to the overall planning mission. This requires awareness of what the ranks should be. The assigning of weights to each theme as well as to the individual factors is conceptually very simple within a mathematically based map overlay operation. The task is difficult manually, however, and we now must use the power of the computer to assign weights and to combine the factors through a simple addition of the factors.

Addition is only one of a large number of mathematical operations that could reasonably be applied to a combination of layers. Multiplication, division, subtraction, squares, square roots, minimum, maximum, average, and many more traditional mathematical operations could be applied as well. The mathematical manipulations that potentially could be employed are quite large and are much more easily performed using digital GIS than they can manually.

A final general category of overlay is still available, even in the analog environment. We'll call this approach to overlay selective rather than mathematical. In fact, **selective overlay** is not necessarily a separate category but rather is an extension of the techniques we have discussed thus far. In many cases we will have a set of rules that allow us to decide which factors we might want to use to cover other factors as we compare two layers. This approach, sometimes called **rules-of-combination overlay** (Chrisman, 1996), allows us to use exclusionary rules, weightings, mathematical manipulations, and Boolean logic in appropriate combinations. Such a set of rules might resemble a formal computer algorithm with its if-then-else command structure. Using our planning and zoning example, we could create a selective overlay-based coverage that follows this simple set of rules:

If moderate or less soil limitations for housing
    and
        moderate to no soil limitations for agriculture
    or
        greater than 200 meters from an archaeological site
then
    minimize coverage values
else
    maximize coverage values

This simple algorithmic structure adds a level of control to the overlay operation by imposing multiple rules to determine how the overlay is to be performed. To develop this routine, our planner may have to know beforehand what the likely combinations are going to be. A tabulation for each layer of the quantities of each value for each factor would be most useful. This requires another form of overlay, called **identity overlay**, which can isolate any of these numbers and preserve all of their attribute information. Because it is normally employed within such overlay procedures, I include identity overlay within the selective or rules-of-combination category. The identity overlay, like the selective overlay operation itself, is far too complex to be performed manually. Few of the operations discussed thus far, beyond the simplest approaches, are realistically possible without some form of automation. Therefore, we will now revisit these basic operations to see how hey might be automated and how automation will enhance their capabilities.

AUTOMATING THE OVERLAY PROCESS

As we have noted, the various overlay processes can become very complex and nearly or totally impossible to perform with analog maps. Variable factor weights, differences in the importance of layers, complex mathematical procedures, and iterative rules of combination all contribute to these difficulties. We will now see how the computer can assist us in performing these often-used techniques. We will examine the basics of computer automation, and we will learn still more techniques for overlay that are seldom undertaken without the assistance of computers.

Automating Point-in-Polygon and Line-in-Polygon Procedures in Raster

Raster overlay is simplicity itself. First, all column and row assignments of all grid cells for all thematic grids must correspond exactly. Under these circumstances all point, line, and polygonal grid cells are comparative on a cell-by-cell basis. For point-in-polygon operations, the grid cells representing points are compared to grid cells of equivalent column and row locations through the logical or mathematical procedures already described. If, for example, one is seeking to determine whether a grid cell in a point theme exists within a set of grid cells in a polygonal theme, one merely needs to perform a comparison of the assigned attribute value of the grid cells in the point theme to those of the polygonal theme (Figure 12.3). For example, let us assume that the point grid cells representing a bird species of interest are assigned a value of 1 and all others are assigned a value of zero. Let us further assume that for the polygon theme there are values of 10 representing forest, 20 representing grassland, 30 indicating agriculture, and 100 representing lakes and ponds. If we multiply these two grids, we get values of 0 where no birds occur, 10 where bird locations are co-located with forest, 20 where they occur in grasslands, 30 where they are found in agriculture, and 100 where they are found in water. This is one of a number of methods (called operators) inherent in what Tomlin (1990) called Map Algebra (Tomlin and Berry, 1979). There are many more methods of comparison of point grid cells with polygon grid cells within raster GIS (De-Mers, 2002).

The process of line-in-polygon proceeds in exactly the same manner except that the grid cell values are strings of cells representing linear objects. Generally, the same operators for point-in-polygon grid overlay are available for line-in polygon methods. Thus we could assign roads different grid cell values and compare them to the grid cell values of the polygons (e.g., neighborhoods). The unique numbers that arise through mathematical comparisons (e.g., multiplication, addition, subtraction, etc.) allow the analyst to ascertain whether certain road types occur in conjunction with selected types of neighborhoods. With most modern raster GIS, the assigned categories follow the grid cell assignments, whereas in the past the analyst had to be aware of the values for each grid cell and what the outcomes would likely be. Thus using a multiplication comparison could result, for example, in gravel roads (assigned value of 2) occurring within Pleasantfield (assigned grid cell values of 25) receiving a value of 50. Because the labels are most often attached to the grid cells, the resulting category of "gravel roads in Pleasantfield" results, making identification easy.

Automating Polygon Overlay in Raster

By extension, using map algebra, we can perform the process of polygon overlay in raster in much the same way, but with considerably more options. (cf. De-Mers, 2002). Not surprisingly, grid cell-based polygon overlay tends to lack spatial accuracy where grid cells are large, but, like its simpler point-in-polygon and line-in-polygon counterparts, it gains an unusually high degree of flexibility and speed because of its simplicity. It is well recognized that raster overlay operations in general are preferred to their vector counterparts because of the computational ease with which they are performed (Burrough, 1983). Because each grid cell in a theme is, by default, coregistered with the same cell for a different theme, there is no need for the computer to calculate composite reference locations for each overlaid theme. Instead, all the computer resources are used to compare the attribute data, thus greatly improving the speed of this operation.

The simplicity of the raster overlay process greatly improves the flexibility of the overlay operation as well. Each grid cell has associated with it a series of numerical values, as well as their attributes and labels. These numerical values can therefore be compared to other coregistered variables through a nearly unlimited set of logical, conditional, or mathematical operations. We have seen how the grid cell values can be reclassified through simple decision rules: measures of size, shape, and contiguity; slope and aspect categorizations; and many more. Each of the resulting neighborhoods is a polygon, and each of these polygons can easily be compared to polygons on other layers through many of the same operations.

Automating Vector Overlay

Having reviewed the simpler raster method of automating the overlay process, we turn to an examination of vector overlay operations. Vector overlay operations offer the same advantage as vector-based automated cartography—they produce map products that resemble the traditional hand-drawn map far more than do raster maps. This similarity brings us back to the hand-drawn overlay process because the maps resulting from vector overlay offer potential solutions to some of the graphical problems associated with manual overlays. We have seen how the process of polygon overlay can become very complex, especially where multiple layers are used to produce a single map product. To perform even a simple polygon overlay using Mylar sheets limits us to the use of very discrete patterns and a minimum of categories. Let us assume, for example, that we have 15 different environmental sensitivity factors in a project such as one originally implemented by Ian McHarg. Using gray-shaded maps means that it is possible to obtain up to 15 classes of environmental sensitivity. Although each of the categories might be important, our visual sense does not allow us to discriminate among more than 8 to 10 categories at a time. The limitations of analog maps also severely restrict our ability to examine corresponding spatial patterns, both because of the length of time it takes to produce them and because categories with many corresponding dark gray shades are likely to look identical to the viewer. Today's vector geographic information systems provide a wide array of tools for performing such an analysis and producing readable maps with the power of multiple colors, shades, and intensities.

Given its utility and the frequency with which vector overlay is employed in commercial GIS packages, it is not a far stretch to question whether a system that cannot produce new layers through, for example, polygon overlay is really a GIS at all. Although such an argument may seem to be purely academic, in reality any limitations in the vector overlay process also limit the utility of the software for the user. To become familiar with all the possibilities, we will look at overlay both as a graphic artifact, where the maps produced cannot be saved as separate layers, and as a method of producing new, retrievable layers. Then we will look at how point-in-polygon, line-in-polygon, and polygon overlays operate in vector. This knowledge will be very useful to those who use GIS as well as to those who act as consultants to potential GIS customers about selecting systems.

TYPES OF VECTOR OVERLAY

CAD-Type Overlay

The first and simplest method of automated vector overlay is very similar to the analog method in that we simply apply the display symbols of precategorized data to a single Cartesian surface. Such data might include a range-graded choropleth map of population, a single value-by-area map showing the locations of wild plant species, land use polygons, line symbols indicating road networks, or point symbols showing archaeological sites. The process of graphic overlay is akin to the method of thematic combination that is used to produce more general maps, like those in a regional atlas. Such a map allows the reader to see a considerable variety of factors laid out on the same graphic at the same time and can be used to show graphically how these different factors might or might not be spatially associated. To perform this relatively simple operation with the computer requires only that the software be able to keep track of the locations of each point, line, and polygon to ensure that all the coordinates can be represented on the screen.

The result of this operation is a visual graphic, not a GIS layer. The software is not responsible for combining the attributes for each object displayed, since the attributes are most often merely labels attached to the graphical elements. There are no tables that associate the labels with other attributes. Neither is there any topology. The image is merely a composite artifact of superimposing the individual images to make a single, unified image. Such products can be printed or plotted as hard-copy output, or, in some cases, saved as graphical computer files. So why aren't these documents GIS layers?

The question is an important one, and it continues to confuse even some experienced GIS users. Let's use an example from CAD software. Say that we have digitized a map of roads using CAD software. On another layer of the software, we digitize the locations of water bodies from the identical map. Next, we combine the two layers into a composite that shows both roads and lakes. Visually we can locate the roads that would get us to the lake so we can go boating. This is useful information, so we save the composite map for another time. Later, another user wants to ask the software which roads enter the area and run parallel to and within 100 yards of the lakefront. Although the solution to this simple query can readily be observed by a person who has found your map and is looking at it, there is no way for the computer to answer the question because it does not know explicitly what names might be associated with what roads. It has no associated database; instead, the names are also graphical devices placed on or near the entities. The computer also lacks the capability to determine any measure of proximity or connectedness because it does not know how to determine these measurements. It is not a GIS map layer. With a little work, it would be possible to import this map to a functional GIS system, to build the database for road names, to create topological structures, and then to answer the question asked. Thus enhanced, our original composite has become more than a graphical device designed specifically for the communication paradigm. It is now a bona fide coverage suitable for analysis under the analytical or holistic paradigm.

CAD-type overlays are very useful within their analytical limits. It is not always necessary to perform computer-intensive queries to determine neighbors or locate the spatial congruity between two objects. These tasks can easily be done by hand. Much of the utility of this type of overlay lies in its simplicity and the rapidity with which it can be produced. Many **AM/FM** (automated mapping/facilities management) systems operate on the same structure and provide exactly what the users want. That is, they produce maps of selected items reasonably fast for users to take into the field or job site, where the graphics can be read and interpreted manually. It is only when the complexity of the databases or the complexity of the queries is very large that the limitations of manual interpretation of automatically produced maps become a problem. Let us say, for example, that you have a map of underground gas lines, overlaid with a map of streets (including street names), overlaid with a map of housing, overlaid with a map of population information. Someone in the area of the coverage reports a smell of gas. You look at your map to see where the leak might be located, but the lines are rather dense because of the high number of symbols representing residences. It is also difficult to separate the lines that represent roads from those that represent gas lines. You are aware that there are several above-ground shutoff valves, but the complexity of the map obscures the tiny symbols that show them. Such uses require overlay operations (as well as others) that can keep track of multiple attributes for each entity and can highlight the shutoff valves, as well as the locations of residences that might need to be evacuated. All this could be done by hand, given enough time. It might, however, prove easier to use an individual map for each theme, to prevent the graphical symbols from interfering with the map interpretation. In short, you need something more for this emergency situation than a basic CAD-type overlay is capable of producing.

Topological Vector Overlay

The idea of topological data structures introduced in Chapter 4 outlined how the modern GIS can explicitly relate points, lines, and areas in a single coverage. We further saw how these relationships allow us to determine functional relationships among the entities, such as defining left and right polygons associated with a specific line, defining the connections among line segments to examine traffic flows, or searching for selected entity combinations on the basis of individual or related attributes. It also established a method for overlaying multiple polygon layers to ensure that the attributes associated with each of the entities could be accounted for and so the resultant polygons with multiple attribute combinations would be maintained. Chrisman and Peucker (1975) introduced this topological result, known as the least common geographic unit **(LCGU)**, specifically to show how the changes in polygonal entities can eventually reach a point at which no further divisions are possible. Associated with these smallest divisions, then, is a specific set of attributes that also cannot be broken down into any further categories. In addition, the attributes associated with the LCGU would effectively emulate the "darkest" or "most sensitive" area polygons found through using the McHarg (1971) method of manual cartographic overlay. For each of the remaining polygons, there will be a number of attribute combinations with lesser "darkness."

Topological Vector Point-in-Polygon and Line-in-Polygon Overlay As we have discovered, most operational vector geographic information systems are designed around the polygon-on-polygon overlay operation. Still, we can take certain steps to automate the point-in- and line-in-polygon operations in vector. We need software that is able to locate the coordinates of each point, each line, and each polygon (and the associated line segments). In addition, the software must be able to calculate the locations of points or lines relative to the polygon segments with which we are to compare them. If, for example, a point is located at coordinates $X = 2.300$ and $Y = 3.450$ in Figure 12.4, the program must be able to determine whether this point is located within the polygon whose outside coordinates match those in the figure. As you can see, in this case they do. In an earlier example, we attempted to show the locations of cowbirds inside forest polygons; we will no doubt want to do this for a number of these points. A polygon overlay search might be performed to create a coverage showing only the polygons that contain cowbird locations. This will require the software to perform a number of calculations to determine which of the polygon line segment coordinates surround each of these points (Figure 12.5).

To fulfill the objectives of our study, we might want to know what types of land cover polygon do contain these point coordinates. In this way, we will be able to determine whether there is a preference for cowbirds in forest polygons as opposed to urban polygons or row crop polygons. Therefore, the software must be able to produce table (in addition to map) output that shows how many points are located in each of the polygon types. The search will require that the computer know not only whether our points occur within polygons but specifically which attributes are associated with each polygon. Software that does not explicitly include topology will be required to check each point in

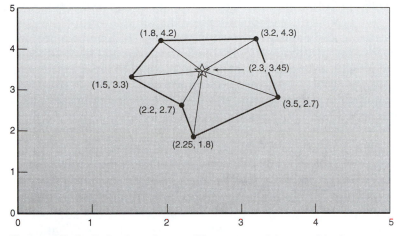

Figure 12.4   Point-in-polygon. Illustration of the graphical nature of point-in-polygon in vector. Example distance calculations are shown for reference.

each polygon line segment to determine the coordinate limits of the polygon. This information will have to be checked against attributes to define what the polygons represent. Then the software must be able to compare these data against the locations of our cowbird point data. As you might guess, for very large databases this can be a time-consuming process. If, instead, our polygons were already defined by their line segments, with each line segment explicitly related to each other line segment, we could save computation time in comparing our points with the polygons and their attributes.

One problem practitioners will encounter with some vector GIS software goes beyond the mere computation time necessary to compare a point coverage with a polygon coverage. Some software, though capable of comparing

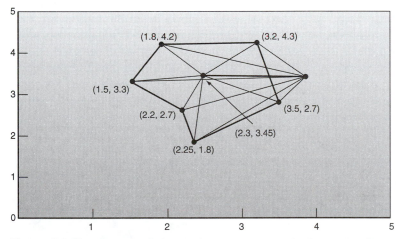

Figure 12.5   Line-in-polygon calculation. Example distance calculations are shown for reference. Note that the addition of numerous line segments (and their associated points) greatly complicates the calculations.

these sets of entities and their attributes, is not capable of preserving the map output as a separate coverage. Although the tabular output is useful, the ability to operate on the final map product is missing. This can be a problem if your intent is, for example, to create variable buffers around the point locations on the basis of a value assigned to each polygon attribute type. This limitation is generally a function of how the program stores its associated table information. Exactly why this occurs is beyond the scope of an introductory course, and often the reasons are of little importance to the user. A simple method of working around this software limitation is to create very tiny buffers (too small to be seen on the output map), which become minute polygons. You can save this coverage, and the points will be represented by the tiny polygons.

By extension, line-in-polygon is a process of comparing the coordinates of lines with those of the polygons to evaluate when the lines occur within the graphical space of the polygons. The computations are somewhat more complex but are essentially comparable because each line segment begins and ends with explicit point coordinate pairs. The only complicating factor in this case crops up when a line begins within a particular polygon and ends outside it. As long as any point coordinate pair falls within the polygon, we can assume that the line also falls within it. This, however, oversimplifies the real situation. It is more reasonable for the computer to calculate where the line segment of each linear feature intersects a line segment for each polygon. Once this is done, a node is placed at this point, and the line attributes and each type of polygon attribute can be kept separate. In this way, if a fencerow encoded as a line feature crosses a polygon of agriculture and a polygon of grassland, we can determine how much of the fencerow occurs in each. We can also create a table showing this relationship.

As you can see, **the point-in- or line-in-polygon overlay** in vector is not simply a matter of showing the graphical relationships; the attribute relationships must be revealed as well. After all, our entities are meant to represent some portion of the real world. You will remember that many real-world features can be encoded as points. And each point could represent not only the location of a single object but also a large set of associated attributes. In the case of our cowbirds, we might have associated with each location such information as gender, tag number (if known), physical condition, size, and whether the bird was observed singly or as one of a mated pair. Each of these attributes could be selected and compared to preselected polygon attributes. Similarly, the lines could show roads and could include road type, road condition, number of accidents per segment, when the road was last surfaced, and many other attributes associated with transportation. Multiple attributes can be compared in numerous ways. And, as we have seen before, they will yield fundamentally different results. The comparisons we use will depend on which attributes we wish to compare, how we want them compared, and ultimately what we are looking for in these comparisons. We'll return to this subject when we discuss vector polygon overlay operations because they are essentially of the same general type. Which of these techniques is available depends largely on the software you are using.

Vector Polygon Overlay   The main problem associated with using point-in-polygon overlay and line-in-polygon overlay is that the lines do not always occur entirely within the polygons. Our most powerful solution to this problem was to let the software determine the points at which each set of lines intersected, so that nodes could be placed at these points. For vector polygon overlay, your task is essentially the same, except that you must now be able to account for the overlapping intersections among lines defining polygons and for the attributes associated with each of these new polygons. Each polygon line segment on one map layer that crosses each polygon line segment on subsequent layers must have an associated node that shows where the segments intersect. The software must also keep track of the attributes for each original polygon as well as for each newly created polygon in the final coverage. Vector polygon overlay can be performed in several ways, and there is an increasing body of how-to literature, but such details are beyond the scope of our discussion. Our concern is to compare mapped variables, not to study the specifics of how the computer program performs the graphics or how it keeps track of the variables. For the following discussion we will assume that the vector-based software incorporates a tabular method of keeping track of variables. This is most often done by using a database management system that is either tied specifically to the graphics or linked externally with them.

Because many vector-based GISs are linked to a database management system, it should be no surprise that database queries based on Boolean logic are used for spatial database queries as well. The **Boolean overlay** is among the more commonly used approaches and is very easy to understand. You remember back in your youth when a teacher drew diagrams showing the relationships between two sets of numbers, two sets of dishes, and so on. Diagrams of that type, called Venn diagrams, are graphic "mappings" of the relationships between two sets. Venn diagrams closely resemble the graphic devices we commonly call maps. Let's take a closer look.

Consider the following. You have two sets of data, each representing land use types. Set A, using the list notation, would be A = row crops, grain cropping, ranching, dairying, forestry, single-family housing, multiple-family housing, retail, manufacturing, and office space. For the second set, B, we have B = dairying, forestry, retail, and office space. To find out what land uses these two sets have in common, we simply perform an intersection on the two sets just as in our nonspatial example earlier.

In the case of Boolean vector overlay, we compare directly not the attributes themselves but rather the space occupied by any two sets of attributes. Thus, if we had a simple map showing only two land use polygons, one might show cropland and the other, an urban setting. Now for the same portion of the earth's surface, we have another coverage showing two other attribute variables, such as owned and rented land. As you can easily see, when we try to find out which portions of land are owned farmland and which are owned urban, we create a new map through the process of intersection that shows the designated portions only. The graphics portion of the program determined the polygon intersections and created specific polygons based on these line intersections. Then the attributes associated with the polygons that fit both criteria are selected from the database, whereas those that do not fit both criteria are not selected. In other words, performing an intersection of the two maps is equivalent to drawing a Venn diagram for the two spatial locations of the two sets of attributes. By doing so, we are now able to locate both the owners of farms and the owners of urban property.

The union operation is the total of all the elements in both sets and can be illustrated with a Venn diagram (Figure 12.6). Extending this idea to our earlier

Figure 12.6   Boolean operators. Venn diagram showing intersection, union, and complement. The same two sets (lists) of land uses could be used in each.

map layers, we could produce an overlay of the two maps to show all locations that were owned or rented or were agriculture or urban, obtaining a map with a single category that included all these attributes. In this case, our map would cover the entire study area (Figure 12.7). A third method would be to use the complement of these two layers, analogous to the Boolean "not" operation, which also can be illustrated by a Venn diagram (Figure 12.6). In this case, we might want to cover these two maps to show all areas that are owned or rented and have some type of human activity, but not human habitation. The effect would be a map showing owned and rented farmland.

Our last example shows how easy it is to overlay variables through the logical complement procedure. It also illustrates, that it is not always necessary to overlay maps to achieve the desired output. To determine where owned and rented farmlands lie would simply be a matter of reclassifying our land use theme ("farmland" would become "owned or rented farmland"), and we would not select from the same coverage the attributes called "urban." This is an important lesson because it shows that simple solutions, requiring no overlay, are often readily available. Because overlay of large polygon layers can be a lengthy process, it is prudent to seek simpler solutions when available. This is especially true when working with Boolean overlay operations, which are relatively simple and offer more cases in which reclassification operations would suffice.

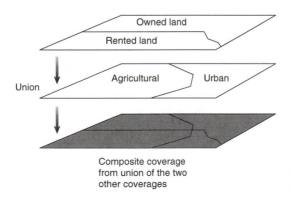

Composite coverage
from union of the two
other coverages

Figure 12.7   Union overlay. The union operation is used in a vector overlay of two layers: the first is a map of landownership versus renting, and the second is a land use map showing agricultural versus urban. The result of this operation is a map of the entire study area, indicating agricultural and urban as well as owned versus rented land.

Still, many other types of overlay operation could be avoided if the reclassification were employed by itself. Or the complexity of the overlay procedure could be greatly reduced by preprocessing the overlays, as we did in raster, by reclassifying the polygons that are not of interest in the subsequent overlay operation. Our purpose is to whittle down the number of categories, thereby reducing the number of polygons. This process makes the results of the overlay simpler and easier to understand. It also has the fortunate side effect of eliminating some computer operations and speeding up the overlay process.

Before we leave this discussion, it is best to revisit the idea of identity overlay. Although it is generally unnecessary to formally define identity overlay as a method, use of the term will allow us to at least have a moniker to hang on it. In identity overlay, we can achieve results similar to our earlier examples except that, although we are using Boolean operators to perform the actual cartographic overlay, we retain the attributes. Thus our resulting polygons show combinations of all those that intersect. In other words, we overlay the graphic entities and tabulate all the category combinations. So, for example, if we have four different categories (say land use, soil type, animal communities, and vegetation type) that intersect in space, all the categories are tabulated. Therefore, we may have a polygon that contains orchards (the land use), mollisols (the soil type), field voles (the animal community), and pecan trees (the vegetation type). By maintaining all these categories, we can easily cross tabulate the amount of each category that corresponds to each other. Many overlay operations are performed so that these statistics can be obtained. Your vector GIS may use identity overlay to perform this operation, thereby requiring you to reclassify the categories once they have been overlaid.

A Note about Error in Overlay

As in our raster examples, the operational use of cartographic overlay tends to involve numerical data more often than simple nominal data. Overlay data may be ordinal, interval, or ratio; they may be range graded to indicate groups of values for a given set of polygons. The values, in turn, place constraints on the use of simple Boolean logic and identity overlay operations that are designed more for handling nominal data than for mathematical manipulation. The more precise mathematical values give us other powerful options for overlay. Instead of being restricted to Boolean logic and set theory, we now have available the same set of mathematical operators we used in mathematically based raster overlay. Any fully operational database management system has the capability of allowing mathematical manipulations of the data in the attribute tables associated or linked with the graphics of the GIS. Because the vector GIS normally contains the attributes and their labels and can link these with the entities, the process is even easier than it would be in the simplest types of raster system, and nearly identical to the raster systems that also contain a database management system toolkit.

The operations of mathematically based overlay in vector are performed much the same on the attributes in the database tables and in each individual georeferenced set of grid cells. Except for computational difficulties due to the graphical procedures, the user will see little difference in the results of such operations. This similarity in operational capabilities will prove very important when we consider cartographic modeling in the next chapter. Because of the nature of the output from vector GIS, the visual results of mathematically based overlay will look quite different in vector and in raster. One would expect the cartographic output to resemble a hand-drawn map. The result, however, can be both surprising and annoying, as dozens or even hundreds of tiny, unexpected polygons of interacting polygons show up, especially around the margins of intersecting polygons. This apparently innocuous visual difference can have quite profound impacts on how we interpret the results of our analysis.

As an example, let's make a simple comparison of multiple land use layers representing separate time periods for identical pieces of land, digitized from aerial photographs classified on the same scale. When we compare a single set of polygons of our urban land use category for just two of these time periods, we notice that the polygons looked identical when mapped separately (Figure 12.8). But when combined, they do not exactly match. Instead, there are a number of sliver polygons at the margins of the overlaid/intersecting polygons (Figure 12.8). Now we must decide: Has there, in fact, been no change from time 1 to time 2? Have there been slight changes in the land use locations? Was there a problem with the digital input of the data in the first place? Instinctively, we know that changes as minor as the slivers in the composite are unlikely, especially because the polygons look so much alike. We can probably eliminate our second possibility as well, again because the polygons are nearly identical in shape, and even minor changes in land use would have imposed greater modifications.

Ignoring the possibility of land use changes of a minuscule percentage, we are left with the answer that the polygons probably are identical, and we try to account for variations caused by digitizing. Or if the polygons are essentially the same shape but one is slightly rotated relative to the other, the slivers might be due to a minor rotation or distortion in one of the aerial photographs from which the polygons were digitized. In this case, we might be able to perform a computerized form of rubber-sheeting. This remedy is readily available in GIS software but rarely gives 100-percent co-occurrence even if done with care. The process also requires that we anchor a number of points that we know should be located at exactly the same place on both layers. Although these are held in place, the rest of the map can be "conflated" or rubber-sheeted to move the polygons until correspondence is achieved. As a result, other polygons that were originally well aligned move out of register. In short, there is no easy solution to this problem, which is much more common in vector GIS overlay operations than one would like to admit. Some users even pre-

Sliver polygons
from an overlay operation

Urban polygon
at time T^1

Urban polygon
at time T^2

Urban polygon
composite

Figure 12.8   Sliver polygons created by vector overlay. These sliver polygons result from minor differences in two marginally intersecting polygons. Do the discrepancies indicate real change or error? Note that most of the slivers occur on the periphery of the two otherwise identical polygons. Care must be taken in interpreting the results of vector overlay, especially if small amounts of change are important.

fer to use raster, thus giving up some locational accuracy, to avoid the tedious adjustments frequently required. Even so, conflation is a valid approach and one that might be considered if a process must be repeated many times or if polygons are very small and the sliver polygons are likely to be tiny.

Problems are compounded when multiple layers are overlaid, especially if the layers encompass fundamentally different themes. For example, if we overlay a coverage of soils with one of vegetation, we no longer have the luxury of assuming that any two polygons might be the same shape. It is not necessarily true that there will be a perfect one-to-one relationship between soils and vegetation. The purpose of the overlay might be to see where the categories deviate. Our dilemma here, then, is to separate the error from the actual polygon differences.

If, as in our last example, each map has its own unique amounts, types, and sources of error for both entities and attributes, the question is this: How do we deal with error in multiple layers? Unfortunately, despite ongoing research attempting to identify how much error is propagated with multiple layers (for a good general survey of this topic, consult Burrough and McDonnell, 1998), there are few general principles and even fewer answers, especially if the layers are from widely different sources (Chrisman, 1987). Your decisions must be based on your own knowledge about the data quality, input methods, and field methods to obtain them. Your response to error is heavily dependent on how precise your output must be for the problem at hand. It is logical to assume that if you have many different layers, the results of overlay will be only as good as the worst coverage employed, but it is very important to note that this is not often the case. This "weakest link" approach, though appealing, is based on the assumption that all layers are equally important. As we have seen, however, we have the capability, in either raster or vector, of weighting each theme through the overlay process. In addition, even if we do not explicitly weight each theme, some will ultimately be more important to our model results than others. Before you overlay your layers, you should become familiar enough with your data to be able to decide what types and amounts of error you can accommodate.

DASYMETRIC MAPPING

Before we leave the topic of polygon overlay, we should revisit an old cartographic technique for detailing polygonal information on the basis of other variables. The technique, called **dasymetric mapping**, is based on the idea of the choropleth map discussed in Chapter 3, and it gives us insights into some of the approaches currently taken inside GIS, as well as others that have not yet been considered by the GIS community. Its use requires that the data be in ordinal, interval, or ratio form and that they exist as a statistical surface (see Chapter 3). The first documented use of dasymetric mapping was for improving the categorization of population densities on Cape Cod through a method called **density zone** outlining (McCleary, 1969). The technique, also called density of parts, was employed to obtain greater detail of individual densities of unresampled areas based on a more detailed knowledge of some smaller subareas that had been resampled (Wright, 1936). An excellent example of the application of density of parts reported in Robinson et al. (1995) closely associates the example with Wright's original 1936 method. The student of GIS should become familiar with this technique because of its power to improve the quality of quantitative data contained in polygons by comparison with more detailed data for another map layer.

We have already employed one method of dasymetric mapping without identifying it as such: pure dasymetry, first used by Hammond (1964) in his work on landforms, entails the delineation or classification of areas of geomorphic type on the basis of a recategorization of topographic data. Topographic surfaces perfectly fit this approach, which calls for continuous distributions of an infinite number of points. When we studied reclassification of continuous surfaces, we created neighborhoods for "intervisibility," "south-facing slopes," "steep slopes," and so on by grouping preselected ranges of our datasets. This was a modified form of pure dasymetry. Once again, we see that the modern techniques we use so often have roots in a time well before the advent of the computer.

Several more forms of dasymetric mapping could prove potentially useful; among the most powerful is the type called "use of other regions with the assumption of correlation," with its strong implication of a form of cartographic overlay. Thus, we might look at our example of density of parts and suggest that this was also a comparison of variables between or among maps. The assumption-of-correlation approach is strikingly different, however. Instead of isolating portions of an area for which detailed study improves the information content, then using this information to improve what we know about the remainder, we use either **limiting variables** or **related variables** contained in other layers.

We used the limiting variables form of dasymetry when we discussed exclusionary factors within overlay. As an example of how this might be applied to improve the quality of our polygonal data and the models we produce from them, let's say that we have a map of Minnesota that shows the population by county. We also have a map giving as a polygonal value the area in square miles for each county. By overlaying the two layers, having divided the population coverage by the area coverage, we obtain a map of population density for each county in the state. However, Minnesota is the "land of 10,000 lakes," which implies a substantial surface on which people do not reside unless they have houseboats. To improve our population density coverage result, we should have "excluded" from the area coverage the amount of water contained in each county. If we had used this "limiting variable" to create a map of land area by county and employed it in producing our population density coverage, the population density of counties with large water bodies would have increased. Such techniques can frequently be used in conjunction with cartographic overlay to isolate and exclude areas that bias quantitative polygonal data.

Our final example of dasymetric mapping, called related variables, corresponds strongly to our use of mathematically based cartographic overlay methods as in both raster and vector GIS. We have seen that variables contained in our polygon layers often interact in ways that are more complex than simple exclusion. Statistical tests such as correlation and regression are often used to show how geographically dispersed variables are related to one another and how these relationships can allow us to predict variations in one based on the changes in the other. This is no less true when using GIS. If we, for example, know that a high correlation exists between percentage of cropland and percentage of slope on the land, we can predict the amount of cropland on the basis of this correlation and its associated regression line. With this information, we can develop a detailed predictive coverage of "% cropland" on the basis of the slope alone. Alternatively, we might have layers of existing percentages of cropland and slope. By cartographically overlaying these, we would create a coverage showing the true relationships between these two variables in a particular area, or we could overlay our composite on a map of "predicted % cropland." The differences between the two would give a visual display, as well as a quantifiable difference between the actual and the predicted, which could be used to evaluate the predictive model. The areas that do not fit the predictive model could then be compared to other layers to develop hypotheses concerning variations in the model. This is an excellent example of how GIS can be used to develop hypotheses for scientific applications of GIS or to create predictive GIS models for decision making in commercial applications of GIS. We will look at the use of this powerful application in the next chapter.

As we have seen, dasymetric mapping has a great deal of potential to improve the use of GIS in both academic and commercial settings. Still, the subject is relatively untapped within the GIS literature. Raster GIS lends itself readily to many forms of dasymetry, and several methods are being employed every day without the knowledge of the users. Vector GIS is also capable of performing some—perhaps all—of these techniques, but few have made any serious, systematic attempt to evaluate its potential (Gerth, 1993). An awareness of dasymetry as a potential set of cartographic overlay methods will likely result in vast improvements in the GIS overlay toolkit currently available to users.

SOME FINAL NOTES ON OVERLAY

Because of the visual appeal and intuitive nature of cartographic overlay, this set of techniques is often considered to be what GIS is. This perception limits the number of potential GIS users and may even result in reduced sales of commercial products. It also keeps many users from evaluating other powerful techniques already available in GIS. The student should be aware that, despite the power of cartographic overlay, there are numerous alternative methods for solving GIS-related problems. Overlay is frequently found to be far more powerful when combined with other methods of spatial analysis than when used in isolation. In the next chapter, we begin to develop complex cartographic models that will force us to select appropriate techniques and to combine them in a rational manner to obtain useful results.

Constantly remind yourself that cartographic overlay is powerful but mindless. No GIS software is capable of evaluating whether the layers you are using are functionally related. For personal applications, you should learn which spatial variables are likely to be functionally related before you begin blindly overlaying them. If you are operating a GIS for a client, it is even more important to be vigilant because the client will turn to you for answers. Think about your variables before you select those that you will overlay. Determine why they might be related. You might even consider pretesting statistically a sample of the variables to help you make this determination.

A useful statistical approach is one that employs the integrated terrain unit mapping approach (Dangermond, 1976). Very much like biophysical mapping units, **integrated terrain units (ITUs)** generally reflect the assumptions that all the data are integrated on the ground and that you employ a method, such as digitizing from aerial photographs, that requires you to extract these interacting variables from the same source of information. This ensures that the variables you include are related spatially, if not in a logical sense. If you do not have the capability of employing ITUs and can't find a logical reason, or at least a statistical reason, that would indicate a relationship among your variables, perhaps you should re-evaluate the solution to your problem. This is no different from testing the relationships between the surrogate radiometric responses gathered from a satellite and the land features they are meant to represent.

Terms

AM/FM	identity overlay	overlay
biophysical mapping	integrated terrain unit (ITU) mapping	point-in-polygon overlay
Boolean overlay		related variables
CAD-type overlay	LCGU	rules-of-combination overlay
dasymetric mapping	limiting variables	
density zone	line-in-polygon overlay	selective overlay
exclusionary variables	mathematically based overlay	

Review Questions

1. What are some obvious limitations of the manual overlay process? What advantages have been offered through automation?

2. What are the advantages of raster-based overlay operations over their vector-based counterparts? What are the disadvantages?

3. Why is overlay still among the strongest operations in most modern commercial GIS?

4. From your own studies, give examples of how point-in-polygon and line-in-polygon could be useful to you.

5. What are the limitations of CAD-type overlay operations? Give an example of how CAD overlays would prohibit subsequent analyses on the product of the overlay process.

6. What is dasymetric mapping? How can it be employed in vector GIS? raster?

7. What is the difference between related variables and limiting variables in dasymetric mapping? Give an example of the use of each variable type.

8. What problems or difficulties might you encounter when comparing multiple layers with vector overlay techniques? Are there any approaches you might consider to solve these problems? What should you do before you overlay to lessen these problems?

9. Give some examples of how portions of dasymetric mapping might readily be applied with existing raster or vector GIS. Why do you suppose so little attention has been paid to the use of dasymetric mapping in GIS overlay?

References

Adams, J. S., 1970. "Residential Structure of Midwestern Cities." *Annals of the Association of American Geographers,* 60(1):37–62.

Burrough, P.A., 1983. *Geographical Information Systems for Natural Resources Assessment.* New York: Oxford University Press.

Burrough, P.A., and R.A. McDonnell, 1998. *Principles of Geographical Information Systems.* New York: Oxford University Press.

Chrisman, N., 1987. "The Accuracy of Map Overlays: A Reassessment." *Landscape and Urban Planning,* 14:427–439.

Chrisman, N., 1996. *Exploring Geographic Information.* New York: John Wiley & Sons.

Chrisman, N., and T.R. Peucker, 1975. "Cartographic Data Structures." *American Cartographer,* 2(1):55–69.

Dangermond, J., 1976. "Integrated Terrain Unit Mapping (ITUM)—An Approach for Automation of Polygon Natural Resource Information." Publications of the Environmental Systems Research Institute, no. 160, 11 pages. Redlands, CA: ESRI.

DeMers, M.N., 2002. *GIS Modeling in Raster.* New York: John Wiley & Sons.

Gerth, J.D., 1993. "Towards Improved Spatial Analysis with Areal Units: The Use of GIS to Facilitate the Creation of Dasymetric Maps." Unpublished MA paper, Ohio State University, Columbus.

Hammond, E.H., 1964. "Analysis of Properties in Landform Geography: An Application to Broad-Scale Landform Mapping." *Annals of the Association of American Geographers,* 54:11–19.

Hills, G.A., P.H. Lewis, and I. McHarg, 1967. *Three Approaches to Environmental Resource Analysis.* Cambridge, MA: Conservation Foundation.

McCleary, G.F., Jr., 1969. "The Dasymetric Method in Thematic Cartography." Ph.D. dissertation, University of Wisconsin, Madison.

McHarg, I.L., 1971. *Design with Nature.* Garden City, NY: Natural History Press.

Monmonier, M., 1991. *How to Lie with Maps.* Chicago: University of Chicago Press.

Robinson, A.H., J.L. Morrison, P.C. Muehrcke, A.J. Kimerling, and S.C. Guptill, 1995. *Elements of Cartography,* 6th ed. New York: John Wiley & Sons.

Sauer, C.O., 1925. "Morphology of Landscape." In *Land & Life,* J. Leighly, Ed. Berkeley: University of California Press, 1963, pp. 315–350.

Simpson, J.W., 1989. "A Conceptual and Historical Basis for Spatial Analysis." *Landscape and Urban Planning,* 17:313–321.

Steinitz, C., P. Parker, and L. Jordan, 1976. "Hand-Drawn Overlays: Their History and Prospective Uses." *Landscape Architecture,* 56(4):146–157.

Tomlin, C.D., and J.K. Berry, 1979. "A Mathematical Structure for Cartographic Modeling in Environmental Analysis." In *Proceedings of the 39th Annual Meeting, American Congress on Surveying and Mapping,* Washington, DC, pp. 269–284.

Tyrwhitt, J., 1950, "Surveys for Planning." In *Town and Country Planning Textbook.* London: Architectural Press.

Wikle, T.A., 1991. "Computers, Maps and Geographic Information Systems." *National Forum,* summer, pp. 37–39.

Wright, J.K., 1936, "A Method of Mapping Densities of Population: With Cape Cod as an Example." *Geographical Review,* 26:104–115.

CHAPTER 13

Cartographic Modeling

In the following pages, we will view the analysis subsystem as a set of interacting, systematic, and ordered map operations that together perform some very complex modeling tasks. These complex models are composed of much smaller, simpler model components that can be solved with relatively few analytical operations. Each one can then be combined with others to create larger modules of more complex models, and these can be combined further until your entire model is constructed. Like any system, a spatial model can almost always be broken down into these component parts as long as you are familiar with how the model works.

This chapter emphasizes the development of appropriate model flowcharts. Flowcharting forces you to think carefully about what data elements you need to obtain the answers you are seeking. It is easier to detect missing themes, or to spot elements that will be used many times, by examining a flowchart during model development than to correct a finished model. Flowcharting of cartographic models should be a required preliminary step to the development of any model. During this process, you may discover that you do not have a clearly defined objective, whereupon you will have to rethink your goals for building the model in the first place. In addition, even if your primary interest is not modeling but database development, you cannot properly develop a database without first specifying the final product that is to be obtained from the modeling process.

When you examine the models and flowcharts given in the text, you will probably find dozens of ways to improve on the approach or to enhance the model by including additional information. The models given should not be considered to be absolutely complete. Nor should you assume that they represent the only solution to the respective problems. Instead, you should envision alternative solutions for each problem—perhaps even creating an alternative flowchart for each. Then make a comparison—it may appear that your approach is more easily accomplished (i.e., a more elegant solution) or more realistic (i.e., more likely to give an accurate representation of what is being modeled). Practice is essential to good flowcharting, as it is to good modeling.

When you are finished with this chapter you should be able to:

1. Understand how commands are assembled to produce a cartographic model.

2. Illustrate the cyclical movement from subsystem to subsystem inside a GIS.

3. Know the difference between deductive and inductive cartographic modeling and the advantage of deductive over inductive approaches.

4. Determine the difference between explicitly and implicitly spatial variables and discuss the role of spatial surrogates for nonspatial variables.

5. Produce and explain a simple model flowchart.

6. Explain the problem of cartographic modeling constraints that are too severe or too loose and define methods of circumventing these problems.

7. Describe the idea of weighting of coverages as they impact the outcome of cartographic models, especially with regard to how these weights might affect the error component of the output.

8. List and discuss three basic areas of cartographic model verification.

9. Describe alternatives to cartographic output for GIS results.

10. Discuss the role of prototyping and preplanning in the production of a successful cartographic model.

MODEL COMPONENTS

In our last few chapters we saw a wide variety of possible individual techniques for working with points, lines, areas, and surfaces. Each of these simple techniques alone can be powerfully applied to the solution of geographic problems. Each has a wide array of possible options and subtle variations that could radically change the result of an analysis. As a practical matter, the number of combinations could be described as nearly infinite.

Fortunately, there are limitations to the ways in which the options can be applied. To develop a neighborhood function that shows the number of land uses within a specified area, for example, you would use a measure of diversity, ignoring all other options because "diversity" correctly answers your question. This constraint on the numbers and types of option used to answer specific questions greatly simplifies GIS analysis. It also focuses your attention on the techniques or options that are absolutely necessary. You will recall that you have many options for overlaying one thematic map with another. You first decide what result you are looking for and what is the best solution. For example, an overlay operation using the logical complement may produce a map that has the same polygonal entity shapes as a map based on union. One approach requires you to perform a second operation to reclassify the result because the attributes are the opposite of what you want. Both achieve the same answer, but one is more efficient than the other. Efficiency is very important in working with large models, both to prevent you from making mistakes and to simplify the overall model as you begin combining many commands.

As with the options for each command, the potential number of command combinations is daunting. Fortunately, the potential for interaction of commands is also limited by the nature of the models you are creating and the types of objects on which you are operating. Although it is common to reclassify polygonal data prior to polygon overlay operations, this is not generally done prior to topographic viewshed analysis. Each command combination is contingent on the nature of the data, the type of operation being performed on the data, and the answer sought. It would be very useful to be able to create a set of do's and don'ts for which commands will be combined with which others and which should not. Unfortunately, such a cookbook would be too large to be practical. Instead, you will have to experiment, learning from your own experiences and those of others around you. Formalized procedures for combining GIS functions to analyze complex spatial systems are now available, and we will examine them in some detail.

THE CARTOGRAPHIC MODEL

The term **cartographic modeling** was coined by C. Dana Tomlin and Joseph K. Berry (1979) to designate the process of using combinations of commands to answer questions about spatial phenomena. More formally, a cartographic model is a set of interacting ordered map operations that act on raw data, as well as derived and intermediate map data, to simulate a spatial decision-making process (Tomlin, 1990). Let's examine this definition in greater detail.

The first condition of the definition is that the map operations must interact with one another. The purpose of each individual operation performed on a theme or element must be to create a result, usually another theme or variable, that can be used by the next operation. The next operation may be a further manipulation of the same output, say to isolate certain map variables, or it may operate on a separate input variable or map layer, perhaps as a map overlay operation.

Let us say that you have a map of "landcover" for a study area, and you want to determine whether there are sufficient large, contiguous, federally owned rangelands to make it worth your while to petition the government for permission to graze your cattle there. Your first operation would be to reclassify the map layer to create a new layer called "landuse," to isolate the land uses related to the land covers, as in reclassifying all the grassland polygons as either pasture or natural grasslands. You also would create a map layer called "whosegrass," to show pastures that are owned by the federal government and those that are privately owned. This may require you to perform another reclassification, although you could have done this work during your initial reclassification. Your next map operation would be another reclassification to create a layer called "bigfed" that isolates federally owned grasslands that contain over 25 hectares of continuous grass. This reclassification will call for a clumping function that isolates polygons of 25 hectares or larger (Figure 13.1);

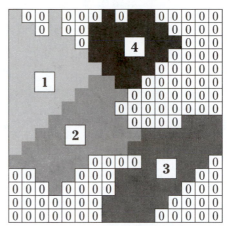

Original values representing different types of rangeland. Each grid cell contains one hectare of land.

New values representing a ranking of contiguous rangeland parcels based on size. 0's represent those under 25 hectares. 1's are for areas with 44 hectares, 2's for areas with 42 hectares, 3's for 41 hectares and 4's for 28 hectares.

Figure 13.1   Reclassifying neighborhoods by size. Clumping to isolate the polygons of 25 hectares or larger.

you will need to combine this clumping operation with a neighborhood function that separates out contiguous polygons.

Thus you have the opportunity to operate either on the raw data, as in our original ("landcover") map, or on layers derived from this initial operation, as in the reclassified "bigfed" coverage used to isolate large, federally owned pastures. A cartographic model employs each analytical operation on either raw or intermediate map data. Each step is combined to add information to the data by giving the data a context about which it relates. Our original layer, "landuse," was not specific enough. It related to all possible land cover in the study area, without regard to what the question might have been. As we manipulated this layer, we created "value-added" layers of specific attributes that could be used to decide whether the layer contained federally owned pastures large enough to graze a certain number of cattle.

This brings us to the last portion of the definition. Our manipulation of raw and intermediate data was designed to simulate the process of deciding whether the coverage in question contained "large enough" contiguous areas owned by the federal government. A negative result will require you to look elsewhere for such grasslands, whereas a positive result will prompt you to begin seeking the permits necessary for your grazing operation. A positive result means simply that there are sites available. But suppose you want the option of increasing the size of your herd. You would then have to perform yet another operation on the derived coverage "bigfed." By asking the computer to rank by size the federally owned pasture polygons that exceed 25 hectares, you will be in a position to determine which combinations would allow you to increase your grazing herd. These steps add information to each preceding data element incrementally.

Now that you have completed your simple decision model, think carefully about the GIS subsystems involved. The first step was to input the "landcover"

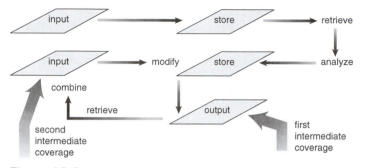

Figure 13.2 Cyclical modeling process. The illustration shows the movement from one geographic information system (GIS) subsystem to another to produce new maps.

coverage, employing the input subsystem of the GIS. Next you had to store the map and edit any mistakes. When you were satisfied that it was an accurate model of present land use, you had to retrieve it for analysis, calling on the storage and editing subsystem of the GIS. Having retrieved the coverage, you reclassified it on the basis of attributes more closely associated with anthropogenic use of the land. The reclassification process used the analysis subsystem. The product of the reclassification is a stored map layer, representing use of the storage and editing subsystem as well as the output subsystem, even though the output may never have been printed as a hard-copy map. You then retrieved the new layer to continue to analyze it. And at any point, you might have wanted to view the map.

Clearly, the foregoing process is not linear—we did not go from input to storage to retrieval to output, as in CAC. Instead, we went from input to storage to retrieval to analysis to storage and output of our first intermediate layer. From there, we retrieved the first intermediate layer and output the map on the monitor to determine what analysis to perform next to create a second intermediate layer, and so on. In short, the process of cartographic modeling is cyclical rather than linear (Figure 13.2). The cyclical nature of cartographic modeling allows you the greatest flexibility to move from subsystem to subsystem in a series of data transformations all designed to produce a final **spatial information product** (SIP).

TYPES OF CARTOGRAPHIC MODELS

There are strong similarities between the categorization of statistical techniques into descriptive and inferential types and the major types of cartographic models. Tomlin (1990) separates cartographic models into descriptive (like the statistical counterpart) and prescriptive (somewhat related to the inferential statistical techniques) types.

Descriptive cartographic models are those that describe and, under certain circumstances, explain particular patterns and pattern associations produced from analysis. The simplest **descriptive models** are those that merely illustrate existing conditions by isolating preselected phenomena and presenting the results in an understandable form. Despite its simplicity, this type of model is still among the more common because it offers a relatively straightforward way of obtaining easily recognized patterns of spatial phenomena. Moreover, the GIS professional should see this as the starting point for more complex models that might be suggested.

It would be natural to go from pure descriptions of existing conditions to a model that predicts how these existing conditions might affect the location of an industry, how these conditions might indicate a change in natural vegetation through time, or where an optimum habitat for waterfowl might already exist. Such a model may, for example, describe the results of damming a stream by showing the areas above the dam that would be inundated. It could also show the areas currently experiencing flooding downstream that would benefit from construction of the proposed dam. A city planner might be able to predict locations of potential urban expansion by using knowledge of where expansion has taken place in the past to identify spatial phenomena that might be used as predictors.

The latter case should be examined more carefully because it shows the predictive potential of descriptive cartographic models. **Predictive models** allow the user to determine what factors are important in the functioning of the study area and how these factors are associated with each other spatially. Prediction on the basis of these associations can be very tricky. It requires that the variables have a clear and verifiable causal relationship. As we saw in Chapter 12, a spatial association among different mapped variables does not, of itself, dictate that there are cause-and-effect relationships, only that the variables occupy more or less the same space. Knowledge of the environment being modeled is as vital in predictive cartographic modeling as it is in inferential statistics such as regression analysis.

Predictive modeling is most often associated with the second major type of modeling defined by Tomlin (1990): prescriptive modeling. As we will soon see, however, there is no clear separation between descriptive and **prescriptive models**. Rather, the two might be thought of as two ends of the cartographic modeling spectrum, with prediction and prescription increasing as we approach the prescriptive end. Take the following example of prediction based purely on descriptive modeling techniques. The owner of a very large ranch wishes to evaluate the carrying capacity (i.e., the number of animals that can survive on the land) of her property holdings with respect to grassland for cattle and habitat suitability for native wildlife such as quail. The model requires her to develop a database showing all the vegetation types on the ranch. In addition, she needs to know the above-ground biomass (the amount of vegetation by weight that occurs above the ground level) for the grassland vegetation, as well as the locations of large patches of noxious weeds (that might be harmful to her stock) and significant clumps of shrubs (needed by the quail for shelter and foraging). From these variables, she produces a cartographic model showing the areas that have the minimal habitat needed to support a reasonable quail population. She also produces a thematic map indicating the minimum necessary carrying capacity for the cattle. The model is descriptive in that it shows where the owner could reasonably expect to put her cattle and where quail might survive as well. At this point, she could simply herd her cattle into the adequate grazing portions of the ranch and prevent the livestock from moving into the quail habitat by fences or other means.

Suppose, however, that she does not wish her cattle merely to graze successfully but rather wants them to have only the very best grazing lands, to ensure that the animals are fattened as quickly as possible. Such a manipulation of the database could reasonably be called predictive because in effect it is predicting that the areas with the highest carrying capacity will result in the most efficient use of the land for grazing. This model is only slightly more complex than the original, but it carries with it an element of prescription because it **"prescribes"** the "best" use of the land.

Prescription also shows the model user how manipulating existing attributes can improve an overall situation. For example, some of the best grazing land may contain many shrubs, making it prime quail habitat as well. If the same location also contains poisonous weeds, the rancher may decide to plant (or transplant) shrubs in areas that are less productive for cattle while eliminating the noxious weeds from the most productive grazing sites.

The GIS could be used to "prescribe" the best places for moving the shrubs as well as targeting for other manipulation locations with noxious weeds. Thus the GIS is now used to suggest the appropriate manipulation of existing holdings to ensure the best solution to the original problem. It is the highest form of predictive model because the user has to determine the interactions of many variables through time, and a much higher degree of predictive capability is called for.

You have probably noticed a common theme of predictive modeling, even if an individual case is not purely prescriptive. All these models require, as a prerequisite, a description of existing conditions. That is, when developing prescriptive models, first you describe and then you prescribe. The change from one to the other is a gradual transition from pure description with no action needed to increased predictive power and increasing prescriptive capabilities. The nature and complexity of the problem will dictate the type of model you employ, no matter what you call it.

INDUCTIVE AND DEDUCTIVE MODELING

Whether you are working with purely descriptive models, prescriptive models, or somewhere along the spectrum, generally models can be formulated in two ways. The first, called the inductive method (from inductive reasoning), moves from specific elements to a general statement. That is, you start with a large collection of spatial data for a particular study area. Such an approach lacks design and often results in failure of the GIS to perform to the users' expectations. Still, it has some validity, especially in a research environment where spatial data have been collected over a long period of time for individual research projects. Under such a setting, the empirical spatial data are often examined (mined) in a search for important patterns. Different map layers are compared and tested for correspondences among the elements of other themes to determine similar spatial patterns or associations that might indicate an ongoing process within the study area.

Trial and error can be used to impose order on items acquired without careful planning. The same can be said of using existing GIS databases. If there are enough of the appropriate variables, and if there are resulting correspondences among variables on different layers, it is entirely possible to create models that will be useful for making some decisions. You will probably also find that map layers you have spent many hundreds of hours creating are never used. Many such projects have contributed to scientific research, but the lack of design in the inductive approach to cartographic modeling is often inefficient for many commercial applications. Although science attempts to explore data, real-world applications are often much more specific in their goals and should therefore be thought out in a more deductive manner.

Deductive models move in the opposite direction from **inductive models**. That is, you begin with a specific recipe or formulation that addresses very specific questions. The advantages of the deductive approach to cartographic modeling should be obvious. You wasted no time or money producing unnecessary map layers. You develop only the layers you need, and you can concentrate on preparing these layers with the care needed to answer your questions.

FACTOR SELECTION

Recall from Chapter 5 that one of the problems we face in creating models is the selection of appropriate variables. In some cases, it is relatively simple to decide which factors you should use. If we are going to model a transportation net, common sense tells us to include all the available road network. It is also obvious that we need to know which are dirt roads and which are superhighways, which are one-way streets and which are two-way, which have traffic signals and which do not, and so on. All these factors are accessible in direct data sources that relate explicitly to what we are trying to model. However, not all data are as easily obtained. Nor are all variables explicitly defined.

Suppose you want to model the potential impact of a hazardous chemical spill along a rail line that runs through your town. You have explicit information of the location of the rail line, and you might have manifest information about the actual materials being transported. You also have explicit spatial information about the number of housing units located in the hazard zone. Because the population in your area is highly mobile, however, you don't know exactly how many people live in each household, nor do you have information about any special-needs populations such as elderly or handicapped people. To develop your model, you will need to use surrogates for these variables. Perhaps you will sample the households in each neighborhood to get an idea of the average number of people living in each unit and the average number of special-needs individuals. By generalizing these data, you can obtain an estimate for each neighborhood. These estimates will be used in developing your potential spill impact model.

MODEL FLOWCHARTING

Whether you are using a deductive or an inductive approach, an extremely useful technique to assist you in formulating your model and determining the appropriate data elements is the **flowchart**. Model flowcharting requires that you isolate each item or element (coverage) that is to be used in your model. Each element should have a very specific, unique theme that is representative of a single factor or group of factors in your model. Flowcharting allows you to determine whether you have all the necessary map elements. It also allows you to evaluate whether each is unique. If you have several redundant data elements representing the same theme, you can eliminate one or more of these from your model, and the input time will be reduced.

Figure 13.3 is a simple deductive cartographic model flowchart that attempts to locate the best site for a mountain cabin. The result we are seeking is very specific, and at least four major factors need to be considered in making the decision. Let's start with infrastructure. You will need to know about the availability of running water, electricity, perhaps gas for cooking or heating, roads for access, and other facilities (sewerage, etc.). Second, there are political or legal factors. You must, for example, know whether a parcel of land of interest is owned by someone willing to sell. You will probably need to obtain a permit for building, or you may find that government mandate prevents you from building your mountain cabin on land zoned as nonresidential. If your cabin is to be located in an earthquake-prone area, design or construction restrictions may be present. Fire-prone areas will require you to use fire-retardant materials. Such restrictions place constraints both on where you can build and on how much the effort will cost you.

A third group of factors to be included in your cartographic model are aesthetic. After all, you probably are building a mountain cabin to use as a retreat from your usual environment. You will want the cabin to be out of sight of the city (while still within driving distance). In addition, you want an unobstructed view of the mountains. You may also want the cabin to face in a particular direction. Perhaps you want the back of the cabin to face west so that you can view the sunsets from your back porch.

The fourth group of siting factors are physical factors. You will need a slope that is not so steep that construction will require expensive measures to prevent the finished cabin from moving downslope. Also to reduce costs, it would be better to choose a site that does not have dense tree cover. A site with a small clearing surrounded by trees would be ideal. Soil engineering properties will also have to be evaluated, because you do not want to locate your cabin on unstable soils, especially soils with a large amount of high shrink-swell clay. You may not have access to sewerage facilities, and you may have to plan for a septic tank system. This will require you to factor the infiltration capabilities of the soil into your calculations.

Let's move to the actual model now, looking closely at the way in which each data element was chosen to represent individual unique factors. Later, we will examine how these elements are combined to complete the model and indicate the best site for our mountain cabin. We begin by looking at the infrastructure portion of our model. For data regarding the first factor—accessibility to running water—we need a map layer that shows where all the municipal water lines are located. The layer "waterline" would be created by digitizing an analog map of water lines. Similarly, we create layers for electric lines ("electricity"), locations of gas lines ("gaslines"), and all available roads in our study area ("roads").

Next, we turn to political and legal factors. First, from cadastral maps (maps showing landownership) we digitize "ownership," the most current landowner-

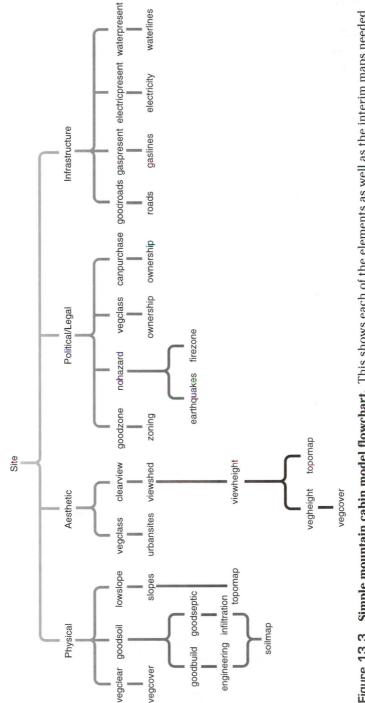

Figure 13.3 Simple mountain cabin model flowchart. This shows each of the elements as well as the interim maps needed to produce the final solution—the site.

ship map. This layer allows us to determine whether there are available un-claimed sites and/or from whom we need purchase the land. Next, we obtain a zoning map and digitize "zoning," a layer that permits us to determine which ar-eas are not zoned for single-family dwellings. We also need to know whether cer-tain locations are earthquake zones or fire hazards. Data on these two potential problems will likely come from different sources. We obtain a map of earth-quake-prone areas and digitize these ("earthquakes"), and we obtain forest fire hazard maps and digitize those to create a theme called "firezone." Although the last two layers are physical parameters rather than political or legal, it is likely that permit restrictions are already in place for such areas.

An evaluation of aesthetic qualities requires maps showing the locations of nearby municipalities ("urbansites"). Topographic maps would provide infor-mation about obstructions ("topomap"). Maps of forest types tend to exclude information about the sizes of trees because of the amount of time required to obtain such information. You might have to create your own forest cover map, either by performing a ground survey of some of your more suitable locations or by using aerial photography of the area and mapping the forest stands your-self to create the needed layer ("vegcover").

Our final group of factors includes physical parameters. Slope, the degree of change in elevation over horizontal distance, can be obtained by means of a layer we already have—topomap. Although slope is a unique element, it is de-rived by modifying existing data. Thus we have not violated our rule about se-lecting unique elements. Another element of our physical parameters that can be obtained from an existing map shows areas with little forest vegetation or with forest vegetation containing an appropriate clearing. The existing layer, called "vegclass," can help us obtain this information.

Our two soil parameters will likely be obtained from a single layer as well. A detailed soil map will normally have both engineering properties and in-filtration properties; either of these will be explicitly encoded to produce separate maps, or they will be associated through a database system and con-nected to each of the soil-type polygons. We will create these maps by first creating the soil layer ("soilmap"). The methods we use to keep track of the various soil properties will largely depend on our GIS software. Alternatively, we could simply encode the soil polygons and retain a paper copy of the soil survey from which we obtained the soil polygon information. These survey reports associate a large variety of soil factors with each soil-type polygon (Table 13.1).

Working Through the Model

Look closely at the flowchart we used for the mountain cabin siting model: all the appropriate elements are included, and they all flow to the final site for our cabin. Something is still missing, however. Ask yourself how the intermediate layers were created from the initial data elements. To complete our flowchart, we need to indicate what functional capabilities of the GIS we use to get from one branch to the next. Not only is this step important in modeling, but it also assists us in determining the needed functionality of any GIS for a specific pro-ject. Let's consider a revised version of the flowchart (Figure 13.4).

TABLE 13.1 Typical Page from a Soil Survey Report Showing Soil Polygons and Small Portions of Tabular Information Available for Each Soil Type

Soil Name and Map Symbol	Dwellings Without Basements	Dwellings with Basements	Small Commercial Buildings	Local Roads and Streets	Septic Tank Absorption Fields	Sewage Lagoon Areas	Playgrounds
Ev: Eudora part	Severe: floods	Severe: floods	Severe: floods	Severe: frost action	Moderate: floods	Moderate: seepage	Moderate: floods
Gravelly land: Ge	Moderate: slope	Moderate: slope	Moderate: slope	Slight	Severe: small stones	Severe: seepage, small stones	Severe: small stones
Judson: Ju	Moderate: shrink/swell	Moderate: shrink/swell	Moderate: shrink/swell	Severe: frost action	Slight	Moderate: seepage	Slight
Kennebec: Kb Kc	Severe: floods	Severe: floods	Severe: floods	Severe: floods, frost action, low strength	Severe: floods, wetness	Severe: floods, wetness	Moderate: floods
Kimo: Km	Severe: floods, shrink/swell	Severe: floods, shrink/swell	Severe: floods, shrink/swell	Severe: shrink/swell, low strength	Severe: percolates slowly	Slight	Severe: wetness
Martin: Mb, Mc, Mh	Moderate: shrink/swell	Severe: shrink/swell	Severe: shrink/swell	Severe: low strength, shrink/swell	Severe: percolates slowly	Moderate: slope	Moderate: too clayey, percolates slowly
Morrill: Mr	Moderate: shrink/swell, low strength	Moderate: shrink/swell, low strength	Moderate: shrink/swell, low strength	Moderate: shrink/swell	Severe: percolates slowly	Moderate: slope	Moderate: percolates slowly
Ms	Moderate: shrink/swell, low strength	Moderate: shrink/swell, low strength	Severe: slope	Moderate: shrink/swell	Severe: percolates slowly	Severe: slope	Severe: slope
Oska: Oe	Severe: shrink/swell	Severe: shrink/swell, depth to rock	Severe: shrink/swell	Severe: shrink/swell	Severe: depth to rock, percolates slowly	Severe: depth to rock	Moderate: percolates slowly, too clayey, depth to rock
Pawnee: Pb, Pc, Ph	Severe: shrink/swell	Severe: shrink/swell	Severe: shrink/swell	Severe: shrink/swell	Severe: percolates slowly	Moderate: slope	Moderate: percolates slowly
Sharpsburg: Sc, Sd	Severe: shrink/swell	Severe: shrink/swell	Severe: shrink/swell	Severe: shrink/swell, low strength	Severe: percolates slowly	Moderate: slope	Moderate: percolates slowly
Stony steep land: Sx	Severe: depth to rock, slope	Severe: depth to rock, slope	Severe: depth to rock, slope	Severe: depth to rock, slope	Severe: depth to rock, slope	Severe: depth to rock, slope	Severe: depth to rock, slope
Thurman: Tc	Slight	Slight	Moderate: slope	Slight	Slight	Severe: seepage	Severe: slope
Vm: Vinland part	Moderate: depth to rock, slope	Moderate: depth to rock	Severe: slope	Moderate: depth to rock	Severe: depth to rock	Severe: depth to rock	Severe: depth to rock
Wabash: Wc Wh	Severe: wetness, floods, shrink/swell	Severe: wetness, floods, shrink/swell	Severe: wetness, floods, shrink/swell	Severe: wetness, floods, shrink/swell	Severe: percolates slowly, floods, wetness	Severe: floods, wetness	Severe: wetness, floods, percolates slowly

Source: U.S. Department of Agriculture.

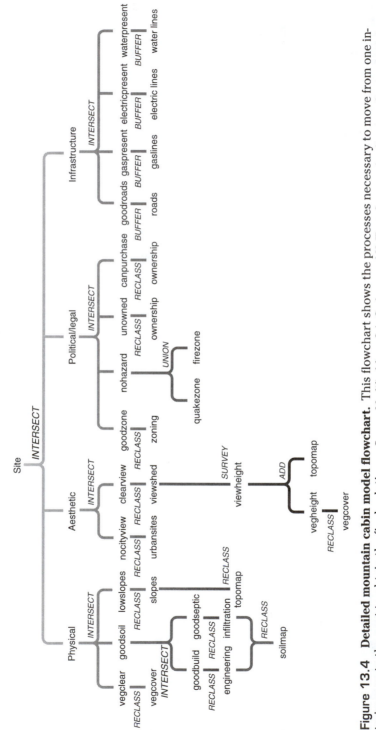

Figure 13.4 Detailed mountain cabin model flowchart. This flowchart shows the processes necessary to move from one interim map to the next to obtain the final solution. *Source:* Modified from S.A. Carlson, and H. Fleet. Systems Applications Geographic Information System (SAGIS) and Linked Analytical/Storage Packages. Workshop, annual meeting of the Association of American Geographers, 1986.

To avoid using specific commands, let's continue to use rather generic functions with mnemonically useful names as has been done so far. To implement the model on your own software, you need only to change my names to the specific commands you would use. We begin by examining the infrastructure parameters, namely, roads, gas lines, water lines, and electric utilities. There is a high probability that none of these will directly serve your cabin site. You need to know whether they are close enough—say, within 500 meters—to make it feasible to build extensions for each. To find out, you begin with one of the existing elements. The digitized map of electric utility lines will tell you where they are located, and it will not be too difficult to create a buffer of 500 meters on each side of the existing electric line elements. The result is a map called "electricpresent," which shows cabin sites that do or could have access to electric power. Next, you might build a buffer around your gas lines coverage extending, say, 250 meters, indicating the likely areas where gas lines could be extended to your cabin—a coverage called "gaspresent." Roads are not nearly as important because you have a four-wheel-drive vehicle. You could travel as much as 750 meters offroad to get to your cabin. Therefore, you build your buffer 750 meters around the existing roads coverage to produce a new coverage, "goodroads."

You now have three separate, intermediate themes, each representing the limits of your requirements for accessibility to roads and electric, water, and gas utilities. Next you must decide how to combine these to produce a final coverage, called "infrastructure," to represent all these factors. Because you need all these elements, rather than just one or two, they must be overlaid in a way that results in a map showing where they intersect. Therefore, the overlay procedure you choose must use some form of "and" search, or intersection, as you learned in Chapter 12. The resulting map should display at least one area that satisfies all three criteria.

At this point you might be wondering, "What happens if there is no area that has all four factors?" This is not an unlikely outcome. If your composite map shows no intersections, you could focus on the factor that is the farthest from intersecting the other two. Let's say that the gas line buffer and the roads buffer intersect, but the electric lines do not. If the distance limits for that factor can be extended—in other words, if the model constraints can be relaxed—you might be able to fulfill the requirements of your infrastructure submodel. For example, you might find that for an additional charge, you could extend the electric lines to 1000 meters. This approach to improving model performance by constraints relaxation is common in cartographic modeling.

If constraints relaxation is impossible—for example, if the electric company is unwilling to extend lines 1000 meters from their existing facilities—you may have to reduce the importance of one or more of your factors. In creating your buffers, you based the size of each on the absolute maximum limits for the corresponding factor. That is, you assumed that all factors were equally important. This is not always the case. Actually, all factors in a model rarely have the same importance. Therefore, unless you are willing to abandon the project, you must weight the four infrastructure factors. Roads are important beyond the 750-meter limit because you will need to get to and from the cabin regularly. This factor must retain a high degree of importance. If gas is available, you could use it to heat your home. You could even have natural gas lighting if you wished. What is left is our electricity, which was the problem in the first place. The solution to your problem is to downgrade your perceived need for electricity. Thus you could reduce the importance of electricity to zero, eliminating it as a model constraint.

Now that we have satisfied our infrastructure submodel constraints, we can proceed to the second submodel, the political and legal factors. Areas not zoned against single-family dwellings ("goodzone") can be obtained by reclassifying the zoning layer to eliminate all areas that do not meet this constraint. Similarly, you can create a map layer of property without a registered owner ("unowned") and another of properties whose owners may be willing to sell or have offered the land for sale ("canpurchase"). Both of these maps use the same initial coverage ("ownership"), and both are produced by simply reclassifying categories to obtain what you want. Your map layers for the remaining political or legal factors, "quakezone" and "firezone," contain the information you need, but your cabin must be located outside both of these areas. You can combine these using an "or" (union) overlay. The resulting output, called "nohazard," shows all areas that do not have seismic and/or fire hazards.

You now have the four intermediate themes needed to complete your political/legal submodel. Once again, to satisfy all these constraints, you combine them through an overlay operation that intersects the factors to produce a map of politically/legally viable locations. Here, you must be careful how you intersect the layers. You must be sure that all four layers indicate good areas for locating your cabin. If one of your themes, for example, the nohazard layer, shows locations that have hazards, your cabin will be located in a rather warm and possibly shaky location.

The third submodel, aesthetic factors, has fewer components than the other three but is somewhat trickier to calculate. Beginning with your theme of urban sites, you may not be able to eliminate all small towns and villages from your study area. In fact, you probably want a nearby community as a source of supplies. You decide that you want no town larger than, say, 1000 people. By reclassifying your map, you can create a theme called "nourban" that shows small towns and villages but no large towns or cities. Although this might be appealing, it is not what you want, because you will need to know where the large towns and cities are so you can determine whether they can be viewed from your potential location. So you reclassify your "urbanzone" map to show where all the cities are located ("cities"). This layer will then be combined with your next one to eliminate such unsightly anthropogenic artifacts from your view.

The second portion of your aesthetic submodel requires that you be able to see around any peaks and any tall vegetation. Remember that the vegetation is growing on the terrain itself. To incorporate this feature into our model, we must find an approximate height for mature forest stands. (The average height at maturity of the various forest types can be found in standard botanical or forestry references.) Then we reclassify our vegcover on the basis of the vegetation classes at hand and call the map "vegheight." The reclassification will produce a theme of estimated tree heights and will be stored as ratio data rather than nominal data. To show what the topographic elevations will be like with tree cover on them, we add the topomap data (also in ratio format) with the vegheight data to produce a layer called "trueheight." This is the intermediate theme we need to determine what can and can't be seen. By using a command that determines the viewshed (see Chapter 10), we produce a viewshed based on our ability to see around peaks of elevation and around the vegetation on those peaks. This coverage can then be reclassified to show areas that offer a clear view of the valley below ("clearview"). Remember, however, we don't want to see cities. The clearview coverage shows areas that are clearly visible from the hillslopes, and the cities map shows all communities exceeding 1000 people. By combining these two layers through intersection overlay, we produce a map that shows viewable areas with cities and viewable areas without cities. The viewable areas without cities will ultimately be combined with the other three submodels.

Completing the model, we next look at the three major factors of the physical parameters: vegetation, soils, and slope. Returning to our vegcover layer, it is easy to reclassify the vegetation categories into those with large, mature tree stands and those with other types of vegetation or with relatively small trees that can be cleared. The resulting layer, called "vegclear," should isolate areas with no mature tree stands. Keep in mind that the scale of most vegetation maps is rather small, offering little detail. Thus any decision that relies on such data should be checked on the ground. Some mature tree stands on very rugged terrain are not very high and hence may not be as difficult to clear as a vegetation map implies. This is another area in which it is very easy to modify constraints to accommodate the decision process. In this case, having mature trees may not be particularly negative because of the ready availability of power saws. In addition, a few trees, properly placed, could provide shade without obstructing the view or interfering with the construction of our cabin.

Among the most severe potential physical constraints in siting a mountain cabin is the steepness of slope. A severe slope will either prohibit the project altogether or radically increase its cost. A slope of 15 degrees or less would be ideal. To find all the areas that fit that constraint, you would begin with the topomap coverage and perform a neighborhood function that separates out the degree of slope throughout the coverage. By reclassifying these slopes so that low slopes are those of 15 degrees or less, you now have a layer ("lowslope") that fits these constraints.

The two soil characteristics that must be considered are engineering properties and infiltration capacity. The first determines the capacity of the soil to withstand a structure like a mountain cabin, and the second classifies a soil according to whether infiltration capacity is sufficient to allow the construction of a septic tank. Both of these factors come directly from the original soils map; either they will be reclassified to fit the appropriate categories, or they will be selected from an associated attribute table. The layers produced ("goodbuild" and "goodseptic") can then be intersected to create a more nearly complete map layer called "goodsoil." This layer completes the three physical factors dictating where we can put our mountain cabin. Through an intersection type of overlay, we can create the final submodel coverage, called "physical."

The final solution is to intersect physical, aesthetic, political/legal, and infrastructure maps because all these constraints must be met. If our model works as planned, we should have at least a single, small portion of land that suits all our needs.

Not so fast. Just when you thought you were finished, yet another problem appears. Earlier we discussed the problem of constraints that were too tight. When we looked at our infrastructure factors, we discovered that insistence on having electricity could limit our choice so severely that the idea of building a mountain cabin would have to be abandoned. There is also the unlikely possibility that our outcome illustrates that we have very good soils with few very steep locations, lots of good mountain views, no major zoning or ownership problems, and plenty of close infrastructure. If this nearly fairy-tale scenario were true, our model would indicate large portions of the study area in which building conditions were appropriate. Under these circumstances, we could build our mountain cabin almost anywhere. Thus an apparently wonderful outcome shows that the model was not very useful in helping us make a decision.

Whereas when modeling constraints were too tight we loosened them to make the model more flexible, now we do the opposite. In this case, our model constraints are too lax, and we will have to tighten them a little so that the GIS user can have the very best mountain cabin location from the model. As before, this can be accomplished in two ways. The first would be to modify the criteria. For example, we could require soils of the absolute best quality for building the cabin and installing a septic tank. We could also require slopes of less than 7 degrees rather than the 15 degrees originally selected. Another factor that could be modified is the size of town or city that would be acceptable within our viewshed. As you can see, quite a number of changes could be made to the model. In fact, it is generally easier to produce a range of values for each factor, whereupon the answers obtained can be ranked, rather than treated simply as "yes" and "no" answers to the facets of the problem.

Again, an alternative approach can be used to tighten our model constraints. Although physical, aesthetic, political/legal, and infrastructure parameters are all important, they do not have equal importance. Physical limitations of the soil and slopes have a substantial impact on your ability to build your cabin. However, if no lands are zoned for single-family dwellings, and the zoning commission is unwilling to give you a variance, the physical parameters are no longer important. Instead, you have one set of factors that can be called absolute limitations, whereas others might be overcome by modification of the constraints or by a willingness to invest more in the project. Determining which factors can be weighted and which are absolute or preemptive will help the cartographic modeler adjust the solution to the individual situation. You must be flexible and very aware of the impacts of these weights on the final outcome. The literature presents a few examples of the use of weighting and reweighting to improve a model and to satisfy the ultimate users (Davis, 1981; DeMers, 1988; Lucky and DeMers, 1987).

Conflict Resolution

It is important to consider factor weights and different thematic layers when you produce a cartographic model. However, weighting will not always produce results that satisfy all possible conditions. Consider the problem of producing a cartographic model that involves two essentially competing views of the world—say, selecting timber harvesting versus the preservation of spotted owls to comply with federal regulations. Even without the federal mandates, there are essentially two antagonistic ideas about the potential use of large portions of land. Both those concerned with preserving spotted owls and those trying to use the forests as an industry that will enable them to feed their families have valid standpoints.

How can we approach such a difficult modeling task? Let's look at one readily available technique suggested by Dana Tomlin. Although not the only approach, it shows that the competing portions of a problem can be conceptually separated out and a reasonable compromise arrived at without employing unusual software, weird logics, or high-priced negotiation teams. The approach, called ORPHEUS, involves creating two separate models, one for each competing side. After evaluating all the factors, you begin to put weights on each of the factors for each party (Tomlin and Johnston, 1988). If you continue to reweight the factors on a pairwise basis, getting, it is hoped, small concessions for each side, your output maps should reveal areas that will allow the competing demands to coexist. The process takes time to perfect, and it seems that there is no substitute for experience. Try this out for yourself by creating a small database, 1 square mile or so, that includes soils data with associated tables of properties. Then select six to eight possible land uses totaling 1 full square mile to place in this area. Now use your GIS to place these land uses, keeping in mind the usual needs for size of some uses, locations of others, proximity to pollution, and so on. Even if you don't immediately achieve positive results, the experience will be very instructive.

Some Example Cartographic Models

To finish our examination of flowcharting models, let's look at three basic examples of models and their flowcharts. Our sample cases have separate themes and use quite different approaches both to modeling and to flowcharting. This should give you an idea of some of the options.

We begin with a descriptive model of deer habitat quality (Carlson and Fleet, 1986) (Figure 13.5). This simplified model is based on the presence or absence of some factors essential to the survival of the deer: availability of water and vegetation for browsing, and shelter. Availability of water is based on a water or hydrology coverage, whereas browsing and shelter are based on the vegetative cover. Browsers normally obtain their food from woody plants, which have their leaves off the ground. Thus the necessary forage and shelter factors are based on a reclassification of the cover vegetation into woody shrubs and trees as useful and grasses and forbs as nonuseful vegetation types. The model is somewhat vague about the precise types of vegetation needed and whether all types of water can be drunk. Still, it should be easy to imagine how these factors could be obtained from the available coverages with basic zoological/behavioral knowledge (i.e., of the animals' requirements for food, cover, and drink).

A second portion of the model is designed to determine habitat quality on the basis of spatial configuration of the landscape units. The primary considerations here are the size, amount of edge for each landscape compartment or patch (i.e., edginess), and integrity (cf. Figures 8.2 and 8.3). This portion should remind you of our earlier discussion about the use of landscape ecological units in cartographic modeling. This is an excellent, easy-to-understand example of landscape spatial variables as they are applied to a single-species habitat. The final portion of the model totals the habitat quality components for the separate landscape components and scores or ranks them through, in this case, averaging, to obtain the composite habitat quality. Once again, it should be fairly obvi-

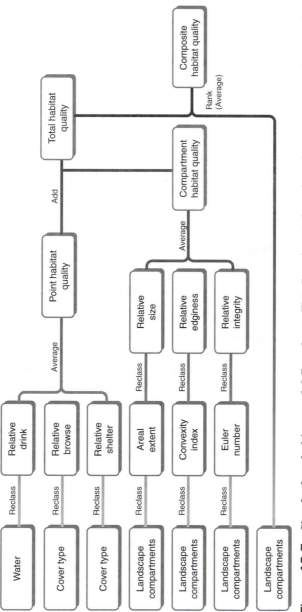

Figure 13.5 Simple deer habitat model flowchart. This flowchart is based on the presence or absence of some primary survival factors: shelter, food and water, and the spatial configuration of the landscape with respect to size, amount of edge habitat, and the integrity or patchiness or the landscape. *Source:* Adapted from Joseph K. Berry (personal teaching notes). Used with permission.

ous how simple submodels can be combined to achieve a relatively complex model of deer habitat. By compartmentalizing each set of operations and each landscape compartment, we can easily produce the individual models.

Our second model is more of a prescriptive model that attempts to define portions of an area that will receive enough exposure to prevailing winds to operate a wind generator (Carlson and Fleet, 1986) (Figure 13.6). There are only two basic coverages: topographic elevation data and vegetation communities. The topographic coverage exposes the orientation or aspect of the topographic slopes, which have been assigned ordinal categories ranging from best to worst (for wind generators). Level ground will, generally be the most useful because there are no obstructions, whereas in our example, northwest is the next best because the prevailing winds come from that direction. As one moves further away from northwest, usefulness declines.

Another factor we need to consider is the distance to obstructions—in this case, trees and slopes—which is easily obtained by using a functional distance operation; this model uses a command called "Radiate." The required generator height, h', based on the heights of obstructions and the distance just calculated, is a simple difference between the obstruction height and the ratio of the obstruction height minus the generator height divided by the distance. The model assumes that towers will be no higher than 80 feet by subtracting that value from the topography, and then it ranks the sites independent of the tower heights, after which it recombines the aspects to obtain sites available for wind generation. The flowchart for this model is considerably different from that of

Steps	Operations
Find direction of slope	**Operations**
level = 1 (best),	Orientation command
NW = 2 (good)	
N,W = 3 (fair)	
S,SW = 4 (poor)	
other = 4 (no good)	
Find distance to obstructions d	Radiate
Find height of obstructions h	Reclass, Topo + Cover
Find needed generator heights:	
ratio $r = (h - h')/d$ for $r = f(d)$	
thus $h' = h - rd$	
select correct r	Subtract $(d - d, f(r))$, Reclass
find rd	Multiply
find h'	Subtract
Eliminate towers	Subtract
Display all ranked sites independent of tower heights	Reclass, Display
Display sites as a function f of (aspect, tower height)	Crosstab, Reclass, Display

Figure 13.6 Wind generator siting flowchart. Model flowchart for the siting of wind generators based on prevailing winds, terrain elevation, and tree heights. This example is designed more as an algorithm than as a flowchart. The mapped data, consisting of elevations and vegetation communities, can be manipulated by means of the commands in the "Operations" column. *Source:* Adapted from C. Data Tomlin (personal teaching notes). Used with permission.

Figure 13.4; it is more of an algorithm than a flowchart. Some may feel more comfortable with this recipe approach, although it is relatively easy to transform both the classic flowchart and the algorithmic approaches.

You might want to try the following exercise, which represents a more statistical method of modeling. If you are unfamiliar with regression techniques, however, you may choose to skip this material and proceed to the model implementation section. If you know something about regression, this model is an excellent example of how dasymetric techniques can be applied using the coefficients of a regression equation to modify the values of polygons.

On the basis of the predictive statistical method known as regression, our last model attempts to predict breakage during timber harvesting (Berry and Tomlin, 1984) (Figure 13.7). Such a value would be useful to the lumber companies as they try to determine their financial bottom line. In this model, the independent variables (maps or coverages) used are percent slope, tree diameter, tree height, tree volume, and percent tree defect. Each variable has a measurable impact (indicated by the regression coefficient) on the predicted breakage. By multiplying each coefficient by each associated independent variable, the model produces weighted maps that, when added to each other and a constant (a necessary portion of a regression equation), yield the total predicted breakage that occurs during tree harvesting. Because this predictive model uses tested, statistically significant predictions for each independent variable cover-

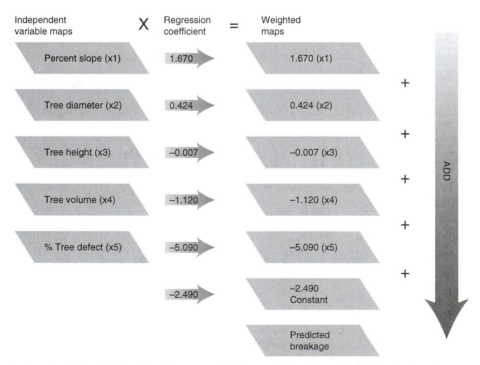

Figure 13.7   Timber breakage predictive model using regression. The flowchart illustrates the use of dasymetric mapping techniques embedded in the following regression equation for conventional felling practices:

$$Y = -2.490 + [1.670 \times (1)] + [0.424 \times (2)] - [0.007 \times (3)] - [1.120 \times (4)] - [5.090 \times 950]$$

Source: Adapted from Joseph K. Berry (personal teaching notes). Used with permission.

age, the results are highly robust and easily tested for validity, a subject we will cover later in the chapter.

MODEL IMPLEMENTATION

A discussion of model implementation might seem redundant, given our flow-chart indicating how the model is to be produced and which elements we need to produce it. But thus far, we have assumed that everything is going to work according to plan and that our results will naturally conform to our expectations. Many GIS practitioners, through trial and error, have seen that it is best to take a conservative approach to model implementation. Although input consumes much more of our time than does any other phase of GIS, we can nevertheless find ourselves spending many hours correcting mistakes if our implementation fails. For this reason alone, it is wise to select a small subset of the study area on which to prototype the model, a technique that will be seen to be very useful in model verification later on.

Using our subset study area and our model flowchart, we begin at the elemental coverage level for each submodel and perform the necessary operations. This activity allows us to examine some of the following basic questions:

1. If we are reclassifying our categories, is our database set up to permit retrieval of the appropriate values? Or if we are reclassifying categories in a nondatabase raster environment, does the order of reclassifying make a difference?

2. If there are alternatives to a single technique or alternative combinations of techniques, which ones are the most likely to give us the correct response with the least amount of work?

3. Are we absolutely sure that the operations selected are representative of how the modeling environment really works?

4. How are we going to deal with missing variables, if we have any, in each layer?

There are many more questions, but this short list makes the point. Be sure you know what you are doing before you do it for the entire database. By performing the operations on a sample of the database, you have a better chance to evaluate what is happening at each step than you would if you used the entire database. In addition, a small sample database makes it possible to test some of the results by hand as they are generated, thereby saving a great deal of time during model verification.

Another important aspect of implementation is determining which intermediate layers will be needed to continue modeling and which should be retained for later model verification. Many GIS practitioners take an extremely conservative approach here, believing that if a layer is created, it probably should be kept, just in case. If you have very large disk space on your computer, this is certainly a reasonable approach. However, if your study area is very large or if you have mounds of detailed data, or even many layers, the result can be a model that quickly exceeds your capacity to store the intermediate layers. This is yet another reason to consider using a small database subset to estimate the amount of data being produced for each intermediate layer. By extending this information, you will be able to decide whether you can systematically eliminate intermediate layers that will no longer be used or whether you need a larger storage device.

Here it is appropriate to state another caution, one that was emphasized when we talked about the input and storage/editing subsystems: back up your data. Whatever media you use for backup, the elemental layers, the intermediate layers, and the final products should be included. During model implementation, you will be creating a large variety of useful output. Some of it will be needed to make refinements to the model during implementation; others you will want to keep in case additional models from the same database are required, and still others will be needed for model verification.

Because the implementation is by far the most exciting part of GIS, it is very tempting to try out many different modifications just to see what they do. To help yourself avoid this temptation, keep in mind that the GIS is capable of producing all sorts of meaningless output. Stick to your flowchart unless you have good reason to deviate from it. Continued model tweaking will frequently result in many layers that cannot be identified later. If you have too large a collection, it can always be purged. Because of the sheer numbers, however, you may wind up discarding needed layers during cleanup operations.

This brings me to my last cautionary note. If you are implementing a GIS model for yourself, be sure to document every layer you produce. Keep records (metadata) of what you have done and of what each coverage represents. Protect your data. This will reduce the danger of throwing out good layers with bad. It will also help to prevent you from unalterably modifying good layers. If you are implementing a GIS model for a client, the same rules apply but are to be emphasized. In most consulting work, the GIS is being operated on by numerous analysts. This means the chance of corrupting finished products is increased. Most modern GIS software has built-in security or administration programs or protocols that will prove very useful during the creation of all those intermediate layers. We will discuss these at greater length in Chapter 15 when we discuss the design issues of GIS.

MODEL VERIFICATION

Cartographic output is an appealing method of communication. The map has a visual impact that allows people to accept the document as a matter of course without questioning either the data from which it was produced or the cartographic or modeling methods by which the product was derived. It should therefore not be surprising that many GIS models are accepted as fact, just as most maps are accepted as fact. This is seldom the case.

The solution is **model verification**. Few examples of model verification can be found either in practical applications of GIS or in the literature. Despite the sparse treatment that cartographic model verification has received, most would agree that it is extremely important. So how do we begin? First, we need to agree on what we mean by verification. The term can loosely describe not just correct models but also useful models. After all, if the results of your analysis are correct but unusable, you have ultimately failed to provide a product. Within this loose definition, three fundamental questions should be asked:

1. Do the data used in the model truly represent the conditions we are attempting to model?

2. Have we combined the model factors correctly, to represent proper factor interactions, thus correctly describing or prescribing the correct decision-making process?

3. Is the final solution acceptable by the users and/or useful to them as a decision-making tool?

Are the data representative of existing conditions? Our first question might appropriately be asked during the input phase of the project, but this is not always possible. Remember that one important reason for flowcharting a model is to determine missing variables. Another is to determine whether appropriate solutions could be derived with the existing data. At times, however, the final models produced are counterintuitive; or as we begin to put the pieces of the model together, we find that the factors we used don't work. Such disappointing results usually mean that something was missing from the original conceptual model. Obviously, it is impossible to evaluate something that wasn't there, even after a flowchart has been produced. Unusual or counterintuitive results might suggest that the model is incomplete or that the variables used were not adequately representative of model constraints. Therefore, it is frequently a good idea to build a small subset of a proposed database and test the model in a small study site that is typical of the overall study area (DeMers, 2002). Such a prototype will allow a quick determination of the model's ability to characterize the overall study area without having to build the entire database.

If the results are positive, we can invest our time in building the complete database and running the model. If, on the other hand, the results are negative, we must begin to dissect the model to ascertain where the problem lies. This is another reason that flowcharting the model is so important. Most often, a problem with a model will appear in one or more of the submodels. In our mountain cabin model, for example, we may experience difficulties with one or more of our physical parameters because very positive mountain cabin site characteristics keep cropping up in areas known to have dense forests. The discrepancy could be due to dated information. That is, our information on vegetation might have been compiled when the stands of trees in question were immature. By updating our information on vegetation stands and running the model again, we should obtain a more realistic outcome.

The foregoing example shows one of the fundamental approaches to evaluating whether data are representative of the important factors—namely, testing them against known conditions. A very good example of this is provided by Duncan et al. (1995), who used bird census data to test whether their Florida scrub-jay habitat model accurately represented the known locations of these birds. The study is particularly useful because it shows that a simple pattern comparison is an insufficient basis for this decision. Instead, these workers used statistical testing to evaluate the results and followed up with a logical discussion of the possible reasons for lack of correspondence.

Another problem that often occurs in determining whether model elements match needed real-world parameters is missing variables. Williams (1985) clearly indicates the possibility of a GIS model with missing variables. Some of these variables may simply be unavailable and hence must be eliminated from the model. In such a scenario, the final model should state which variables are unavailable and acknowledge explicitly that the model is not a complete depiction of the whole. In some cases, the variables are not missing but rather are either too vague or ill defined to be used or are by definition aspatial variables that cannot properly be placed in an explicitly spatial modeling context in the absence of a spatial surrogate (DeMers, 1995).

Our next verification parameters involve the twofold question of whether the factors have been properly combined. The first part asks whether the GIS analysis functions are producing the proper results on the basis of the input parameters. Many GIS operate somewhat blindly in their analysis functions because the software vendor fails to provide specific information about how the algorithms work. Research on a variety of GIS software has examined the differences in performance using the same functionality on an identical database (Fisher, 1993). The results suggest that the user might want to test these different software systems in much the same way. This is seldom possible in an operational setting, however, because of the high cost of purchasing and maintaining multiple GIS software and hardware configurations. An alternative is to produce an artificial database with known parameters—a sort of control database—against which to test each analysis operation for validity of results. This approach, like the former, is time-consuming, and you should consider carefully the importance of embarking on it. If your results, especially for highly complex operations, are extremely sensitive, it might prove to be a useful endeavor. There are few if any guidelines for producing control databases. You must rely on your own knowledge of the available data and your notion of how you would expect the software to perform.

A second approach assumes that you are comfortable with the software algorithms but are not entirely sure that all the correct steps were performed in just the right sequence. To test the validity of the modeling in this case you might subset a small portion of the model and its elemental database. You then reverse the model by following the flowchart from the trunk to the branches. This should produce the same map elements you started with (Tomlin, 1990). Any step that does not produce the same map elements will immediately become obvious, and you will also know which portions of the overall model need to be recalculated. If you have kept your intermediate layers, you can use them instead of starting the model at the very beginning.

The second part of our model verification task is to determine whether the factors are combined correctly. This is much more difficult. Now we must ask whether the factors themselves interact in a manner that actually simulates or models reality. Because of the nearly infinite number of combinations of factors and model types, there is no simple recipe. Here you must rely on your knowledge of the environment you are attempting to model and on your intuition. In most cases, the modeler becomes so familiar with both the environment and the data representing the environment that any discrepancies between the data and the model are immediately apparent. This is identical to an evaluation of whether we are using the correct parameters. If the model seems to be counterintuitive, there is a good chance that something is wrong. If you have eliminated bad data as a possibility, and you know that the algorithms were operating correctly, you are left with the possibility that the model itself is incorrect. In some cases, incorrect or counterintuitive results may, as in statistical analysis, result in a further investigation of how the environment actually operates. In this case, the GIS becomes a useful tool in advancing science as well as implementing it.

Finally, the third major category of model verification concerns the usefulness of the model as a decision tool. There are many forms of output, not just the cartographic output. In some cases, a tabulation of the results might be more useful to the client than a map. Say, for example, that the client is the publisher of a newspaper that is trying to increase circulation. The analysis might produce a list of local residents who are not current subscribers, together with their telephone numbers. The tabulated results could, of course, be used for telemarketing purposes. In this case, a list would be far more useful to the client than a map showing a distribution of circulation.

Or suppose that your analysis yielded a range of numbers of people infected with a virus for each county of a state or region. Some clients would prefer to see a tabulation of these results from high to low rather than a choropleth map. This might be so for at least two reasons. First, users who are uncomfortable with map output, or unfamiliar with choropleth maps, may find the results difficult to interpret. Second, users may want to see the raw data, at least prior to cartographic classification, with a view to determining the critical values for classification.

Still other forms of output are possible. In a few communities, GIS is being used for emergency services. In addition to producing a map showing the shortest route to a fire or other emergency, the system can send an electronic alarm signal directly to the nearest emergency service, therefore shortening response time and, it is hoped, saving lives or property. Although this is perhaps an unusual example of GIS output today, as the technology changes and as more and different clients begin using GIS, such output will become commonplace.

One other aspect of model verification that needs to be considered is the acceptability of the results to the user. This is especially important if your model is not for your personal use. Clients may be searching for particular results to coincide with their intuitive or preconceived idea about what a correct solution is or because they intend it to serve their political or legal mandates. If a client has provided data to be included in the model and has indicated how the model should perform, it might be assumed that the model will meet these expectations. If, for example, you have included values from the users, such as the quality of a particular parcel of land for a preselected land use, you are going to be limited to using those values. The client will probably also expect the handpicked values to produce model output that will either favor or discourage using the land for the designated purposes. In such a case, the GIS is being used to validate the client's existing attitude and to document its derivation. If the results are not what the client wanted, you may have to backtrack through the modeling steps, as before, to see where the model may have departed from expectations. In some cases, you will be asked to modify how you combined variables (often a form of constraints relaxation), and in others, you may be able to teach the client that the initial view was incorrect (DeMers, 1995).

In another situation, clients may require you to manipulate the model to obtain a particular answer in support of a legal mandate to, for example, protect a particular piece of land. Again, the client is looking for a means of validating a decision that has already been made. In any event, you should remember that regardless of whether your model exactly coincides with the client's view, the client is not obliged to accept the results you have produced. These realities, though not always pleasant, should be considered when you decide on your clientele. You are also encouraged to keep abreast of the ever-increasing literature devoted to ethical questions.

Terms

cartographic modeling	inductive models	prescriptive models
deductive models	model verification	spatial information product (SIP)
descriptive models	predictive models	
flowchart	prescribe	

Review Questions

1. With so many commands and so many options to each command available in a GIS, how can anyone possibly decide how to build a model?

2. What do we mean when we say that the process of cartographic modeling is cyclical? Within your own area of interest, create a scenario that illustrates this process.

3. What advantage does deductive cartographic modeling have over inductive cartographic modeling? When might you find inductive cartographic modeling most useful?

4. What is the difference between explicit and implicit variables? What is the difference between implicitly spatial and explicitly spatial variables? What are surrogate variables and how are they used in cartographic modeling?

5. What is the purpose of flowcharting cartographic models? What advantages does this process have over creating cartographic models without them?

6. What do you do if your cartographic modeling constraints are so severe that the model cannot reach a reasonable solution? What are two possible ways around this problem?

7. What do you do if your cartographic modeling constraints are so loose they give results that are too broad to help you make your decision?

8. Why are weights so important to the creation of cartographic models? Why are some factors preemptive rather than simply weighted higher than others? Can you give examples?

9. What are the three basic areas of cartographic model verification? Suggest some methods of model verification for each.

10. Give some examples of cartographic output that might not be as useful as alternative methods or output from cartographic analysis.

11. Why is it important to prototype models before they are implemented on the entire database?

12. What should we keep in mind as we begin model implementation? Why are experimentation and reliance on serendipity not good practices when we are implementing a model?

References

Berry, J.K., and C.D. Tomlin, 1984. *Geographic Information Analysis Workshop Workbook.* New Haven, CT: Yale School of Forestry.

Carlson, S.A., and H. Fleet, 1986. Systems Applications Geographic Information Systems (SAGIS) and Linked Analytical/Storage Packages Including Map Analysis Package (MAP), Image Processing Package (IPP) and Manage, A Relational Database Management System. Workshop Materials, Annual Meeting, Association of American Geographers, Minneapolis, MN.

Davis, J.R., 1981. "Weighting and Reweighting in SIRO-PLAN." Canberra: CSIRO, Institute of Earth Resources, Division of Land Use Research, Technical Memorandum 81/2.

DeMers, M.N., 1988. "Policy Implications of LESA Factor and Weight Determination in Douglas County, Kansas." *Land Use Policy,* 5(4):408–418.

DeMers, M.N., 1995. "Requirements Analysis for GIS LESA Modeling." In *A Decade with LESA: The Evolution of Land Evaluation and Site Assessment,* F.R. Steiner, J.R. Pease, and R.E. Coughlin, Eds. Ankney, IA: Soil and Water Conservation Society, pp. 243–259.

DeMers, M.N., 2002. *GIS Modeling in Raster.* New York: John Wiley & Sons.

Duncan, B.W., D.R. Breininger, P.A. Schmalzer, and V.L. Larson, 1995. "Validating a Florida Scrub Jay Habitat Suitability Model, Using Demography Data on Kennedy Space Center." *Photogrammetric Engineering and Remote Sensing,* 61(11):1361–1370.

Fisher, P.F., 1993. "Algorithm and Implementation Uncertainty in Viewshed Analysis." *International Journal of Geographical Information Systems,* 7(4):331–347.

Forman, R.T.T., and M. Godron, 1987. *Landscape Ecology.* New York: John Wiley & Sons.

Lucky, D., and M.N. DeMers, 1987. "A Comparative Analysis of Land Evaluation Systems for Douglas County." *Journal of Environmental Systems,* 16(4):259–277.

Tomlin, C.D., 1990. *Geographic Information Systems and Cartographic Modeling.* Englewood Cliffs, NJ: Prentice-Hall.

Tomlin, C.D., and J.K. Berry, 1979. "A Mathematical Structure for Cartographic Modelling in Environmental Analysis." In *Proceedings of the 39th Symposium of the American Conference on Surveying and Mapping,* pp. 269–283.

Tomlin, C.D., and K.M. Johnston, 1988. "An Experiment in Land-Use Allocation with a Geographic Information System." In *Technical Papers, ACSM-ASPRS,* St. Louis, MO, pp. 23–34.

Williams, T.H. Lee, 1985. "Implementing LESA on a Geographic Information System—A Case Study." *Photogrammetric Engineering and Remote Sensing,* 51(12):1923–1932.

UNIT 5

GIS OUTPUT

CHAPTER 14

The Output from Analysis

O utput is the final product of any analysis. If our output is unintelligible, we have failed in our mission. After all, our purpose is not just to analyze geographical data but also to provide a tool for decision makers. This requires us to effectively communicate the results. We will examine the technology of GIS output as well as some basic design criteria for producing good quality, understandable output. Both of these subjects are important because both affect the way we communicate to the user. The technological aspects of GIS output are important because the limits of technology impose physical constraints on our ability to produce the kinds of output we need. Beyond the technology, however, we must be aware that users of GIS output have psychological and biological limits and biases that affect the way they interpret the results.

When you are finished with this chapter you should be able to:

1. Understand the primary concern for map design in GIS work and why it is important.

2. Describe the conflict between intellectual and aesthetic map design considerations.

3. Understand the use of the three basic design principles and be able to use them in designing and organizing maps.

4. Show examples of the possible map design constraints and considerations for two- and three-dimensional GIS output.

5. Suggest possible uses of cartograms as GIS output.

6. Identify some types of noncartographic output and understand the design considerations for them as GIS output from analysis.

7. Explain the impact of modern technology on GIS output design.

8. Discuss the role of the Internet with regard to the types, immediacy, and methods of delivering GIS output.

OUTPUT: THE DISPLAY OF ANALYSIS

The final step in analysis is to display the results so that the user can quickly and effectively understand what has been done and what the results mean. Output from GIS analysis is produced as either permanent or ephemeral products, largely depending on the nature of the output devices. Permanent implies that the output is printed or plotted on a tangible medium such as paper or Mylar, or even saved on magnetic or laser disk media for storage. These media are used to preserve the output for an extended period of time, and the form selected depends on whether the results are to be displayed or archived and stored. Ephemeral is applied to text and image output that most often will not be displayed for long periods of time but rather will be viewed, perhaps to demonstrate the results of analysis or, alternatively, as a method of reviewing files in the course of deciding whether they will be used for further analysis or permanent output.

GIS output can also be classified according to whether it is produced in human- or machine-compatible form (Burrough, 1986). Machine-compatible forms of output are most often used as a means of storing material—in which case we move from the output subsystem back to the storage and editing subsystem. Human-compatible output is most often meant to be more permanent and, more importantly, to be viewed and understood by an audience. Although machine-compatible forms of output are important and will require decisions about media, data structure, and volume compatibility between and among computer systems and their storage devices, the human-compatible forms most often cause difficulties. This is primarily because the types of software and hardware are relatively simple in comparison to the variety of human users, with their dissimilar backgrounds and unequal levels of graphic understanding. Therefore, we will concentrate primarily on human-compatible forms of output. We will include permanent and ephemeral types of output because both can be used as devices for communicating results to human users.

At the outset, you should be reminded that GIS output is primarily but not exclusively cartographic. Because the traditional input to a GIS is mapped or mappable data, it is not unexpected that the primary output consists of maps. Indeed, for the next few years, or even decades, this trend is likely to continue. Thus we will spend a large portion of this chapter on cartographic output, but we will also look at some alternative methods, and we will discuss reasons for using these alternative output methods instead of, or in addition to, the map.

CARTOGRAPHIC OUTPUT

As we have seen many times, the map is still the most compact form of communicating spatial information. When we discussed the paradigm shift from pure communication to a combination of communication and analysis, we made it clear that analysis without communication is a serious mistake. The GIS allows us to create many maps in many different forms based on a wide variety of analyses and on an increasingly large set of spatial data. The advent of readily available databases is, in large part, a response to the increasing availability of GIS software. This power makes an understanding of map design more important than ever, especially because the majority of GIS analysts today have limited experience with cartographic production and design. More important, because there are now far more users of maps of many different types, the requirements for effective communication become even more necessary.

The object of creating a map is to elicit from viewers a response that allows them to understand the appearance of the mapped environment (Robinson et al., 1995). This objective is, in large part, a function of the **purpose** of the map and is linked closely to the intended **audience**. The vast numbers of data types, map styles, and available symbols, scales, typography, and other iconographic devices are meant to operate together. Because most GIS output maps are not of the **general reference map** type, which strives to display, on a single map, a wide variety of different geographical phenomena, we will focus on the second basic category of maps—thematic maps. These maps usually concentrate on the representation of the structural relationships of a selected theme or subject. In the case of GIS output, we could easily replace theme with solution because the map output will most often be an answer to a predefined question or the key to the completion of a decision-making process.

The general process of thematic map design involves the selection, creation, and placement of appropriate symbols and graphic objects to show explicitly the important features and spatial relationships of the objects under study. Most often, however, the thematic map will also need a reference system as a comparative framework for locating these thematic objects in geographic space (Robinson et al., 1995). In some cases—for example, when satellite data are the primary graphical objects—a framework becomes especially important because the reader may be relatively unfamiliar with the color representation of earth features through this medium and will not be able to interpret the data without a reference system that includes notations of well-known objects. In most other cases, the reference system provides a method of demonstrating the areal extent of the thematic objects and their relationships to other spatial locations.

Beyond the inclusion of a reference system, perhaps the first rule you should remember about designing your GIS thematic map output is that despite the inherent beauty of the cartographic form, you are designing the map first and foremost to be read, analyzed, and interpreted (Muehrcke and Muehrcke, 1992). Simply put: eliminate needless objects. Any object placed on the map should say something about the spatial distributions and arrangements. An object should not be placed on the map if it has any objective beyond serving the function of the documents. There should be none of the extraneous objects or fancy flourishes typical of maps showing the location of hidden pirate treasure.

Even if you eliminate extraneous objects from your thematic map design, many decisions remain to be made. Map design involves selecting, processing, and generalizing your output data purposefully and with forethought, and then using appropriate symbols to portray them in a way the user can understand easily (Robinson et al., 1995). The decision process entails a sometimes difficult mix of art and science, impression and logic. To make the best use of these components, you should produce a number of preliminary sketches of how your map will look before beginning the actual design. Although you may have access to map-composing software, either separately or as part of the GIS, it is a good idea to begin with pencil and paper, outlining which objects might go where and how they will be represented. This manual operation will save time and effort later, when you are using your software, since sketching is generally easier and very often faster than placing, moving, and relocating complex graphic objects on a screen. Doing the work by hand also focuses your attention on the choices that have to be made and the look you are trying to achieve for the map, rather than on the commands or operations needed to perform the various tasks.

As you might have gathered, map design can be very complex, especially with the large number of options available using the computer. You will also find that the logical or intellectual objectives and the graphic or visual objectives of map design will sometimes be in conflict. These conflicting demands on map production indicate quite clearly that design considerations are nearly always resolved by compromise (Robinson et al., 1995). For example, if you are trying to place house symbols on your map, both logic and aesthetics dictate a consistent approach. For example, you might want to place each house symbol in its exact (analog) position. But if you have a line symbol representing a major stream or river that runs very near the house symbols, either the stream symbol or the house symbols must be adjusted. Under the circumstances, you might prefer to move all the house symbols as well as the road symbol to the right of their actual locations to changing the location of the stream (Figure 14.1).

The choice made in Figure 14.1 reflects the fundamental belief that the location of a physical landscape object takes precedence over a typographical object that does not precisely indicate the actual location of a physical object. Over the generations, cartographers have dealt with many hundreds of thousands of similar conflicts and have therefore established a rather exhaustive set of **conventions** and traditions. They are a result of trial and error and, perhaps more importantly, testing among map users. Their results show the most effective means of compromise. These should act as guidelines for your own cartographic representations. Deviations from established norms will almost always result in a map that is less effective than it could be at communicating the results of your analysis. We will now look at the basic processes, perceptual

True locations of objects   Displaced locations of objects

Figure 14.1   Symbol placement compromise.
Closeup of a portion of a map that illustrates the compromise necessary to allow placement of the stream symbol adjacent to house symbols. One set of cartographic objects must be physically moved to accommodate the other.

considerations, and design principles needed to produce effective map products. You are also encouraged to consult frequently with textbooks on general cartography or more specific texts on map design.

THE DESIGN PROCESS

A time-honored approach to any graphic design process is to begin with knowledge of the basic cartographic elements and symbology as discussed in Chapter 3, combined with a mental image of maps you have seen that communicated the kinds of information with which you are currently working. The first step is to visualize the type of map you are going to create, the objects to be placed on the map, and a basic layout. This initial stage is an intuitive one and results in a rather general plan for your map. As stated earlier, it is probably best to sketch this out freehand on a piece of paper (Figure 14.2a) rather than using the computer, but this is not absolutely necessary.

As your plan develops, you begin to consider which symbols will be used to represent the objects, and you decide on class limits, colors, line weights, and other graphical elements. Again, it might be best to note these ideas on a pencil sketch before you begin using your software, or at least while you are placing objects on screen. Remember, many changes are likely during this second stage, and having notes available frequently proves useful. Even experienced GIS practitioners generally have a pencil sketch nearby at this stage. It is customary to note measurements among objects on the sketch to ensure that when the software is used for their placement, exact numbers can be input to the software (Figure 14.2b). Even if the objects are later moved, the original numbers will likely be referred to while adjustments are being made.

Figure 14.2 Preliminary map composition. *(a)* Hand-drawn and *(b)* computer-generated preliminary sketches showing the general object types, their approximate placement, and a general view of how the map will be composed. The computer-generated sketch includes measurements that indicate distances provided for objects; these notations are useful when the design is to be finalized with the geographic information system (GIS) software.

Figure 14.3 Detailed map compilation. *(a)* The map as it might have appeared on the graphics monitor. *(b)* Initial plotter output with text overprinting and a font shift. These problems are most often caused by differences in graphic language between the software for the geographic information system (GIS) and for the output devices.

The final stage in the design process consists of fine-tuning the work you did in the preceding stage. This stage should employ only minor modifications to the graphical plan developed earlier. Among the more important considerations here, however, is the development of design prototypes on your monitor prior to hard-copy output. Printers and plotters are often many times slower than the graphics monitor and proportionately slow the process of prototyping. Another frequently ignored factor to be kept in mind in the final stage is the visual relationship between your monitor and your output device. Most hard-copy output devices have their own graphical language that can result in a map that looks somewhat different from what you see on your monitor. It is not uncommon to find that a text font used on your monitor is not available on either the printer or the plotter. The plotter may give you text that looks totally different from what you had anticipated or runs one set of text over another. Colors that appear to be easily separable on your monitor may look nearly identical when output on the hard-copy device. Or fonts that you selected with your software may produce unexpected results because the codes for fonts available for the hard-copy output device are different (Figure 14.3). Most software and hardware vendors provide printed versions of symbols, fonts, and color palettes. You should consult these before you finalize your design, but remember that the mix of sizes, shapes, and proximities, not to mention output material responses to inks and wax output material, will all have an impact on the appearance of your map. When you are finished with your design prototype, it is wise to test the output by plotting or printing a small version, or perhaps a small portion of the map, to examine it for these possible pitfalls. Final adjustments can then be made before you print the entire map.

MAP DESIGN CONTROLS

Design principles are dependent on or influenced by a number of external forces known as **controls**. These controls determine the nature of the map produced, the types of graphical element used, and the design principles that will come into play as you draft your final output. To prevent design problems and conflicts later on, you need to keep all these considerations in mind from the outset.

The first and most essential control on map design is the purpose for which the map is being prepared. The **substantive objective** relates to the nature of the data and information you are attempting to display. Thematic maps produced by GIS analysis are meant to present the distributional form of a specific set of analytical outcomes. Here, simplicity is very important. The more focused your graphical output, the more easily understood. With the lower cost of producing maps from a GIS as opposed to traditional manual cartography, the tendency to produce highly complex maps is considerably less prevalent. Still, the large volume of data, combined with the possibility of multiple analyses from a single database, might create a tendency to put too much information into a single map. Your results will be much easier to understand if you present the results of one analysis on each map you produce. If multiple analyses have been performed, you can always combine several maps on a single output, thus maintaining the purposeful identity of each map.

The other aspect of map purpose, called **affective objective**, relates less to what is to be presented than to how the data are presented. Once you have decided what each map is to represent, you must decide on a form of presentation that will convey the appropriate message. If, for example, the map output is meant to illustrate potential hazards, your map design should make it clear that the hazardous areas are important. The output should be stated boldly and decisively in a way that does not disguise the dangers at hand—pleasant appearance might result in an artistically beautiful map product, but it would send the wrong message to the map reader.

A second control on map design is **reality**, meaning that every area mapped has its own character that will place limits on how well you can employ your design criteria. Selected study areas for GIS analysis are frequently locations with highly complex physical, transportation, social, or economic structures. Indeed, these are the areas that need spatial data analysis most. The complexity of soils, road networks, distribution of socioeconomic phenomena, or even the shape of the study area may place serious constraints on name placement, symbol sizes and styles, shading patterns, and the like. Important objects may not always appear in prominent locations on the map, or the numerical differences among variable measurements may be very subtle. Because of the inevitability of such restrictions on the display of your graphical hierarchy, you should become familiar with the physical layout of your study area as well as the nature of the data before you begin the design process. Knowledge of the restrictions placed on your design capabilities will allow you to plan ahead and anticipate solutions before you reach the final stage, when change is difficult.

The nature of the **available data** is a particularly difficult problem in GIS output because the GIS is a spatial analysis tool rather than merely an aid for cartographic output. Most GIS databases produce very large datasets, often with equally large ranges of variables. There are at least two reasons for this enormity of output: first, the purpose of GIS is to analyze data; and second, such analysis is frequently performed for large territories. It is not uncommon to see a single map of land use that contains over 100 categories, even though the sheer amount of data creates difficulties. Selecting 100 color categories that are separable to the map reader is a particularly nasty design task. You may have to modify your design procedure to include a subdivisional hierarchy with different color or shading pattern sets for each subdivision (Figure 14.4).

The opposite problem may be encountered with available data as well. That is, data either are not sufficient to be portrayed (or analyzed) easily, or they are old and were not collected according to rigorous protocols. They may also have been collected in such a coarse sampling scheme that their positions may not accurately portray where they actually occur on the ground. Take the example of a temporal comparison of vegetation maps from century-old surveyor's notes versus a 1990s vegetation map compiled from LANDSAT Thematic Mapper (TM) data. The historical surveyor's maps are likely to contain a few broad categories of vegetation associations, frequently biased by the surveyor's experience. In addition, these associations are often based on a combination of the structure of the vegetation and species type. By contrast, the TM data will reflect a high number of classes developed through digital classification algorithms and based mostly on vegetation structure as observed by the satellite. Performing an analysis of vegetation change between the two is difficult enough, but displaying the results in map form presents an entirely new set of problems. Because TM data are limited to 30-meter grid cells, the spatial accuracy will be limited to what the satellite is capable of viewing. In the case of the surveyor, the spatial accuracy may be quite good, but the classification of the vegetation will likely be much coarser than that obtained from satellite data

Figure 14.4   Map category confusion. Geographic information system (GIS) output showing a large number of land use categories. The visual confusion results from a lack of hierarchy separating the categories. Because people can seldom visually separate more than 8 or 10 shades in a given color, a map like this prevents users from determining what category name belongs to which polygon.

interpretation. In addition, because the TM data are classified **pixels**, the final vegetation change map is going to be rather blocky as well. Awareness of such limitations before you begin to design will prove invaluable.

Presentation is also, in part, dictated by the **scale** at which the map is to be produced. As map scale decreases (i.e., as you take a more regional view), so too will the amount of detail you can place on the map. Decreasing map scale does not simply mean that you make all the objects proportionately smaller. Were you to do so you would soon reach a point at which none of the symbols would be observable. Instead, you need to select the data you will represent based on the planned output scale. Selection, simplification, and generalization will become important as you design a map for differences in scale of output.

Although scale is a factor in allowing the viewer to perceive the map distributions, the audience for the map is also important (Robinson et al., 1995). Because map reading skills vary widely, the map should generally be as easy to understand as possible, with an emphasis on important shapes, names, and pictographic symbols. Alternatively, if you know that you are dealing with experienced map users, you may choose to employ a higher density cartographic representation with more abstract symbols (Robinson et al., 1995). Age is also a factor in presentation because older people often have more difficulty viewing small symbols and text. Some younger readers, too, have partial sight impairment and need larger symbols and text, and people who are colorblind can process black-and-white formats best.

Conditions of use will play an important role in how the map is designed. This category is more far-reaching than whether the map will be displayed as a poster-sized document or put into service as a small, desktop piece, and the full set of viewing conditions anticipated should be accommodated before the design is finalized. If your map is to be used under conditions of poor lighting, for example, the shades, colors, and symbol sizes should be more prominent than would be necessary for good lighting conditions. In addition, maps that will be used in the field, where outdoor conditions would destroy traditional paper, must be laminated with a plastic coating. During the Gulf War in 1991, for example, the National Geographic Society provided the military with hundreds of laminated maps of the region. Each set of viewing conditions should be considered before the final design is developed.

Finally, **technical limits** will affect the approach to map design. In GIS output, the manner in which the map is to be produced is most often controlled by hardware and software capabilities. For example, if you are using a low-end color ink-jet printer, your color combinations may be limited to as few as 16 colors. Or perhaps you have only black-and-white output devices, restricting your design possibilities to the gray-shade environment. In addition, the resolution capabilities, often measured in dots per inch or lines per inch, will affect the size of symbols and text alike. All these factors should be considered prior to commencing map design..

NONTRADITIONAL CARTOGRAPHIC OUTPUT

Traditionally, thematic maps have been seen as orthographic forms that present the display as if the map reader were observing the map from directly overhead. These maps have also concentrated on the production of graphical forms on the basis of their relative shapes and sizes as seen in the real world. Although these forms of thematic cartography are still among the most frequently used, they are far from the only ones available, especially with the computer graphics devices available with most commercial GIS.

Among the first vector graphics display options used for presenting somewhat nontraditional thematic maps were **fishnet maps** or **wire-frame diagrams**. This now prevalent form of cartographic output produces an impression of three dimensions by changing the viewer's perspective from vertical to oblique. Although these outputs are not often useful as analytical devices, they are extremely powerful at communicating the output of analysis, especially where topography is involved. Three additional design factors must be kept in mind when designing these oblique views: **angle of view**, **viewing azimuth**, and **viewing distance**. There is ample literature from which to obtain details concerning these three aspects (Imhof, 1982; Kraak, 1993), as well as other factors such as sun angle (Moellering and Kimerling, 1990) and the impact of DEM resolution. In general, the importance of each factor depends on the perception of size, vertical relief, and the portion of the study area that is most important to the map reader. Today's graphical software allows us to visually overlay other coregistered map data on these apparent three-dimensional maps. Thus the viewer can readily identify relationships between topographic surfaces and other factors such as land use or vegetation.

Today, some vendors also offer an alternative to static viewing of these and many other types of maps. For two-dimensional maps, animation allows the viewer to inspect dynamic systems as they are seen in real life (Moellering, n.d.) or to illustrate phenomena that would otherwise not be perceived because their natural motion is too slow (Töbler, 1970). In addition, displays can be stopped, slowed, or even reversed to obtain additional insights. Three-dimensional maps have for some time been viewable as animations as well (Moellering, 1980). With the advent of the **virtual reality markup language (VRML)**, it is now commonplace to visualize 3-D terrain in quite realistic settings (http://www.dragonbbs. com/members/db7641/vrml [accessed July 11, 2003]). The effect is one of an airplane flying over terrain that is covered with both natural and anthropogenic patterns, giving the viewer an added dimension from which to analyze possible data relationships. The power of this technique of visualization should not be diminished. Because the human eye was designed to notice moving objects more readily than static objects, the use of computer animation of all kinds can only enhance a viewer's ability to see pattern interactions and to formulate both questions and answers about the phenomena studied.

If your system lacks the computing resources to manage computer animation, there are still a number of useful variations on thematic maps that can enhance the viewer's understanding of your results. Shaded relief maps are particularly good tools for visualizing topographic surfaces, without the requirement of creating three-dimensional maps. The shading patterns themselves give an impression of shadow produced by the interaction of the sun with the topographic surface, thus making the map look more like a photograph. Digital orthophotographs can also be used as background to illustrate the relationships of mapped variables to the easily perceived aerial photographic image. For topographic as well as subsurface information, cross sections can be produced along any transect the user wishes. Even block diagrams are easily produced and provide a visually interesting output while displaying both surface and subsurface features simultaneously. The list of options and combinations of these approaches to nontraditional cartographic output is nearly endless.

Finally, the GIS analyst should consider a set of nontraditional cartographic forms that has been relatively neglected among the community of users. The graphic devices called **cartograms** have the appearance of maps, but the spatial arrangements of the depicted objects have been modified not by their actual locations in space but rather by the changes in value of the variable being measured. Thus the distances, directions, and other spatial relationships are relative rather than absolute. Cartograms are seen frequently and give evidence of their usefulness. For example, many urban mass-transit networks are posted inside the municipal buses and trains. These maps plot the relative sequential locations of each connecting stop on a straight line, as opposed to showing the actual route taken. In addition, the spacing between stops is not necessarily a true depiction of the scaled distance between each stop. These conventions simplify the map and emphasize the most important aspect of the spatial relationship—in this case sequence rather than distance. While riding in a bus or subway car, your most important concern is where you want to get off, perhaps followed by where the transit line you are traveling on will connect with the next (Figure 14.5). Such **routed line cartograms** are also found in many road atlases that indicate, by using straight lines rather than road networks, both the distances between major cities on the map and the estimated travel times.

Cartograms modify geographic space to produce an easily understandable model of reality. Another form of cartogram, the **central point linear cartogram**, could easily be used to modify the output from a functional distance model used by your GIS (Bunge, 1962) (Figure 14.6). Among the most used forms of cartogram are **area cartograms**, which vary the sizes of each mapped study area on the basis of the value being examined. A classic example of the use of this technique was presented by de Blij and Muller (1994), who modified the physical size of each nation to illustrate the amount of population rather than the physical land area. An area cartogram can be either contiguous, where all the areas are touching, or noncontiguous, where the areas are not touching (Campbell, 1991) (Figure 14.7). The latter type is sometimes called an exploded cartogram and has the advantage of not requiring the lines to be connecting, thus simplifying the production of the output.

The interpretation of cartograms may require a minor rethinking by users unfamiliar with this form of output. Still, the time spent should be minimal. And once users become fluent in the reading of cartograms, the results generally are well received. However, many users will want to have traditional cartographic forms against which to compare this radically different device of graphic communication.

You may not see many examples of cartograms as you look over the output from GIS analysis at conferences and meetings. This is partly because most members of the GIS community either are not familiar with the form or are not used to producing it. A second—and perhaps more important—reason for the relative lack of cartogram output is that few vendors offer software that readily performs this task in an automated environment. Tobler (1967) developed a number of FORTRAN programs designed to automate the process. As you become more familiar with your software and as you gain confidence in your abil-

Figure 14.5 Linear cartogram. Linear cartogram for the Chicago transit system. *Source:* Chicago Transit Authority, used with permission.

ity to create macro programs for unique tasks, you may want to refer to Töbler's programs, modifying them to suit your own needs or the macro language used by your software (Jackel, 1997). In any event, the use of cartograms as an alternative to traditional map output is certain to increase as both GIS operators and users become more familiar with them.

GIS ON THE INTERNET

Cartographic output, whether traditional or nontraditional, has proven quite useful for many decision-making situations. Its limitation is one of distribution and

Figure 14.6   Central point cartogram. (From Bunge, 1962.) *(a)* The illustration shows a conventional representation of travel times from central Seattle in five-minute intervals. *(b)* This is a cartogram showing functional distances in five-minute intervals. *Source:* Adapted from W. Bunge, in *Theoretical Geography,* 1962. Lund Studies in Geography. Series C. General and Mathematical Geography, No. 1 (Lund, Sweden: C.W.K. Gleerup, Publishers for the Royal University of Lund, Sweden, Dept. of Geography).

access. Modern digital technology allows maps to be shared over great distances through the Internet. One of the many examples of this sharing is the Topozone web site (www.topozone.com), which allows users to specify specific locations, scales, and formats of data to be retrieved (Figure 14.8). Another example that has proven quite useful is Terrafly (www.Terrafly.com), which allows the user to not only peruse digital imagery data, but to scroll or "fly" through the data from place to place, and change the spatial scale as well. These two commercial examples demonstrate a distributed environment, sometimes called a distributed GIS, consisting of a client (the map user) and a server (the map or spatial data provider). Many of these applications are in the first stages of development, and a few are relatively well developed, some of which allow the user to download GIS compatible data for insertion into existing databases. Web sites that allow this ready access are available through university web sites, commercial operations like GEOWeb where clients and data distributors share their data, and ESRI's Geographic Data Community where both academics and commercial vendors allow ready access to their archived data in multiple formats.

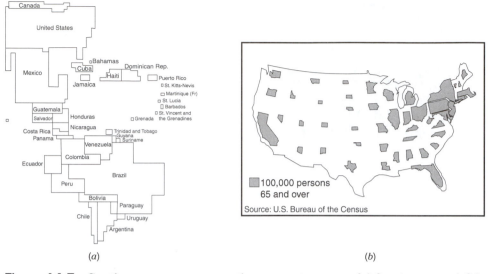

(a)   (b)

Figure 14.7   Contiguous versus noncontiguous cartograms. *(a)* Contiguous and *(b)* noncontiguous value by area cartograms. (The requirements for contiguity apply only to areas on the same landmass; hence, the Caribbean islands are shown separate from each other and from the mainland of the Americas.)   *Source:* Part (b) J.M. Olson, "Noncontiguous Area Cartograms." *The Professional Geographer,* 28(4), p. 377. Blackwell Publishers. Used with permission.

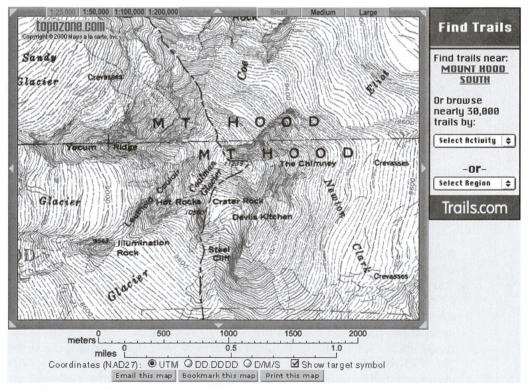

Figure 14.8   Maps on the Internet. A map of Mount Hood retrieved from the Internet.

Beyond data viewing, browsing, and sharing, the GIS output subsystem has been greatly enhanced with the development of GIS application software capable of not only distributing maps in digital form over the Internet, but also allowing an increasing amount of query and analytical capabilities over the Internet. This means that the client is no longer required to own the GIS software at all unless substantially more sophisticated analyses are necessary. Such dynamic maps are evident in the MapQuest® searches which are so commonplace on the Internet. In such simple applications, users are able to locate places of interest, addresses, and even produce travel directions from place to place. In this way the output of GIS is not just the maps but also the added capability to allow the client to search, pan, zoom, and analyze the data contained within them as well, in a dynamic, distributed, interactive environment. As more datasets become available, and as the software becomes more sophisticated, the trend toward distributed GIS will increase. While at present few fully functional distributed GIS operations are available, there will likely soon be a time when much of GIS will be performed at a considerable distance from the server. This is particularly true with the advent of increasingly fast Internet access routes via DSL and Digital Cable. As this activity increases, the already fuzzy lines among the four GIS subsystems will become even fuzzier.

NONCARTOGRAPHIC OUTPUT

Despite the prevalence of maps as a form of GIS output, certain data are better presented in noncartographic form or in addition to the map product. There are two general reasons for this relatively frequent occurrence. Either the map provides output that is less immediately understood because of the audience, or the map is not the desired output from analysis. In the former case, the map usually is supplemented with other forms of output, whereas in the latter, it most often is replaced. A few examples of enhanced or alternative output should prove useful here.

Interactive Output

Perhaps the classic example of replacing map output with alternative types has already been alluded to: namely, the enhancement of emergency services such as the 911 system by means of GIS. In the standard implementation of this system, residents of large metropolitan areas may not get a speedy response when they dial the 911 network to report a fire because there are many fire stations throughout the city that could respond. The 911 operator wants to make certain that the closest station will always be the one to take the call. In the enhanced implementation, information about, say, the location of a fire, is transmitted electronically to an automated GIS system while the operator is still on the phone. The GIS matches the address of the actual road network inside the database. In turn, the program determines which fire station should respond and reacts by immediately sending an alarm signal to that unit, alerting the dispatcher to the current emergency. In this case, the result of the GIS analysis is an electronic response rather than a map.

The program could also be designed to determine the quickest path from the fire station to the fire, and this information could be electronically transmitted to an output device at the appropriate fire station. The output could be in the form of a text-based route, perhaps including a map. Although computer networking and GIS implementation today are seldom so sophisticated, the scenario is not beyond current capabilities. It is much more likely that multiple output would be available under less pressing (i.e., nonemergency) circumstances.

Suppose that you are using the network functions of your GIS to route a moving van from one location to another. This is an analytical task commonly performed by auto clubs for individual motorists as well. If you have used these services, you know that there are generally two types of output. The first is a map that highlights the roads that lead most efficiently from one place to another. The second is a text-based road log that shows, in miles, the distance traveled on each road, the names of the roads, the intersections at which you are to turn and the direction of the turn, as well as the number of interchanges when available. Although the correct route could be discerned by careful attention to any of the individual components, together they provide additional information that makes navigating along the highways a simpler task. Even though many of us are comfortable with a regular map, having the road log along makes us more confident, particularly when the available map is highly generalized. Thus road logs offer a good example of how cartographic and noncartographic output can complement each other to provide a more comprehensive solution to a problem.

Some new automobiles are now equipped with GIS units linked to digital maps that allow for onboard navigation. With these systems, it is now possible to interact with the map to determine your current position and to make decisions as to where you go next. Such systems are becoming more commonplace and indicate an effective use of interactive computer output within everyday use.

Even less sophisticated uses of interactive output from GIS could easily be envisioned with the introduction of GIS in, for example, county land records offices. By keeping current records of landownership records, for example, it would be easy to determine who currently owns a single property. In such a case, realtors or neighborhood associations could search for ownership records by selecting individual properties rather than examining endless text-based records. The system might also increase the speed of data entry and the currency of the data for the records officials as well.

Tables and Charts

There are many other alternative forms of output from GIS (Garson and Biggs, 1992), but among the most common are tables and charts. Earlier we considered a GIS client in the newspaper business and saw that the most useful output would be a tabulation of potential customers. Here, a simple noncartographic output designed to increase circulation could be produced without the use of expensive equipment.

In other situations, the use of tables and charts could greatly improve the understanding of cartographic results. Assume that you are analyzing land use patterns as they change through time in an area that is losing natural habitat and agricultural lands as a result of population pressures. A series of maps might show where the changes from wildlife habitat and farming to urban land uses are most severe. The map legend may include tabulations of total land uses by type for each time period. However, much more specific information is available in the database. The tabulation could be enhanced by simply producing a bar graph under each map to show the total amounts or percentages of each land class. In addition, you might produce a bar graph showing the percentage change for each land class for each subsequent time period. You could improve this simple graphic technique by creating a matrix showing "from" and "to" changes for each time period. Such a table would prove very useful in understanding which types of land use were responsible for the loss of the largest proportion of natural habitat and which were most responsible for the loss of agricultural lands.

The foregoing example illustrates that for any GIS analysis, there are many options for noncartographic output, both in type and in form. Just as cartographic display involves design issues, so too are design issues a consideration for noncartographic output. One simple example is frequently encountered among users of computerized spreadsheets. Nearly all of these software programs allow the user to produce a wide array of different types of tabular output, graphs, and diagrams showing the results of analysis. The design questions that arise are numerous and encompass complexities associated with color perceptions, personal preferences, and the capabilities of output devices. Let us look at just a few of the more important considerations.

Tabular output from GIS analysis is of two general types. The first is contained in the textual material of the map legend and can often be treated as part of the map design process because it is used to connect the attribute data to the entities presented on the map. A second type of tabular output employs lists of attribute data, tables showing relationships between raw attribute data and map entities, text showing extended-data dictionary entities for improved understanding of what the legend means, correspondence matrices showing relationships among sets of attributes, and so on. These outputs can be placed inside the exterior map boundary (the neat line) if they are to be presented as a combined finished product, or they can be output separately for the user to peruse at will (i.e., alone or in conjunction with a separate map).

Design considerations for text-based output involve the basic concepts of purpose, readability, and audience, familiar from our introduction to map design. Before producing text and tabular data, consider the needs of the user. First, will the text-based output effectively replace the map as a communication device? Or will it sufficiently enhance the readability of the map output to warrant its production or inclusion? In our example of the newspaper publisher needing lists of potential subscribers together with their telephone numbers, the answer to the first question is an obvious yes. This is exactly what would serve this client best. Many other situations will be far more subtle, and you will have to determine the value of the tables themselves or the value added to the map by inclusion of tables and text.

If you intend to include tables and text, the next item you should consider is readability. Text and tables are fairly commonplace and more easily understood by the general public than maps are; hence, they tend to be easier to design. Still, a few things should be considered to ensure basic readability. First, the font should be simple in style and large enough to be seen at the desired combination of viewing distance and lighting conditions. Fancy fonts may by themselves seem particularly attractive, but simpler is usually better. Keep in mind the need for contrast when you select the color of your text output. If the tables or additional text are to be included inside the map boundary, be sure that their colors contrast sufficiently with the map background. To further ensure separation, add a border, taking care that it is not so heavy that it conflicts with graphics devices on the map proper. Avoid acronyms and abbreviations whenever possible. The less explanation your tables require, the more useful they will be. For large tables, you may decide to outline major categories, or even all the cells in the table, since these options are familiar to many from word processing or spreadsheet programs. Text or tables lying inside the map boundaries should be separated with a border, which should be functional but not so heavy that it conflicts with other graphics devices on the map. Finally, employ the same common sense in depicting your data that you used in designing your map.

As you produce text and tables, you should also remember the nature of your audience, using terms that will be familiar to the client. In addition, the type of table you create should be useful as well as understandable. Returning to our example of land use change, a set of tables indicating the amount of change in each category for each time period may not contain as much information as a change matrix table or to–from matrix. However, if the client is interested only in the amount of increase or decrease in a select few land use types, the additional information from a matrix may be both uninteresting and generally confusing. The text or table should say something about the phenomena studied, but no more than what the reader needs. Alternatively, if you or your client really need to know both the amounts and the directions of the land use changes, a change matrix might be entirely appropriate. Before you produce your tabular output, be specific about what you or your client needs to know and how best to present it.

Tables can easily become too large or cumbersome to be useful. Monstrosities with 40 rows and 40 columns are not likely to be useful because they require far too much time to examine. An easy alternative is to present the results not as tables but rather as graphic representations of tabular data. Most common microcomputer spreadsheet programs illustrate a wide variety of traditional business graphics devices for the presentation of data: line graphs, pie charts, bar graphs, and the like. There are, however, far more options available to represent the phenomena that can be output from GIS analysis. Among these options are triangular graphs representative of, for example, soil textures; climographs indicating the relationships among soil moisture, temperature, and precipitation variables for a region; and wind roses or star diagrams indicating predominant directions of wind, glacial flow, or other phenomena for selected points (Monkhouse and Wilkinson, 1971). It would be futile to try to examine each in detail, or even to consider all the possibilities. Instead we'll look again at some basic considerations for the more general types of line graph, trusting in your own background and experience to select the appropriate specimens for the data with which you are working.

The most common line graphs employ a Cartesian coordinate system with the standard ordinate (vertical scale) and abscissa (horizontal scale). Another, less common type of graph is the polar or circular graph, where the points are plotted on the basis of a combination of angular bearing (the vectorial angle) and distance from a point of origin (the radius vector). In one variation, the points of the line graph are plotted on a triangular grid. Such curves, as mentioned earlier, are common when showing the relationships of three variables; in a soil texture triangle, for example, the percentages of sand, silt, and clay are represented on the same graph. For all line graphs, in general, the observed points for the variables, usually classified as dependent and independent, are connected either by individual straight-line segments between pairs of observations or by a smooth curve indicating the general trend along the points. A modification of the standard Cartesian line graph is the histogram or bar graph, which displays the response of each observation by means of a wide band extending either vertically from the abscissa or horizontally from the ordinate. The histogram is more often used when the number of observations is low; it is more generally representative of the noncontinuous nature of many observations. For data that represent continuous variables, such as changes in barometric pressure, the lines themselves replace the points (Monkhouse and Wilkinson, 1971). You are not likely to encounter these true line graphs in GIS analysis, where the spatial dimension is normally stronger than the temporal dimension.

Line graphs can be simple line graphs, displaying a single series of values connected by a line, or they can resemble a polygraph, producing multiple lines, each representing a selected variable. Simple histograms may have a single series of values rising like ribbons, or multiple values, each represented by a ribbon of unique color or pattern. These multiple line graphs are produced primarily to allow the direct comparison of numerous variables under the same circumstances. Multiple line graphs can be produced in two general fashions: as multiple sets of adjacent lines or bar symbols, or as stacked graphics. For stacked graphics versions, the values of each observation are physically combined with those of the preceding values, to give an impression of area. Such plots are especially useful if the values themselves are meant to indicate an additive relationship. In the case of stacked histograms, for example, the height of the bar is based on a combination of all the different variables at a single observation point. Under this circumstance, if the bar was meant to represent the variables %sand, %silt, and %clay, the total bar would be at 100 percent because each of the values is additive.

Designing line graphs has become relatively simple with the power of the computer. Some software, however, reduces our options for output or preselects the ways in which the output will appear. These default settings can, and often should, be overridden by the user. Good line graph design choices for GIS output can be indicated in general with a few simple comments. Although most software offers very colorful and stylish options, such as the appearance of three dimensions for histograms and ribbons for lines, these tend to be unnecessary and generally are less readable. Try to avoid using large, garish symbols as data points because these outsized symbols often interfere with the lines. Unwieldy data points can easily be replaced by numerical values, which are more informative and less confusing. Avoid radical differences in direction of the internal marks used to produce shading patterns, such as diagonal lines next to crossed lines next to diagonal lines in the other direction. Such inelegant graphical design will often result in unnecessary eyestrain. (The same can be said for adjacent choropleth patterns on maps.)

Keep the important lines prominent. Although most software allows you to place grids on your graph, the fewer you use and the lighter the line shade and line width, the more prominently the important lines will show. Make the size of the vertical and horizontal scale extents as close to the maximum value as possible. This is not always easy with available software. You may want to ignore this rule if you have multiple graphs that show relationships between or among different places. In this case, all the graphs should carry the same maximum sizes to avoid confusion, or the false impression that small values are larger than they really are. Finally, remember what you are seeking: simplicity and legibility. Simple, clear fonts, and lines and symbols with good contrast should remain the most important criteria.

The sheer numbers of graph types and options can be somewhat overwhelming. There are a number of good references, based primarily on business graphics and graphics presentations. You should consult these or others like them when you intend to include business-type graphics in your work (Cleveland and McGill, 1988; Holmes, 1984; Meilach, 1990; Robertson, 1988; Sutton, 1988; Tufte, 1983; Zelazny, 1985).

A final note should be made about another form of noncartographic output that is becoming more prevalent in GIS output. Many commercial vendors now include the option of incorporating digital photographs showing, for example, buildings, site characteristics, and specimens of plants and animals found in certain areas. These digital images, often explicitly linked to the attribute database, can add a great deal to the understanding of what is represented on the map by giving the reader a set of familiar examples. The images can either be scanned into appropriate software or photographed directly with the use of digital still cameras or even videotape. More of these images are likely to appear as output from a GIS. Like all graphical and nongraphical devices, they should be used sparingly and only to achieve a desired effect. Because of the graphical complexity typical of these images, it is easy to overwhelm the associated map. And, as always, make sure the images are appropriate and useful and that they add value to the output.

Terms

affective objective	controls	substantive objective
angle of view	conventions	technical limits
area cartograms	fishnet maps	viewing azimuth
audience	general reference map	viewing distance
available data	pixels	wire-frame diagrams
cartograms	purpose	virtual reality markup
central point linear cartograms	reality	language (VRML)
conditions of use	routed line cartograms	
	scale	

Review Questions

1. Why is it important to be familiar with cartographic design in GIS? Why is it even more important with GIS than with traditional, hand-drawn cartography?

2. What is our primary concern when producing a map from GIS analysis? How does this matter relate to the map as an art form?

3. Give a concrete example, other than what is in your text, of how intellectual and aesthetic design considerations can conflict. Show how the problem you describe can be solved.

4. Given the ready availability of computer software for designing and organizing graphics, why should you begin the map design process with pencil and paper?

5. Indicate, through concrete examples, the three basic design principles.

6. Show, with examples, what some of the possible design constraints might be in GIS map output.

7. Give an example of how too many data can produce problems in map design. Give an example of map design problems due to insufficient data.

8. What are some design considerations involved in producing good quality, three-dimensional map output?

9. Give as many examples as you can think of to describe how you might find animation useful as a form of cartographic output.

10. What are cartograms? Why are they not in general use in GIS today? When might they prove useful as output from GIS analysis? Give some examples, perhaps from your own laboratory exercises.

11. List some types of noncartographic output. What are some of the more unique types? Think about how you might decide whether these should replace or augment an existing map.

12. What are the basic design considerations for the production and use of line graphs as GIS output?

13. How does technology impact the design considerations for GIS output?

References

Bunge, W., 1962. *Theoretical Geography.* No. 1 in Lund Studies in Geography, Series C, General and Mathematical Geography. Lund, Sweden: C.W.K. Gleerup, Publishers, for the Royal University of Lund, Sweden, Department of Geography.

Burrough, P.A., 1986. *Geographical Information Systems for Land Resources Assessment.* New York: Oxford University Press.

Campbell, J., 1991. *Map Use and Analysis.* Dubuque, IA: Wm. C. Brown, Publishers.

Cleveland, W.S., and M.E. McGill, 1988. *Dynamic Graphics for Statistics.* Belmont, CA: Wadsworth Publishers.

de Blij, H.J, and P.O. Muller, 1994. *Geography: Realms, Regions and Concepts,* 7th ed. New York: John Wiley & Sons.

Garson, G.D., and R.S. Biggs, 1992. *Quantitative Applications in the Social Sciences.* No. 87 in Analytic Mapping and Geographic Databases Series: Newbury Park, CA: Sage Publications.

Holmes, N., 1984. *Designer's Guide to Creative Charts and Diagrams.* New York: Watson-Guptill.

Imhof, E., 1982. *Cartographic Relief Presentation,* edited by H.J. Steward. Berlin and New York: Walter de Gruyter.

Jackel, C.B., 1997. "Using Arc View to Create Contiguous and Noncontiguous Area Cartograms." *Computers and GIS,* 24(2):101–109.

Kraak, M.J., 1993. "Cartographic Terrain Modeling in a Three-Dimensional GIS Environment." *Cartography and Geographic Information Systems,* 20:13–18.

Meilach, D.Z., 1990. *Dynamics of Presentation Graphics,* 2nd ed. Homewood, IL: Dow Jones-Irwin.

Moellering, H., 1980. "The Real-Time Animation of Three-Dimensional Maps." *American Cartographer,* 7:67–75.

Moellering, H., n.d., "Traffic Crashes in Washtenaw County Michigan, 1968–1970." Highway Safety Research Institute, University of Michigan, Ann Arbor.

Moellering, H., and A.J. Kimerling, 1990. "A New Digital Slope–Aspect Display Process." *Cartography and Geographic Information Systems,* 17:151–159.

Monkhouse, F.J., and H.R. Wilkinson, 1971. *Maps and Diagrams: Their Compilation and Construction,* 3rd ed. London: Methuen & Co.

Muehrcke, P.C., and J.O. Muehrcke, 1992. *Map Use: Reading, Analysis, Interpretation.* Madison, WI: J.P. Publications.

Robertson, B., 1988. *How to Draw Charts and Diagrams.* Cincinnati, OH: North Light.

Robinson, A.H., J.L. Morrison, P.C. Muehrcke, A.J. Kimerling, and S.C. Guptill, 1995. *Elements of Cartography,* 6th ed. New York: John Wiley & Sons.

Sutton, J., 1988. *Lotus Focus on Graphics* (5 vols.). Cambridge, MA: Lotus Development Corporation.

Töbler, W., 1970. "A Computer Movie Simulating Urban Growth in the Detroit Region." *Economic Geography,* 46:234–240.

Töbler, W.A., 1967. *Automated Cartograms.* Department of Geography, University of Michigan, Ann Arbor.

Tufte, E.R., 1983. *The Visual Display of Quantitative Information.* Cheshire, CT: Graphics.

Zelazny, G., 1985. *Say It with Charts: The Executive's Guide to Successful Presentations.* Homewood, IL: Dow Jones-Irwin.

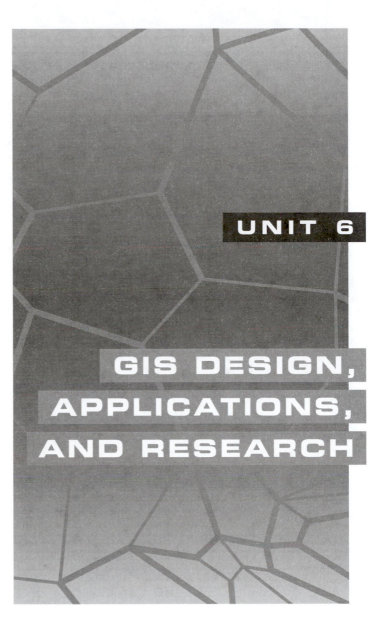

GIS DESIGN, APPLICATIONS, AND RESEARCH

CHAPTER 15

GIS Design

As you become more experienced with GIS, you will see how design flaws contribute to system failures. Some GIS operations are incapable of producing the desired results because of poorly selected or badly organized data, incorrect data models, or software with limited capabilities. In some cases, system managers have underestimated the time involved in developing the databases. Conversely, you also see well-designed GIS that are underutilized, either because system complexity prohibits easy use or because of lack of understanding of the overall capabilities of the software. In still other cases, systems that were once highly useful have lost that utility because of aging software and hardware configurations, aging data, or other shortcomings that rob the system of flexibility required to adapt to changing demands. Finally, some failed systems will be seen as direct reflections of personnel problems that arise when organizational structures are changed forever by the introduction of new technologies into previously manual settings. Much of what we know of successful systems can be learned by examining failed systems, just as successful businesses have learned from disasters experienced by competitors.

In this chapter we focus on the business side of GIS for two reasons. First, the vast majority of GIS clients are institutional: consulting firms; service and utilities organizations, businesses; and governments. In short, most users provide a product or service requiring the application of GIS. Our second reason for concentrating on the business side of GIS is that the success or failure of the system, regardless of user or application, depends on much the same set of criteria used in the business community. If you can make a system work in the budget-conscious business world, it is likely to be successful for other investigations.

Ultimately, your business is the business of digital geography. Although the current subject matter may seem to deviate considerably from what you have learned thus far, success depends on being able to combine business-oriented systems theory and operations research with the basic data models, abstractions, and analytical capabilities of the practicing geographer. The utility of geography is what makes GIS a successful enabling technology. Companies use GIS because they need to answer geographic questions. They employ GIS design experts—liaisons between the organization and the geography they need to solve spatial and spatio-temporal problems.

Design activities come in three interacting classes: analytical model design, spatial and spatio-temporal database design, and system design. Analytical model design attempts to develop effective, realistic mathematical models of spatial settings, situations, and related processes. It endeavors to represent the elements of real-world settings required for decision making. To be effective, we must consider the real world in the context of existing data models and structures. Spatial and spatio-temporal database design seeks to optimize the organization of the necessary spatial data within a computing environment to allow both effective and sustained system functioning. These latter two portions of GIS design are nearly inseparable, while system design, the development or selection of conceptual GIS models that link current geographic information science theory and available technology with spatial problems, more often has an institutional context. This chapter examines all three of these general classes of design activity within two larger categories: (1) analytical and database design and (2) institutional (system) design.

LEARNING OBJECTIVES

When you are finished with this chapter you should be able to:

1. Define what we mean by GIS design as it applies to the success or failure of modern systems, as opposed to the GIS setups of the 1960s.

2. List and define the components of analytical and database design.

3. Explain the close relationship between analytical GIS design and GIS database design.

4. Describe the process of isolating the spatial components of a GIS problem in general terms. Provide a concrete example of how this works within a specific knowledge domain (e.g., designing a wildlife preserve or analyzing a crime pattern).

5. Define system design and discuss why it is important for GIS.

6. List and explain the functions of the internal and external players in the GIS institutional setting.

7. Describe the idea of conceptual design as it applies to GIS and have a feel for how people problems are a primary concern in this area.

8. Discuss what design creep is and enumerate its possible negative effects. Explain how it can be avoided.

9. Define spatial information products (SIPs) and discuss how they can drive the design of a GIS and how they are integrated from local views to global views of the GIS in an organization.

10. Discuss the idea of cost-benefit analysis as it applies to GIS in an institutional setting. In particular, describe how costs and benefits are determined.

11. Understand the major database design considerations, especially with a view toward system verification and validation.

ANALYTICAL MODEL AND DATABASE DESIGN

Spatial Processes and the Scope and Structure of the Model

Fundamental to effective GIS modeling is the recognition of the explicit spatial nature of both the geographic objects and their underlying processes. Failure to do so is a primary reason that high-powered geographic information systems are often underutilized. This is particularly true with regard to processes, but this, in turn, is often caused by a failure to perceive a spatial pattern of existing conditions in the first place. Once a pattern is perceived, it is at least possible to research the likelihood that the pattern is purely happenstance or that a process or set of processes is responsible for it. Although this may be simple in some cases, in others it may require substantial research. In both cases, the first step is to perceive or recognize the pattern and then to describe it.

DeMers (2002) provides some detail on the variety of tools that allow us to perceive spatial patterns, but a synopsis is valuable here. Maps, aerial photographs, and satellite imagery that show a planimetric view of our world are obvious tools for acquiring pattern descriptions. These graphical forms explicitly display the spatial locations, associations, and juxtapositions of objects in geographic space. Frequent exposure to such tools sensitizes the viewer to the spatial nature of the world around us. Alternative tools for recognizing spatial patterns include statistical analysis, charts, and graphs. These tools are particularly useful for relating one set of variables to another. The scattergram, for example, is a commonly recognized method of discovering relationships among variables.

We have already seen how patterns, especially nonrandom patterns, typically both develop from underlying spatial processes and produce changes in these processes. Not everybody is aware of this relationship, however. Many businesses fail because they do not recognize the power of the spatial proximity of their potential customers. Customers that are nearby are generally far more likely to frequent stores that require less travel time than they are to travel great distances for the same products or services, even when the cost is less at a more distant store. This is particularly true when the distance incurs a cost of both fuel and time or where nearby stores cater to the particular needs of its proximate clientele. If you are a GIS consultant working for someone else, you may need to explain this to your client.

Establishing the Effective Spatial Domain of the Model

Our last example showed a relationship between distance and clients. This may also be true between animal predators and prey in wildlife modeling, between criminals and victims in crime mapping, and many other situations. Depending on your specific need, the relationships among spatial variables may not always be this obvious. Before starting a GIS project, it is important to identify which factors are most likely to be important to the system being modeled. In short, we need to be aware of the effective spatial domain for each model we build. An effective and relatively simple way of doing so is to begin with the expected outcome from the GIS and working backward. DeMers (2002) often compares this to envisioning a GIS modeling process as a tree with the trunk as expected outcome, the branches as the general categories of thematic elements, and the leaves as the more specific data elements that are input to the GIS and from which the model is built. Precisely for this reason the development of a GIS flowchart is used to identify possible missing data elements, aspatial components, and poorly understood relationships. The mountain cabin example in Chapter 13 is designed specifically to illustrate the process. Although it explicitly illustrates the flows among data elements, this example does not provide much guidance as to what the computational linkages might be. This requires us to take a more detailed look at the role of the powerful GIS analytics in connecting the data elements.

GIS Tools for Solving Problems

No single volume provides a comprehensive set of recipes or best practices for applying the GIS toolkit to each myriad task to which it could potentially be applied. Such a book would likely be impossible to write because of all the possibilities. However, Mitchell (1999) has created a set of useful tables outlining some of the more common tasks the GIS practitioner might encounter, their conditions of use, types of output resulting from their implementation, and the tradeoffs encountered through their application. (These are summarized on the inside covers of the book.)

Study Area

Study areas are most often driven by a formal decision about why the area is under study. Thus, for a county-level study, the entire county will be the study area. For a study based on a natural phenomenon, such as pollution in a stream, the entire stream drainage basin may be selected as the study area because the pollution will come from all upland areas that drain into all the stream's tributaries. For a study of a lumber company contemplating use of a GIS to keep track of its logging operations and landholdings, the landownership will be the deciding factor for selecting the study area. In short, the study area may be selected on the basis of political boundaries, physical boundaries, or landownership, or it may reflect primarily the economic and data constraints that are found in the early stages of the design process; in any event, its choice is project driven.

If there are detailed data for a small portion of a prospective study area but more general data for the rest, the detailed data need not dictate the extent of the study area. Instead, the larger area of interest can be considered to be the overall study area, and the subarea for which there is more detailed information can act as a prototype for a detailed analysis of the larger study when data become available or affordable. In addition, the detailed data for the subarea might be useful for improving the knowledge of the larger area through the use of density of parts or another dasymetric mapping technique.

Whenever any form of interpolation is scheduled, the study area boundary, at least for the elevational coverage, must be extended sufficiently to ensure that the interpolation will produce the correct results. Finally, the larger the study area, the more expensive in time and money will be the production of the database. Often, cost awareness is a major constraint in defining a study area in the first place. In addition, the larger the study area, the greater the chance of a lack of data for all the coverages needed to perform the analysis. Selecting a smaller study as an initial prototype is a good choice if it will allow the organization to demonstrate the utility of the system. Once that has been effectively done, the system sponsor is much more likely to supply more resources to increase the size of the study area.

Scale, Resolution, and Level of Detail

Related to the extent of the study area is the scale at which you want your data input. This is seldom a straightforward matter because of the disparate scales of cartographic data available for building cartographic databases. As you remember, the smaller the scale of the maps, the larger the amount of error and the greater the generalization. A good rule of thumb is to take the largest-scale maps available for the task. The importance of the aerial coverage to the models being built will also dictate what scales will be acceptable. Although there are no formal guidelines for determining an acceptable scale, most GIS professionals apply the "best available data" approach, recognizing that more detail is better than less detail.

When working with raster data models, you will also need to know the acceptable size of grid cells within the larger map area. Again, smaller provides better detail but also drastically increases the data volume (i.e., the smaller the grid cells, the more you will need a single coverage). Although storage may not be a problem for most modern computer systems, functions that require extensive neighborhood searches will be slowed by massive increases in the grid array. In some cases, the grid cell size may be dictated by the smallest item you wish to represent, in others by the requirements of the model (DeMers, 1992), and in still others by issues of compatibility with other digital data such as input from a satellite.

Classification

Prior to analytical GIS design, we need to consider the available data sources as well as the classification systems appropriate for our models. It might prove useful to examine the kinds of input data that are available prior to deciding on the classification system. Use of a more detailed classification is often preferred for two reasons. First, it gives the user the largest amount of data; second, if one theme can be compared to another using a low level of detail, it is relatively easy to aggregate classes, whereas disaggregation may be difficult or even impossible. Of course, the most detailed classifications also present two possible problems. First, the classifications may be designed for a particular task that has little to do with the project at hand; and second, the more detail, the greater the chance for errors in classification in the first place.

But classification is more than just selecting the correct level of classification detail. It is important to remember that in GIS we frequently are making comparisons among maps. This means that we need to be able to correctly compare the classifications from one coverage to another. We also need to be able to produce consistent classifications within a single coverage, often from different sources. Suppose, for example, that you require a regional coverage of soil classes based on different county-level soil surveys. When surveys have been developed in widely differing time periods, it will be necessary to cross-walk them to ensure a uniform classification. Burrough (1986) provides a more detailed discussion of the problems of classification, especially as they relate to potential error in the GIS.

Coordinate System and Projection

The decision as to which coordinate systems and projections to use is often dictated by the regional extent of the study area and by the availability of data. Again, there are no hard-and-fast rules, just some basic commonsense guidelines that can be applied to the problem. For study areas covering many states, it is generally advisable to avoid the use of any coordinate system that has a limited geographic range of accuracy. As for projections, we have seen that the properties needing to be preserved (i.e., the properties that will be most important to analysis) are those that will dictate the correct projection. In general, for both coordinate system and projection, spatial and temporal compatibility are vital to the decision-making process.

Selecting the Software

Whether for an individual or an organization, it is not always simple to select the correct software system, particularly since many software packages are beginning to look very much alike. This is particularly true of the commercial systems, but several open-source GIS packages are viable solutions even for complex modeling situations because they conform to a rigorous set of standards developed by the Open GIS consortium (www.opengis.org [last accessed October 12, 2003]). For individuals and even some organizations, the choice is often an economic one. For others, the choice may require a specific data model, although this is much less common today because the software can move from one data model to another almost seamlessly. Finally, the choice of system may be driven by the expected **spatial information products (SIPs)** for the project. The choice of hardware platforms and peripherals is dictated by economic constraints, compatibility with software, accuracy requirements, learning curves, and personal preference.

So how do we decide for ourselves or for our clients which system to purchase? A reasonable approach is to prepare a document indicating analytical needs, cost limitations, accuracy needs, and training needs. Then these specifications are submitted to a large number of vendors for competitive bidding. You may wish to append a request that each bidder provide a list of current customers, both satisfied users and clients who have experienced problems. The responses you receive will tell you about the kinds of organization that use the system (giving you an idea of their modeling needs) and will allow you to learn from firsthand responses about the personal satisfaction levels of various users. The final selection of the GIS should be among the last decisions to be made, but unfortunately, it is often one of the first.

INSTITUTIONAL/SYSTEM DESIGN

System design is concerned with the interactions of individual people, groups of people, and computers as they function within organizations (Yourdon, 1989). The introduction of such new technology as GIS requires different or additional training for people doing the work, requires that funds be allocated for software and hardware, changes the flow of information within the organization, and therefore modifies the organization's structure. And now that we have large databases to work with, we must be more concerned about the integrity and quality of the data that are accessible by many different individuals. Wide accessibility, in turn, increases the need for security and quality control measures within the organization.

GIS system design can be subdivided into two highly interactive parts: technical design (internal) issues and institutional design (external) issues. The internal issues deal most often with the system functionality and the database. Will the system work the way we need it to? Can we answer the questions we need answered? Have we got the correct data in the right format? Do we have people with the right training to run the system? Can our system adapt to changing demands? These are some of the general questions that accompany the process of internal or technical design. But although we need to be sure that our system functions properly, we also need to understand the relationship between the GIS operation and the organizational setting. Do we have the funding necessary to permit continued operations? Can we obtain data at reasonable cost? Do we need to employ applications programmers to customize the software? Will we have adequate software support from our GIS vendor? Will we be legally responsible for errors made through our analyses? And are we meeting the larger goals of the organization beyond the immediate end of performing GIS analyses? All of these questions are important institutional considerations.

As you can probably see, the technical design issues cannot be separated from the institutional issues. Even GIS operations that are brilliantly successful from the technical design standpoint will fail if we lose the support of our organization or external sponsor. For example, even if your GIS-based efforts to enhance delivery of newspapers within the *New York Times* organization are successful, if upper management sees the added expense of operating the GIS as excessive; your operation will likely be suspended. Or if your government organization fails to obtain funding for the use of GIS, you will have to abandon such efforts. And, of course, if your system fails to provide the needed answers to the questions at hand, you are far more likely to lose your institutional support. Internal and external forces impact the overall system in which the GIS operates. As with any system, proper planning can make all the parts work together more effectively. We will look at a modern structured approach to successful GIS system design in the following pages.

System Design Principles

Among the first concepts developed in the field of system design was the **project life cycle**. In most operational GIS settings, many projects are conducted simultaneously. For each project, we must make basic decisions about what needs to be done, when it needs to be done, and who will be responsible for doing it. Because each project has a beginning, a middle, and an end (we hope), we can say that it has a life cycle, which, in turn, dictates the operations and organizational structure that will bring the project to a successful conclusion. If you are working on a single project, the project life cycle will focus on that one undertaking. The principles remain the same, but the operational details will differ.

Whether your project is to be performed by you, within a small group, or within a larger organizational structure, a methodology must be in place to give a framework for finishing the job. You might think of an assigned term paper as a simple example of a project having a life cycle. The paper may require field work, computations and analysis, review of the literature, compilation of results, and writing. You don't (I hope) begin writing the paper until the preparatory operations are complete. There is a need for a structure that enables you to perform the correct tasks in the right order to get to the final result. Because you are the only one involved, defining the tasks and establishing the proper order of execution probably will suffice. In a setting characterized by many interconnected projects designed to meet the needs of management, however, two other objectives of the project life cycle must be met. There is a need to ensure consistency among all the different projects so that all will have the same level of reliability. In addition, there must be built-in points at which management can determine whether it is time for the project to go forward. We can formalize these three objectives of the project life cycle as follows:

1. Define the activities of the project and the order in which they are to be performed.

2. Ensure consistency among many projects in the same organization.

3. Provide points for management decision making regarding starting and stopping individual phases of the larger project.

It is important to note, however, that the project life cycle is only a guide to management, a framework for the project; the manager still makes the fundamental decisions. Many important aspects of the organization (supporting the workers, fighting political battles, ensuring good morale, etc.) are still fundamentally up to the manager. The life cycle provides guidelines to facilitate the making of the right decisions at the correct time (Yourdon, 1989).

System Development Waterfall Model

Among the first approaches to the project life cycle was a technique called **bottom-up implementation**. This method, though flexible in terms of the number of steps and the exact steps taken, is highly structured and progresses linearly. The steps cover the definition of user requirements, the specification of functional needs, systems analysis, detailed design, the testing of individual modules, the testing of subsystems, and finally system testing. In the computer industry, this approach is often called the **waterfall model** of system design (Boehm, 1981; Royce, 1970) (Figure 15.1).

The waterfall model was developed to provide a structure for the systematic movement from requirements analysis through testing and final operation of an information system. It generally cascades from conceptual design through program detail design, code creation, and code testing to program implementation. Using the waterfall model as a tool for developing a GIS presents several problems. First, because the model requires each step to be completed properly before the next proceeds, any delay in a single step will slow the entire system (Yourdon, 1989). Within a GIS context, then, if we know most but not all of the user requirements, we may not proceed to the input phase until the rest of the requirements have been received. Another problem with the waterfall model is its linear structure. Although we may want our GIS system to progress in a linear fashion, nearly every implementation is going to encounter problems. As it turns out, people are better at making improvements to an imperfect system than at anticipating all possible problems before the initial implementation. Clients may also neglect important details at the outset or discover new uses for the GIS as

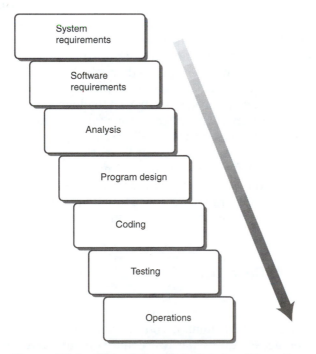

Figure 15.1 Waterfall model of system life cycle.

they see the system implemented. Or a client may find the economic situation changing resulting in either more uses being included or costly portions of the system being cut. If we followed the waterfall approach to system development, we are likely to have a completed or nearly completed system before we discover that other things need to be added or mistakes need to be fixed.

Some General Systems Characteristics

The more specialized a GIS is, the less adaptable it is for new tasks. If, for example, you are developing a GIS system specifically oriented toward digital remote sensing applications, you are most likely to incorporate data models and techniques designed to enhance and classify the pixels obtained from a satellite system. If later you decide to incorporate a gravity model designed for use with a vector data structure, your tools will likely be inadequate for the new tasks. You will have to add vector data modeling capabilities to the system. Then you will have to train the users. This is nearly like creating a new system and placing it on top of the existing one.

The size of the system has much to do with the amount of resources that must be applied. In general, the larger the system, the more resources are needed to run it. For example, if you are creating a GIS for a very large region, with many possible tasks and hundreds of themes, the input alone is going to consume more time for digitizing, more costs for obtaining data, more digitizing tablets, and probably more copies of the software than in a smaller system. Once the system is implemented, you will need more people to analyze data and to keep the data secure, you will need more output devices, and you will have to hire more managerial staff to keep track of the many parts of the analysis.

Fortunately, systems can be broken into component parts, each of which can be managed separately. This is another basic principle of systems, and one that is of paramount importance for larger systems. Each of the separate analyses can be envisioned as a subsystem that can be operated much like a single project. Strongly related to the ability to break GIS tasks into smaller components is the concept that many systems, perhaps even most, have a tendency to grow. So even if your original GIS operations performed only a few analytical tasks, as the knowledge and skill of the personnel grow, and as the utility of the system becomes better known both to management and to the client base, the system will have to grow to adapt. The more successful the operation, the more likely it is to grow. But success is often partly a function of communicating success to the institution and its sponsors. Most commercial operations serve many people. Therefore, knowledge of the institutional setting of the operation becomes important.

The Institutional Setting for GIS Operations

Whether a GIS is operating in a university environment, a business, or a government organization, it is not in a vacuum. If you are doing your own work, it is likely that you will need access to the hardware and software, but you should be sensitive to the requirements of other users so that your project does not interfere with theirs. If you are working on a larger project, you will need to coordinate your activities with the other users while keeping the goals and objectives of the overall project in mind. In turn, a project funded by external sources will require the system manager to meet the goals and objectives of that outside sponsor. The hardware and software most often will come from outside sources as well. As you can plainly see, there are many players in most GIS operations, and these players are both internal and external.

The System and the Outside World

Figure 15.2 shows a simplified model of the relation between most GIS operations (the system) and the world at large. Inside the system are interactions with the people responsible for day-to-day operations of the GIS, with those who operate the system, and with those responsible for project management and oversight. This is illustrated by the two-way internal arrow in Figure 15.2. The second set of arrows shows inputs and outputs to and from the outside world. This is the larger framework within which the GIS must operate. Now let's identify the internal (system) players and the external (world) players and see how they interact.

Internal Players

The system comprises three basic sets of players, each with different objectives. The **system users**—those who will use the GIS to solve spatial problems—are most often people who are well trained in GIS, perhaps in a specific GIS. Digitizing, error checking, editing, analysis of the raw data, and output of the final solutions to queries are the primary tasks of the system users. Whether trained in university courses or on the job, the system users will require additional training because of constant changes both in software and in the demands of the system for new analytical techniques. In many GIS operations, this group is

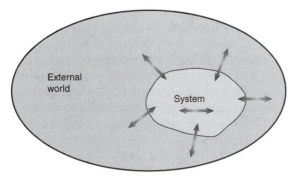

Figure 15.2   Internal and external players. The geographic information system (GIS) in relation to the outside world.

likely to be the most transitory, with new employees replacing those who have moved on to different positions.

System operators are responsible for day-to-day system operations, more often performing tasks that allow the system users to function efficiently. Their work involves troubleshooting when programs hang up or the complexity of the analysis requires additional insights. They are also responsible, in most cases, for training the users. This suggests that in many cases system operators have experience as system users. They also know enough about the hardware and software configurations to be able to make adjustments for upgrades. In addition to acting as system administrators, they frequently act as database administrators, keeping track of the security and integrity of the database to prevent possible loss or corruption of data.

The security and integrity of GIS databases are integral to the long-term operability of a GIS, particularly within large organizations. System operators are responsible for ensuring that the most current themes are not corrupted by system users who are not familiar with the overall database contents, structure, and possible uses. They are responsible for granting access to the databases through password protection, workspace allocation, and other security measures. This is especially important where the database is to be modified through data input, editing, or analysis where old themes are to be replaced by new. Beyond providing secured access, it is often necessary for the system operator to archive one or more copies of the database, so that mistakes do not corrupt the existing data and so that computer malfunctions do not destroy valuable information.

Beyond providing for system and database integrity, the system operator is also responsible for data security involving sensitive GIS data. Such situations might arise where absolute locations of, for example, endangered species should be hidden through data aggregation to avoid exploitation of that information. Another might arise in defense operations where only portions of a database might be provided to each individual system user whose security clearance might limit their access to the overall database, or even in research situations where field researchers have shared valuable data with GIS scientists on a cooperative basis. The system operator should examine the nature of each security issue and outline specific, formal procedures prior to starting a project. In most cases, these procedures should be explained to both the internal and external players.

Both system users and system operators function within a larger organization often called the **system sponsor**—the institutional parent that provides the funding for the software, hardware, operating expenses, and salaries. It also provides the political viability of the system. In other words, if there is no larger organization, there is probably no GIS. The system sponsor can be a university research group that obtained grant funding for a project, a government body whose work requires the use of GIS, or a commercial organization that performs GIS operations for paying clients. Funding and the political viability of a GIS system also require that the system sponsor engage in the long-term planning needed to ensure a steady supply of projects and continued funding.

External Players

Because most GIS operations today do not use homemade software, they rely on the first of our external players, the **GIS supplier**, a company that develops and markets GIS software systems of one or more types. Suppliers also are responsible for providing software support and updates of the software as new and improved methods are put into the system; they may work in concert with hardware development companies to provide bundled packages of software and hardware. In many cases, training is provided for GIS users either on site or at this supplier's facility, through contacts with the system administrator.

The **data suppliers**, the second external player in the GIS world, tend to be either private or public. The private company may provide internally generated data or data obtained from public agencies, modified to better fit needs expressed by the user community. Public agencies, primarily federal government agencies, provide data for large portions of the country. In many cases, these data are designed for use by the agencies themselves, but the data are available to external users as well. Private companies are most often for-profit organizations that supply data at a cost-plus-profit level. In some cases, the data can be provided for specific regions and in the formats needed for a specific use. Some public agencies offer data at cost, or in some cases at no cost at all, provided the use is for nonprofit GIS operations. Appendix A gives a list of some public sources of data.

Another group of external players, a group that is becoming increasingly important as GIS software becomes more sophisticated and more object oriented, is the **application developers**. Application developers provide user interfaces to reduce the reliance on specialized GIS professionals to perform common tasks. In many cases, the programming is done in macro languages provided by the GIS supplier to support applications development without the need for interfacing with traditional computer languages. Although in the past most application developers were internal personnel who had years of experience with the analytical capabilities of the GIS software, today there are many small consulting firms specializing in the development of applications for third-party clients. Application programmers can limit the training costs incurred as new GIS users enter the organization. They provide a "point-and-click" environment that allows the system to be used by analysts with only a limited amount of knowledge of the internal functionality of the GIS software.

Finally, we have the **GIS systems analysts**. Members of this group of external players specialize in the study of systems design, the primary focus of this chapter. Most often systems analysts are part of a team of professionals responsible for determining the goals and objectives of the GIS within an organization, fine-tuning the system so that it provides the appropriate analytical techniques, and ensuring the successful integration of the system into the organizational framework. In short, the systems analysts act as navigators for organizations using GIS. Commonly, the systems analysts are part of a larger consulting firm specializing in the implementation of GIS for third-party users, but in some cases they are employees of the GIS suppliers.

Cost–Benefit Issues

Even if we assume that the employees are interested in the use of the technology because they see that it will make their tasks easier to perform and might even ensure them of a more secure employment situation, we need to convince the system sponsor that the benefits of introducing GIS into the organization will outweigh the costs. By and large, corporate CEOs feel that there is generally no totally reliable measure of what the return on investment is. Cost–benefit analysis, developed shortly after World War II, is a common technique for examining the return in real dollars on the basis of the dollar amounts invested in a system. A breakeven benefit-to-cost ratio (B:C) of 1:1 indicates that there is 100-percent recovery of costs, whereas if the numerator is larger than one, for example, a ratio of 2:1, it indicates that the financial benefits are twice those of the costs. The most difficult issue of such analysis is the identification of the dollar values of both costs and benefits. Little is known about the exact costs of implementing a GIS, however, and even less about the economic benefits from such an implementation. This is partly because for each GIS implementation, the numbers and complexity of the maps that are input to the GIS change radically. In addition, because of the long time periods necessary to develop the initial databases, short-term returns on this investment frequently are minimal. To begin a GIS project, an organization's management must be convinced that there will even be long-term benefits.

Recent research has indicated that there are workable methods of evaluating the costs and benefits of implementing a GIS (http://www.nysgis.state.ny.us/costanal.htm [accessed October 13, 2003]). Gillespie (2000) provided a description of a 10-step process, resulting from an empirical study of 62 federal government GIS implementations performed by the U.S. Geological Survey in 1990 and 1991. Subsequent implementation of this model in the state of Montana proved that it was a useful approach. Like most such systems, it requires the users to enumerate the cost components such as hardware, software, training, data conversion (often a very large component of the costs), application development, and many more (Silva, 1998) as well as the benefits. The benefits can be divided into four groups (Wilcox, 1996). (http://gis.esri.com/library/userconf/proc96/TO300/PAP282/P282.HTM) [accessed October 13, 2003]:

1. Direct—improvements over manual methods.
2. Agency—increased productivity and quality.
3. Government—benefits to other agencies.
4. External—both qualitative and quantitative.

Direct benefits usually include such operating efficiencies as reduced staff costs, more products from existing staff, reduced outsourcing, and even reduced map purchases. Agency and government benefits include better outcomes from more timely, more complete, or more accurate data. These benefits usually result in better customer satisfaction and/or greater sales. External benefits are the hardest to quantify and may require proxy or surrogate measures. Wilcox (1996) suggests the example of using GIS to destroy or to avoid destruction of archaeological sites, which might result in increased tourism in the former case and/or reduced fines to oil or gas companies who wish to explore areas in and around the sites. Beyond providing the list of possible costs and benefits, Wilcox (1996) also presents a long-term view of cost–benefit analysis, showing that in one case she examined the benefit-to-cost ratio increased from 0.5:1 in its first year of operation to over 6:1 in its sixth year, for a cumulative ratio of nearly 3:1 over the six-year period. This clearly indicates that the benefits of a GIS are more often than not long-term, while the costs, especially the initial costs, are often borne in the early stages of GIS operations.

GIS INFORMATION PRODUCTS

GIS information products are largely the result of analysis within the software. What the specific information products are depends on the nature of the organization, its goals, and its experience with the system. Although many organizations have a fairly structured set of products—for example, Rand McNally Corporation makes a relatively definable number of products (Calkins and Marble, 1987)—others, especially scientific research organizations, have a large and ever-changing set of potential products that may be difficult to define (Rhind and Green, 1988; Wellar, 1994). But even in scientific organizations, there are most often general ideas of what the GIS is capable of doing and what the technology might offer the individual institution.

How Information Products Drive the GIS

Organizations that have a particular goal in mind when they consider implementing a GIS are quite likely to have some ideas about what the output of analysis might look like, even if they have only a vague sense of how it might be derived. They most likely know roughly what general data might be part of the system. Many organizations recognize the value of the data in an automated form even before they are able to see the potential utility for analysis. To extract particular spatial information products from the users, we must recognize the close linkage between what goes into the system (data availability) and what comes out.

Because members of the user community cannot be expected to be instant GIS experts, it is our job to help them define their needs. You should note that it is our job to *help* define their needs! We are present not to tell them what they need, but to *help* them tell us what they need. Thus the systems analyst must play the role of educator to elicit descriptions of needed products from the users. Perhaps this will clarify my insistence on the need to understand the analytical capabilities of the GIS as well as the database creation and management tasks.

To act as educators, we first spend a great deal of time determining how a GIS could best serve each user. This is best done in person, rather than through a review of questionnaires, because interactive question-and-answer sessions are most likely to lead to an understanding of the tasks each user normally performs and to enhance explanations of functions the GIS might be able to perform to match those tasks. Any documentation indicating what the user produces (sample maps, written responses to questions or reports of decisions, applicable legal mandates, etc.) will prove useful in making a good match between the existing products and the SIPs obtainable through GIS. As you perform this review, you should keep in mind that the users will be more interested in what can be obtained than in how it is derived. As the process moves toward implementation, the how questions can be addressed.

Organizing the Local Views

Because we most often will be working with many potential users of differing individual needs, it can be useful to be able to keep track of the relationships between each user and each SIP. Tomlinson (n.d.) suggested the use of a **decision system matrix**, listing the users along the side of and the products along the top; Figure 15.3 shows a highly simplified version. Not only will this device allow us to keep track of the SIPs for each user, but later, when we must integrate each of the local (individual user) views into a larger global view, data from this matrix will provide much of what we need. An alternative to the decision matrix is an **organizational diagram** such as that shown in Figure 15.4. The global view at the bottom of the diagram illustrates the overall organization's needs; the next tier isolates the local views of individual users, and the products, differentiated by user, appear at the top. This type of diagram, like the matrix format, is useful for integrating the local views into a more general global view. Both can also be used to decide which products are the most important, based on which are most often called for by individual users.

Avoiding Design Creep

Design creep, which occurs in the absence of organizational learning, can be defined in terms of its results: at one extreme is a system with more functionality than is necessary, and at the other is a system with an incorrect functionality. Figure 15.5a shows a model for a rather structured approach to GIS design, proceeding from the feasibility study at the left through the design phase and finally into the implementation phase. The two curves indicate the relationship

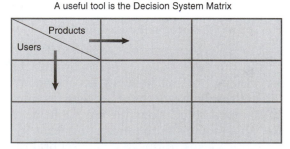

Figure 15.3   Decision system matrix. Simplified decision system matrix used to organize individual user views of the geographic information system (GIS).

Figure 15.4 View integration. Organizational chart used to display individual user views of the geographic information system (GIS). *Source:* Duane F. Marble, Department of Geography, Ohio State University, Columbus, Ohio. Used with permission.

between the technical design process (light curve) and organizational learning. As you can see, organizational learning begins late in the process, when the system is almost completely designed and is nearly ready for implementation. Thus users are forced to learn and operate a system that may not meet their needs.

Figure 15.5*b* shows the more flexible spiral model (Marble, 1994), in which organizational learning drives system development. As users become famil-

Figure 15.5 Avoiding design creep. Two approaches to system design: *(a)* a structured, linear approach that leads to design creep and *(b)* the Marble spiral model, which avoids design creep.

iar with what a GIS is capable of, they can describe their system needs at a point well in advance of the run-up to implementation. You can also see that the more sophisticated the organizational learning, the more sophisticated the system becomes, but the learning curve is always above the design curve. In short, the users are driving the design of the GIS rather than being driven to learn by an in-place system. The latter approach avoids design creep and allows the system to grow in complexity as the organization's needs increase.

VIEW INTEGRATION

As we have seen, most organizations will have many GIS users, and it is vital that the needs of each be ascertained and addressed. The individual sets of requirements are called **local views** (of what the GIS is going to do and how it is to do it), and as a preintegration task we often find it useful to list a set of preferences for each of these local views, based on management needs. We also need to decide on a methodology—perhaps a decision matrix or an organizational chart—for later integration of the local views (called **view integration**). One useful technique is to produce a pairwise (dyadic) grouping of similar individuals or groups. This can be performed easily with either the decision matrix or the organizational chart approach. The most common needs, those being met for the most people, will provide us with one method of identifying a set of priorities that can be compared to the larger organizational goals.

When conflicts exist between and among local views, additional discussion will be necessary to pinpoint the problems. Conflict resolution may require arbitration, modification, and even redefinition of the local views to achieve correspondence to the overall organizational goals. If isolated needs are found to be incompatible with the global view, several iterations of intermediate views may be required to combine individual local views into larger groups. This merger of local views will ensure that the vast majority of individual user requirements are being met. Final decisions for determining the global view will often rest with management, where individual user needs are considered as part of the overall needs.

Terms

application developers	GIS systems analysts	system design
bottom-up implementation	local views	system operators
data suppliers	organizational diagram	system sponsors
decision system matrix	project life cycle	system users
design creep	spatial information	view integration
GIS supplier	product (SIP)	waterfall model

Review Questions

1. What accounted for most system failures in GIS in the 1960s? What causes most GIS system failures today?

2. Why do we need to consider system design for GIS? Why aren't some users' needs met? Why is it important to know what GIS analytical capabilities are when we are designing a system for third-party users?

3. What is a system? How does this design construct expand our definition of GIS? What is the difference between technical design and institutional design?

4. What is the concept of a system life cycle? What are its objectives? Why is the waterfall model of the system life cycle inappropriate for designing and implementing GIS in organizations?

5. What is a system sponsor? Why is it important in the overall, long-term functioning of a GIS in the institutional context?

6. Who are some internal players in the institutional setting? What important tasks do they perform?

7. Who are some external players in the institutional setting? What role do they play in organizational implementations of GIS?

8. What is conceptual GIS design? Why is it not enough for designing a GIS for an organization? What are some of the people problems of GIS?

9. What is a cost–benefit analysis, and what role does it play in designing a GIS? Discuss some of the methods used to measure GIS benefits?

10. What is design creep? Explain how it can be avoided.

11. Describe the importance of a flexible, open-ended GIS design process that emphasizes educating the users prior to implementation. How does this help eliminate design creep?

12. List and describe the potential benefits of GIS in an organizational framework. Create a fictitious organization and discuss the potential benefits for it.

13. What are SIPs? How are they related to the input to a GIS? How do they drive the implementation of a GIS?

14. Describe and diagram the decision system matrix approach to organizing local views of a GIS. Do the same for the organizational chart approach. Suggest how these methods might be used for later view integration.

15. What major database design considerations do we need to look at for GIS design? Define some general rules for each.

References

Boehm, B., 1981. *Software Engineering Economics*. Englewood Cliffs, NJ: Prentice Hall.

Boehm, B., and P.N. Papccio, 1988. "Understanding and Controlling Software Costs." *IEEE Transactions on Software Engineering,* SE-M(10):1462–1477.

Boehm, B.W., 1986. "A Spiral Model of Software Development and Enhancement." *ACM Software Engineering Notes,* 11(4).

Brooks, F.P., 1975. *The Mythical Man-Month.* Reading, MA: Addison-Wesley.

Burrough, P.A., 1986. *Geographical Information Systems for Natural Resources Assessment.* New York: Oxford University Press.

Calkins, H.W., and D.F. Marble, 1987. "The Transition to Automated Production Cartography: Design of the Master Cartographic Database." *American Cartographer,* 14(2):105–119.

DeMers, M.N., 1992. "Resolution Tolerance in a Forest Land Evaluation System." *Computers, Environment and Urban Systems,* 16:389–401.

DeMers, M.N., 2002. *GIS Modeling in Raster.* New York: John Wiley & Sons.

DeMers, M.N., and P.F. Fisher, 1991. "Comparative Evolution of Statewide Geographic Information Systems in Ohio." *International Journal of Geographical Information Systems,* 5(4):469–485.

Gillespie, S.R., 2000. "An Empirical Approach to Estimating GIS Benefits." *URISA Journal,* 12, No. 1.

Marble, D.F. 1994. *An Introduction to the Structured Design of Geographic Information Systems* (Source Book, Association for Geographic Information). London: John Wiley & Sons.

Mitchell, A., 1999. *The ESRI Guide to GIS Analysis:* Volume I: *Geographic Patterns & Relationships.* Redlands, CA: ESRI Press.

Rhind, D.W., and N.P.A. Green, 1988. "Design of a Geographical Information System for a Heterogeneous Scientific Community." *International Journal of Geographical Information Systems,* 2(2):171–189.

Royce, W.W., 1970. "Managing the Development of Large Software Systems." *Proceedings, IEEE Wescon,* pp. 1–9.

Wellar, B., 1994. "Progress in Building Linkages Between GIS and Methods and Techniques of Scientific Inquiry." *Computers, Environment and Urban Systems,* 18(2):67–80.

Wilcox, D. 1996. (http://gis.esri.com/library/userconf/proc96/To 300/PAP282/P282.HTM). Proceedings, Arc/Info User Conference, paper #282.

Yourdon, E., 1989. *Modern Structured Analysis.* Englewood Cliffs, NJ: Prentice Hall.

CHAPTER 16

GIS Applications

The power of GIS both as a technology and as a science has resulted in its adoption by an increasing array of new disciplines. Although most of us will not be directly involved in all of these disciplinary applications, we will typically be exposed to many of them. The following pages provide a range of GIS implementation experiences, both applied and theoretical. This exposure will expand our view of the scope of GIS, providing us with insights and innovative uses of the technology, and demonstrating the limitations of the software. Although GIS implementations share many functional similarities, each of the following describes some different aspects of the software, data, and design considerations.

No single text can demonstrate all the possible uses of GIS. As such, this chapter is meant to be representative, not exhaustive. The scope of applications presented is limited by personal experience and exposure, by the availability of literature, and by time and space. The examples chosen are not necessarily the newest, nor always the most innovative, but are instead designed to be more generally instructive and informative of the general applications areas adaptable to the GIS environment. These are presented in order of increasing sophistication, beginning with GIS as archive and **spatial data query**, then moving on to **pattern detection** and **characterization**, **pattern exploitation**, **pattern comparison**, space-time modeling, and finally predictive modeling implementations. Each topical heading employs the word "emphasizing" to illustrate that no GIS implementation fits purely into any of these categories.

As you go through these examples, a number of "tips" are presented that may prove useful in your own GIS experiences. Some of these tips will indicate things you should do, while others will be things to avoid. The lessons learned from these example applications can save time and effort, not to mention money as we implement our own geographic information systems. They should also be augmented by attendance at regional, national, and international GIS research and software applications conferences. Use the following pages to gain insights from others and to critique their work. Then attend your own user conferences to apply these skills throughout your career.

LEARNING OBJECTIVES

When you are finished with this chapter you should be able to:

1. Categorize GIS applications in hierarchical fashion in increasing order of complexity.

2. Describe at least one specific example application to each category of the hierarchy.

3. Identify at least three alternative existing examples or subject matter applications for each category of GIS application.

4. Suggest and provide some detail of at least one potential application not mentioned in the text, or readily available in the literature.

5. List at least five sources of information for GIS applications.

6. Explain the utility and importance of each application identified.

7. Demonstrate some basic tips necessary to implement each of the applications.

8. Link the applications to previous chapters from Unit 3 on spatial analysis.

9. Within the individual application context explain the tips provided.

10. In the absence of specialized GIS software, describe a method of using existing GIS software to create a space-time model.

11. Explain the differences between correlative and explanatory predictive GIS models.

12. Explain why it is sometimes necessary to disaggregate both dependent and independent GIS variables.

13. Provide an explicit example of multivariate GIS models where the differences among independent variables cause confusion in the model results.

14. Suggest solutions to the example provided in number 13 above.

ARCHIVE EMPHASIZING APPLICATIONS

It might be easy to discount the purely archival functions of the GIS, but to do so would be a serious mistake. The GIS-as-storage paradigm illustrates that the storage and retrieval subsystems of the GIS do more than just act as a computer repository of spatial data. Its more important archival function lies in its ability to organize data spatially, most often in a georeferenced format, thus allowing for eventual retrieval through appropriate spatial queries. This is equivalent to converting the address data contained in your local telephone directory to geospatial data where you could eventually ask questions about who lives within a particular region. Although all the data are essentially contained within both archival methods, the GIS, like the maps on which it is often based, provides a spatially explicit storage medium absent from the telephone directory.

Such inventory and archival tasks are the primary responsibilities associated with parcel-based land information systems. These systems comprise explicit definitions of landownership and land value. Because such data change so rapidly, the primary functions of the GIS in this case are to allow the retrieval of specific land parcels, reconcile the correctness of the parcel descriptions and ownership, and make appropriate changes as needed.

Traditionally, such landownership systems have been created around a set of analog maps called plat maps and/or text-based land records whose information could only be updated or corrected manually and whose users were required to be at the physical location where they were stored. This limits the speed at which records can be added, deleted, and corrected, thereby limiting the overall effectiveness of the system. It further limits access to the data to those who are able to physically visit the records. GIS is increasingly being applied to the automation of such land records. One interesting example of such an application is the Idaho Department of Land Records which has not only developed a digital database of the State's land records but has also implemented it on-line to allow access by users not in the locale. Their manual is available on-line as well. The URL is (http://www2.state.id.us/lands/Bureau/MIS/GIS/LandRecs/users_manual.htm). Recent research indicates that what often appears to be a simple analog-to-digital transformation process is often replete with difficult problems ranging from missing data, incorrectly entered land records codes, incompatible coding schemes, seriously out-of-date records, and conflicts over landownership (Rich, 2001). Many of these problems require the development of detailed protocols for decision making just for database development.

✓ Tip: Close examination of the land records prior to GIS database development is a necessity. Many problems can be avoided by creating an appropriate logical data model early in the process rather than waiting until the project is under way. In particular, entity-relationship diagrams would prove very useful in examining the potential queries and associated data availability so that effective data retrieval can be ensured.

These protocols normally require interaction with land records specialists and government officials especially because they are often legally responsible for using these land records for decision making, including tax assessment. Disputes over landownership, acreages, and tax valuation are often easier to solve if the actual parcels in question can be retrieved spatially. Text-based documents can also be **hot-linked** (i.e., linked to the maps and able to be displayed either singly or within the map themes) to the GIS software to illustrate legal descriptions of land parcels appearing on the map (Figure 16.1).

Land disputes are not limited to small parcels at the county or local level but may even extend to questions of national sovereignty. There are, for example, a growing number of Native American tribes who either explore or actually employ GIS software to archive and evaluate their landholdings (Doucette, 1994). Beyond using these datasets for estimating their potential for economic development and ultimately self-determination, they are also being compared to his-

Figure 16.1 **A digital land records map with a hot link showing how legal documents can be included as part of the database.**

torical records and treaties for negotiating or renegotiating their conditions. Another successful application related to Native Americans and **nation building** requires not just archiving the land records themselves, but also preserving the native language and history (Hartman, 2001). One excellent example of this comes from the Coeur d'Alene Indian Nation GIS, which includes hot-links to digital videos of tribal elders describing, in their native tongue, the history and traditions of individual locations within the tribal landholdings (Figure 16.2). The database, digital images of places owned by the tribe, and verbal descriptions by the tribal elders are all made available to future generations of Coeur d'Alene tribal members. Although this is, on the surface, a simple storage and retrieval operation requiring no substantial analysis, its utility is obvious, and its potential as a tool for nation building and cultural preservation is profound. Although many of the techniques used in this multidimensional GIS application are available only to tribal members, the web site itself (retrieved [September 1, 2002] from the Coeur d'Alene tribal GIS; http://www.cdatribe.org/gis/) is very nice, and contact information is available from the GIS personnel.

Similar tasks related to the use of GIS for preservation involves recording and documenting a wide variety of alternative types of information. Today, both libraries tasked with archiving and maintaining old maps of all kinds, and a large number of museums, particularly anthropology and natural history museums, are adopting GIS technology as the solution. Although many map libraries contain large volumes of maps, these maps take up substantial space and are easily damaged when they get older and more brittle. Many of these cartographic documents are now being committed to digital form through readily available scanning technology as a method of preserving them. Among the

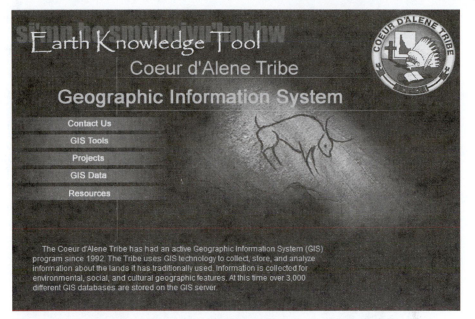

Figure 16.2   Image of the Coeur d'Alene Indian Nation GIS website.
(http://www.cdatribe.org/gis/) Accessed 11/16/2003.

many examples is the Perry-Castañeda Library map collection at the University of Texas (http://www.lib.utexas.edu/maps/index.html). Beyond historically significant maps, much extant map data are now being converted to active GIS datasets that can be directly incorporated into software for analysis. These collections, particularly the GIS datasets, can often be searched directly or over the World Wide Web by theme, by geographic extent, and even by scale. Although like most GIS applications, substantial time and energy are required to create these datasets and their catalogs, doing so ultimately simplifies the map librarian's tasks and provides a much more utilitarian map library. Soon, in addition to the growing number of government and private map archives, libraries will play a major role as a repository for GIS datasets.

Museums are among the best-known repositories of important data and information, much of it being inherently spatial. The need to store and retrieve museum data quickly is driving the development of many large-scale, interactive, easily accessible databases composed of anthropological, plant, and animal specimen data—essentially virtual anthropology and natural history museums. As collections grow and as our need increases to link newly collected natural resources datasets with archived anthropological and biological data, especially in a geographic sense, the greater the need to incorporate GIS technology to manage them becomes. Anthropological and biological surveys and museums, national and international conservation organizations, scientific research bodies, and others are currently endeavoring to provide these needed services. One substantial effort of this nature includes the Illinois State Museum, which has been a functional part of the state of Illinois GIS since 1983. Their efforts also include the FAUNMAP project—an effort to create an electronic database for the late quaternary distribution of mammal species in the United States (retrieved [May, 15, 2002], from the Illinois Natural History Museum http://museum.state.il.us/research/faunmap/). The database can be purchased as GIS datasets or converted to individual maps on-line.

Among the most comprehensive of natural history applications (retrieved [September, 1, 2002] is from the Integrated Taxonomic Information System [ITIS] http://www.itis.usda.gov). Other efforts related to the creation, sharing, and analysis of biological data include those of the National Biological Information Infrastructure (NBII), the Species Analyst activities at Kansas University, and the activities of Arizona State University. All of these efforts encounter another major problem related to the sharing of information from unit to unit. A set of data standards for museum databases called the Darwin Core (retrieved [September, 1, 2002], from the Darwin Core http://tsadev.speciesanalyst.net/ DarwinCore/darwin_core.asp) has been established to illustrate the types of data that should be incorporated into the databases, as well as general information on organizational structure and procedures. Beyond the necessary aspatial information provided, it also includes data on spatial locators. The Darwin Core, is a specialized version of metadata.

> ✓ Tip: Although museums have substantial specimen resources, traditionally the museum curators have been more concerned with the quality of the specimens than with their geospatial locations. This is particularly true of older specimens, which, in some cases, may be vague descriptions and in others are simply incorrect. Care should be taken in creating GIS data layers from museum records. In many situations, it is necessary to build predictive GIS layers to determine the feasibility of specimens existing where the museum records indicate.

PATTERN DETECTION AND CHARACTERIZATION EMPHASIZING APPLICATIONS

A growing number of GIS practitioners are discovering the enhanced utility of GIS as a method of visualizing spatial and spatio-temporal data (DeMers, 2002). Although other methods of visualization exist, there is a growing recognition of the increasing capability of GIS to display geographic patterns of data not found in other types of software, but inherent in the nature of digital mapped databases. The robust nature of the computer display makes the analog crime "pin map" (Figure 16.3), for example, much easier to create and update. In many cases, the maps themselves are much more informative than their analog counterparts. By converting from discrete point data to statistical surfaces, the pin map, for example, could display crime as a probability surface showing predictive regions of crime (hot spots) rather than requiring the map user to identify these regions unsystematically from the pin locations.

A similar pattern identification approach uses a range of remote sensing techniques linked to GPS devices on-board tractors and other farm machinery.

Figure 16.3   Typical pin map. http://www.greverandward.com/images/pinmapb.jpg

As the sensors relay electromagnetic signals designed to detect, for example, deficiencies in nutrients, areas of pest damage, or locations requiring additional drainage, the GPS provides detailed information about exactly where, within the fields, these locations occur. This **precision agriculture** methodology allows the modern high-tech farmer to create a nearly real-time map of crop conditions. This application has become important enough to warrant its own journal *Precision Agriculture: The International Journal on Advances in the Science of Precision Agriculture.* This journal is available on-line at http://www. kluweronline.com/issn/1385-2256/contents (retrieved September 1, 2002 from Kluwer Online).

PATTERN EXPLOITATION EMPHASIZING APPLICATIONS

Both the crime pin map and precision agriculture examples provide, as a next step, the ability to exploit the patterns discovered. In the case of crime hot-spot analysis, police departments can use these hot-spot maps to deploy additional police resources where crime is most likely to occur. Thus the output of one set of analytical operations goes through the cyclical process of GIS to become input to another more exploitive application, which could itself become a map of police deployment. With precision agriculture, the response to identification of pattern can be nearly instantaneous as the farm machinery can be computerized to apply necessary levels of fertilizer to parts of the field based on computer models of the selective need. In this way, only those portions of the crop requiring additional fertilizer will receive it, thus reducing the overall cost of fertilizer. Once again, this illustrates the cyclical nature of GIS modeling we visited before.

Among the more common forms of pattern exploitation are route finding and site location. Route finding assumes an exhaustive level of knowledge about linear patterns of road networks and is therefore among the best examples of pattern exploitation. We have seen that route finding is not necessarily a simple matter of finding the "shortest" path from one place to another, which itself is not always a simple task. If you live in a large city, you encounter some of the more relevant issues regarding route finding such as one-way versus two-way streets, speed limits, stop signs, left-versus right-hand turns, construction activities, time of day and related traffic flows, season, and many more. Because of the extensive use of networks, your GIS software may include a specialized package for route finding, or you may be using a GIS that is specialized for these tasks. In either case, the interactive, cyclical nature of spatial analysis with GIS is very well illustrated by route-finding tasks, particularly because the parameters of the model change depending on time of day.

✓ Tip: Route finding tasks go under the heading of time-critical GIS applications because they rely so heavily on temporal changes in traffic patterns. Traffic changes radically as accidents occur, as rush hour approaches, and as construction zones change. Weather can also have a profound impact on road conditions. For example, heavy rains will often make travel on some streets impossible. In Duluth, Minnesota, many very steep streets that are commonly used in the summer are closed during winter months because they become impassable. This means that change detection and constant updating of the database become critical in such GIS applications. On-board navigation systems are examples of GIS applications that attempt to keep track of changing conditions in real time.

Another typical application designed to exploit observed patterns is that of site location. We have already examined the mountain cabin example, so let's take a somewhat different case here that includes viewshed analysis as a primary component in its decision-making process. Let's say that you are trying to locate a 182-foot water tower in such a way that it will be viewed from as little territory (i.e., as few locations) as possible. Your first step is to perform the viewshed analysis based on the topography, the expected height of the tower, and the proposed location. The viewshed (Figure 16.4) provides a set of polygons within which the tower can be seen, thus giving the planners ample evidence of the possible impact its placement will have on the surrounding area. Such a technique was applied in the Murphy Tower viewshed analysis (retrieved September 8, 2002 at http://www.harpersferry.org/Issues/mf.html). The outcome was a decision to place the tower where the planners had originally planned.

Figure 16.4   Murphy tower viewshed. http://www.harpersferry.org/Issues/ murphytower-viewshed.jpg

✓ Tip: You might remember here that the height of the tower (182 feet) made the viewshed analysis somewhat more complex than its simpler alternative. A major problem with this technique is that one typically decides on a location for the tower—and then proceeds to develop the viewshed from that location. It would be more interesting and perhaps even more useful, if the viewshed analysis were performed on all possible location, thus providing a number of alternative locations for placement. The viewshed analysis performed for an arboretum on the University of Minnesota Campus did this by selecting a number of possible observation points rather than just one (retrieved August 8, 2002: http://www.mnplan.state.mn.us/EPPL7/ applications/arboretu.htm).

PATTERN COMPARISON EMPHASIZING APPLICATIONS

In the previous example, some implicit data layers were employed in the analysis such as topography, and the point location and height of the proposed tower. This is typical of how viewshed analysis operates, but it would have been desirable if we could have introduced additional layers through some form of map overlay to make the viewshed more meaningful as compared to other factors than simple visibility. As noted in previous chapters, map overlay was one of the primary driving forces behind the development of GIS in the first place. The primary reason for this is that a large portion of the decisions we make with maps is derived by comparing more than one spatial factor.

Typically, the map overlay approach is used to either eliminate or exclude areas that are not suitable for some use, or to rank the layers and their individual areas based on some absolute or relative applicability to the problem. Such applications can actually get very complex and include more than map overlay, but other forms of data integration as well. Among the more sophisticated of these is the use of GIS to determine where powerline corridors are to be located across large tracts of land (Helner, 2001; Vann, 2000). In such circumstances we find a wide array of criteria and factors, including:

1. Exclusionary factors (such as parcels of land that simply cannot support the creation of powerlines).

2. Weighted factors (such as the physical capability of the land to support such activity).

3. Aspatial criteria (such as rights of eminent domain) that must first be assigned a spatial dimension or used as layer multipliers or weights.

4. Conflicting factors (e.g., the well know "not-in-my-backyard or NIMBY" problem) that require negotiation and conflict resolution strategies that are not inherent in the GIS software itself.

This powerline corridor allocation problem often presents among the widest array of analytical operations contained in a single set of circumstances (retrieved [September 3, 2002] from http://www.gisdevelopment.net/application/utility/power/utilityp0020pf.htm). Not only are there a vast array of problems to overcome for the single powerline corridor selected, but there may be an infinite number of possible locations in which to place the corridor. As we have already seen repeatedly, it is best to begin by flowcharting the problem—thus creating a set of identifiable and measurable criteria and associated thematic maps and data layers. But, again, because we have no limits (except, of course, those related to cost and expediency) we are forced to use more than flowcharting in our model building. This type of model requires us to narrow down the study area by using the exclusionary factors first, perhaps building an exclusionary model and associated flowchart first. In this way, we may find that our potential areas for placing the corridor are much more limited than what we started with. Even this approach may require us to either loosen or tighten our model constraints before we begin working on the nonexclusionary portions of our model.

SPACE-TIME EMPHASIZING APPLICATIONS

Among the more interesting, and more complex, applications of GIS is the modeling of space change through time. Two primary examples of space-time emphasizing applications are used here: an agricultural land use change model that is fairly typical of how spatio-temporal modeling takes place and a more complex forest fire model that illustrates how change of state can be incorporated into the modeling process. Without using highly specialized GIS software such as the raster GIS package PC-Raster® (Wesseling et al., 1996), non-GIS modeling techniques (e.g., Stella), and programming tools, the GIS is often somewhat limited in its ability to perform space-time modeling. Still, even in the absence of these external tools, it is still possible to at least simulate the effects of time on spatially explicit models.

In the first scenario, we have a case study of the changing land use patterns of a small part of the Northern Virginia Military District (NVMD) in central Ohio (Simpson et al., 1994). As with our last example, we will see how map overlay can be applied to this problem by creating four different vector digital map layers, each with the same basic thematic data but at different times. Each map layer, acquired from digital aerial photography, represents a single point in time. Because we use the same categories for each time slice, we can legitimately overlay the datasets without concern over their comparability as overlaid themes.

In our NVMD case, we see time snapshots of 1940, 1957, 1971, and 1988. We chose readily identifiable land cover themes that anyone obtaining the same aerial photography could easily reproduce. The process of modeling time was to examine the differences in each land cover category by employing an intersection overlay. This can be done by pairwise comparison (overlay). If a particular category of land increased in size through time, its polygons would reflect that, and the intersection overlay would show polygons of land cover increase. If the category shrank, the resulting polygons would demonstrate decreases in land cover. In the case where one or more categories of land cover is absent from one time frame, the results will be readily visible. The result of this is to demonstrate both the overall changes from 1940 to 1988 and to illustrate any trends.

Improvements and enhancements to GIS software are made all the time, some of which are designed specifically to address the linkage between spatial modeling where time is not explicitly addressed with temporal modeling where space is not explicitly addressed. The previous discussion showed how, in the absence of such tools, one can still create a space-time model. This topic will be discussed in more detail in Chapter 17 on GIS research.

✓ Tip: As you might surmise, this technique contains several possible pitfalls. First, the use of a vector data model increases the necessary decision making the modeler has to make, particularly with regard to very small polygons. Are they real changes, or are they just a function of the digitizing errors? One way to eliminate the decision-making difficulties is to use a raster data model. It won't necessarily improve the overall accuracy, but it will make it easier to tabulate the results. A second problem is that the choice of land cover categories places restrictions on how much information can be derived from the results. The coarser the categories, the less information you will be able to extrapolate. A third problem is in the selection of time slices, which also places restrictions on how trends can be recognized. The more time slices you have, the more trend data can be extracted and the more exact your model will be. Do you see any others? How about the truncated patterns on the edges of the maps?

PREDICTIVE MODELING APPLICATIONS

We have already seen one simple example of predictive modeling where we created maps of predicted crime activity and exploited that knowledge for the deployment of police. We have also seen how we can obtain nearly instantaneous response to such predictions through the combination of near-real-time remote sensing, GPS, and computer-directed hardware in precision agriculture. In both of these cases, however, the predictions were rather simplistic and contained limited temporal aspects. Instead, we employed spatial interpolation to predict the outcome from one place to another. The true test of our ability to do prediction is to be able to predict changes through time, as well as to extrapolate spatial data where two bounding data points are not available for interpolation.

Such tasks are particularly difficult in situations where little data are available—situations often encountered in archaeological research. Among the more challenging and interesting predictive uses of GIS are those involving the search for archaeological sites based on data obtained from existing sites. One such study in Italy, characterized as landscape archaeology, took advantage of known relationships among such factors as landforms, soils, vegetation, and proximity to water to predict the locations of Paleolithic camp sites (retrieved [September 8, 2002] at http://deis158.deis.unibo.it/archeo/chapt_1.html). The success of this approach presupposes first that the recommended factors are both **correlative** and **explanatory**, and second

Figure 16.5   The combination of models to create a more accurate predictive map.
An example from the Chesapeake Bay study.

that they are available as data layers in sufficient detail to be able to be put into a GIS.

Similar research in the Chesapeake Bay area of the United States revealed a clear relationship between distance to water and the location of archaeological sites, but also showed that other relationships are not always as clearly defined as one might wish (Wescott and Kuiper, 2000). Wescott and Kuiper (2000) took a number of unique workarounds to refine the initial distance-to-water based model. First, they divided the archaeological sites into two distinct types (see tip), each with rather unique requirements. In addition, they separated the water sources into unique types. As an example, they subdivided shorelines into bay, river, creek, and lake/pond types. In this way a single GIS data layer is effectively disaggregated to yield more useful information. Characterization of the topographic setting into floodplain/beach, terrace, hillslope, hilltop, bluff, and interior flat areas proved more useful than slope and aspect because of the subtle nature of the topographic variability. By effectively combining these parameters for each of the two archaeological types first, then combining them, these researchers we able to produce some effective predictive models for both types (Figure 16.5): http://www01.giscafe.com/technical_papers/Papers/paper057/.

✓ Tip: The separation of the archaeological sites into two types is an important lesson for GIS modelers. Although we often spend a fair amount of time examining the predictive parameters, we forget to separate out the independent variables (those whose distributions we are trying to predict). When we have several independent variables, the dependent variables often cancel each other out, especially if the items we are trying to predict operate very differently from each other. Gender differences (as, for example, with wild animal species) often require individual models as well.

Despite the apparent quality of the results from studies like that in Chesapeake Bay, Church et al. (2000) suggest that such models are limited in quality because they are "bereft of theory" and that the data used to develop them are often "of doubtful reliability." They recommend that the field of archaeology adopt a replacement of the traditional correlative predictive models—those based on known correlations among variables, such as those already outlined above, with explanatory models based on a through knowledge of landscape dynamics. As they explain, and as stated by Kohler (1988), such models attempt to bridge the dynamics of living systems and observed outputs. As we saw earlier, the temporal dimension is not currently one of the strengths of GIS. It is, however, becoming more important as a modeling tool, and it is, therefore, likely to become more commonplace. We will leave such cutting-edge approaches to our final chapter.

Terms

correlative predictive models	hot-spots	pattern detection
	nation building	pattern exploitation
explanatory predictive models	pattern characterization	precision agriculture
hot-linked	pattern comparison	spatial data query

Review Questions

1. Provide a list of the hierarchy of uses as outlined in this chapter. What is the primary reason for this particular structure?

2. Next to each hierarchical category of use discuss one application in depth. Now, using a formulation flowchart, describe the general approach to its modeling (assuming all necessary elements).

3. List at least two applications for each category of applications that are not explicitly described in the text but are available in the literature and/or web.

4. List at least five specific trade magazines, research journals, and/or web locations where you might find GIS applications for each of the following subject matter categories:

   a. Terrestrial ecology

   b. Oceanography

   c. Criminology

   d. Urban or regional planning

   e. Emergency services

   f. Homeland security

5. Why do we say that GIS modeling is a cyclical process? What example was used in this chapter to illustrate that? Provide another example from models you have seen that show how pattern identification is only the first output from simple analysis and is later incorporated into pattern exploitation models.

6. In previous chapters we learned about the process of flowcharting. What makes the flowcharting different in the powerline corridor siting problem above? Why is the problem itself so difficult? How do we systematically attack this problem?

7. What are the limitations of using a traditional, static GIS to examine spatial temporal data? Why might raster data be more useful for such modeling tasks than vector?

8. Why do we need to spend as much effort to examine independent variables as well as dependent variables when developing GIS models? Pro-

vide an example not given in your text where several independent variables may very well introduce serious problems for GIS model development.

9. What is the difference between correlative predictive GIS models and explanatory predictive models? Why is this distinction important?

10. Beyond spatial interpolation, what other GIS predictive methods are available to us?

References

Church, T., R. J. Brandon, and G.R. Burgett, 2000. "GIS Applications in Archaeology: Method in Search of Theory." In K. L. Westcott and R. J. Brandon, Eds., *Practical Applications of GIS for Archaeologists: A Predictive Modeling Kit.* London: Taylor and Francis, pg. 135–155.

DeMers, M.N., 2002. *GIS Modeling in Raster.* New York: John Wiley & Sons.

Doucette, M., 1994. "GIS Addresses Land Issue of Indigenous Peoples." *GIS World,* 7(1):28.

Hartman, J., 2001. "GIS and Preservation of Tribal Culture," *Native Geography: Guide to the ESRI Native American/First Nations Program,* Vol. 1, pp. 20–21.

Helner, T., 2001. "ArcGIS 8 Enabling Energy Delivery Resource Planning (EDRP)," ESRI International User Conference.

Kohler, T.A., 1988. Predictive locational modeling: history and current practice. In W.J. Judge and L. Sebastian, Eds., *Quantifying the Present and Predicting the Past,* Denver: Bureau of Land Management, pp. 19–96.

Longley, P., G. Higgs, and D. Martin, 1994. "The Predictive Use of GIS to Model Property Valuations." *International Journal of Geographic Information Systems,* 8(2):217–235.

Rich, M., 2001. *A Model Cadastral Geographic Information System Process: The Torrance County Case Study.* Master's Thesis, Unpublished, New Mexico State University, Las Cruces.

Simpson, J. W., R. E. J. Boerner, M. N. DeMers, L. A. Berns, F. J. Artigas, and A. Silva, 1994. "48 Years of Landscape Change on Two Contiguous Ohio Landscapes." *Landscape Ecology,* 9(4):261–270.

Vann, R., 2000. "Populating a GIS of Utility Corridor Assets Using an Integrated Airborne Data Acquisition System." ESRI International User Conference.

Westcott, K. L., and R. J. Brandon, Eds., 2000. *Practical Applications of GIS for Archaeologists: A Predictive Modeling Kit.* London: Taylor and Francis.

Wescott, K. L., and J. R. Kuiper, 2000. Using a GIS to Model Prehistoric Site Distributions in the Upper Chesapeake Bay, in K. L. Westcott and R. J. Brandon, Eds., *Practical Applications of GIS for Archaeologists: A Predictive Modeling Kit.* London: Taylor and Francis, pp 59–72.

Wesseling, C.G., D. Karssenberg, W.P.A. Van Deursen, and P.A. Burrough, 1996. "Integrating Dynamic Environmental Models in GIS: The Development of a Dynamic Modeling Language." *Transactions in GIS,* 1:40–48.

CHAPTER 17

GIS Research

Geographic information systems provide a robust set of tools for the input, storage and editing, retrieval, manipulation, and output of geospatial data and information. It would be easy to become complacent about the state of the art of both the tools as they are currently implemented inside the computer and the theoretical foundations on which they are based. This is especially true because the software has become so commonplace and its user interfaces so elegant that nearly everything imaginable seems to have already been considered and is comfortably at our fingertips. In many of the previous chapters, however, particularly those in Unit 3, we have encountered indications of the limitations of both the theory and the technology. You might remember, for example, the discussion of how difficult it can be to explicitly define the numerical values representing friction surfaces. You might also recall the research going on concerning different methods of modeling viewsheds and how new algorithms may be needed to more accurately represent the real conditions of viewshed analysis.

Until recently, relatively few data models and associated data structures were used to represent geographic reality inside the computer. Software vendors are attempting to improve these data representations, and, in some cases these new methods have already been operationalized. Among the many changes are a continuing move from UNIX-based workstation environments to PC-based implementations, adoption of new computational norms involving visual-based computer languages, and object-oriented models and database management systems. The graphical user interfaces for GIS software have already improved the utility of GIS software immensely over their earlier command-line counterparts.

GIS digitizing software has become more sophisticated and more accurate. Large-format scanners have improved in quality while dropping markedly in price. Hard-drive space has grown from megabytes through gigabytes and now to terabytes, and the price per unit of disk space has plummeted. At the same time, the speed of the computer processor has led to both faster input—especially scanned raster input—and more sophisticated compression and decompression algorithms.

More algorithms are being added to the GIS toolkit every day—both those designed as enhancements or options for existing analytical techniques, and those that might be considered totally new methods of analysis. Previous research has examined the validity of a minimal set of existing algorithms, and the different theoretical frameworks and constraints upon which they are based. Some researchers have also attempted to examine the efficiency of algorithms from both a computational and an implementation viewpoint.

Beyond data models and computational geography, there is a growing interest in enhancing the output subsystem of the GIS. In Chapter 14 we examined both traditional and nontraditional methods of output, but much remains to be done. One of the more intriguing themes for future GIS research is that of geographic visualization. Some of this research is being applied to make the output from analysis more meaningful and more telling. Some is also being linked to methods of geographic data mining wherein the visualization of spatial data contained in the GIS database is being employed through the cyclical GIS analysis process to allow interactive examination of data. By developing both statistical and visualization techniques for detecting patterns within large volumes of interacting and fused spatial data, in the future the analyst will have a much more robust set of tools with which to work.

Although definitions vary, what we have loosely identified in these past few paragraphs is an emerging science known as **geographic information science** (GIScience). For those who are not satisfied with using canned packages but would rather enhance both the contents and the package itself, this is a very exciting new discipline. Moreover, the new science is growing rapidly and is likely to result in new, more powerful, and more realistic models of the spatiotemporal nature of reality. Researchers have many different views of the specific directions that GIScience should take. These are often related to perceived levels of urgency, which in turn may depend on the degree to which they might have encountered them through applications or to the particular focus of their own research. The following pages attempt to provide a balanced survey, but keep in mind that your author is just as susceptible to experiential bias as those whose work he cites. In addition, neither the order in which these emerging research themes are presented, nor the amount of material devoted to each, is meant to imply any level of importance.

Some readers might be asking why emerging research themes are presented in an introductory text. Because of time constraints or other reasons, some instructors may choose to ignore this final chapter entirely. The answer to this question is that, although many of you may be more interested in the day-to-day operations and applications of GIS, others of you may wish to pursue GIScience and related technologies as a full-time career. With luck, many of you might pursue the very research themes described here. The intent is more to pique your interest than to burden you with detail. As such, the pages that follow are more likely to have many more questions than answers. If you have gotten this far, you are probably among those who wish to pursue these exciting and challenging subjects.

LEARNING OBJECTIVES

When you are finished with this chapter you should be able to:

1. List and describe at least five major areas of research likely to push GIS into the future.

2. Describe the fundamental difference between research with GIS technology and research on GIS technology.

3. Define the difference between geographic information systems as enabling technology and geographic information science as a focal point for research.

4. Describe some of the differences between cartography and geographic visualization as well as some of the similarities.

5. Define and discuss data mining and knowledge discovery and compare them to spatial cognition, data representation, and geographic visualization.

6. Describe the term *ontology* and discuss its importance to the future of GIScience.

7. Review the potential impact of new satellite imagery and other remote methods of data capture on GIScience.

8. Discuss the problems with and the role of both error and uncertainty as a yet unformalized methodology within GIS software and its uses.

9. Describe why scale is still such a problem in GIScience.

10. Discuss why algorithm development is so important in GIScience despite the apparent sophistication of existing GIS software and the ready availability of so many talented programmers.

11. List and describe three areas where GIS applications are likely to become increasingly important.

12. Describe the need to examine the sociological impacts both on and by GIS technologies.

GEOGRAPHIC REPRESENTATION (INCLUDING ONTOLOGY)

Among the first problems addressed in the early days of GIS development was how we represent geographic data in the digital environment. The problem is far more complex than finding ways to represent points, lines, polygons, and surfaces and their associated attributes, however. If all geographic phenomena, including objects, processes, relations, were universally understood under all circumstances by all possible users at all different scales, the task of converting these to a form that could be encoded and analyzed in digital form would be considerably reduced. This speaks volumes about the importance of ontology as a fundamental GIScience topic. Not only do we often not have a thorough knowledge of these topics, many of which are situation specific, but we also lack compiled data from human subjects (both experts and nonexperts), as well as methods and tools for describing, accessing, comparing, and integrating them. This means that many GIS software packages use different methods of representation and different methods of analyzing data for the same conditions or set of circumstances.

So what don't we know? A few examples might be useful rather than going into an extended, and mostly philosophical, discussion. We have already seen that friction surfaces are time, circumstance, and organism dependent. How can we, for example model the potential migration patterns of humans (in an anthropological sense) or animals (in the biogeographic sense) if we don't know what the relationships are between the organism under study and their physiological conditions related to specific terrain or travel conditions? How can we effectively model crime if we don't understand how criminals act and interact with their environment? Landscape ecologists, for example, have been able to quantify many aspects of the physical environment, such as fragmentation and isolation, but little is known about the impact of such landscapes on gene flow, migration patterns, desiccation rates, and many other ecological and biological conditions. How are some land-based organisms able to find their way around when others are not? More importantly, how do we represent such conditional contexts within the GIS software?

SCALE

Many of the **ontologies** that are lacking also have associated with them the problem of scale and its representation. Biologists have recognized for some time that both space and time are scale dependent. This means that one cannot represent organisms and their environment at a detailed resolution and then simply scale up to look at the larger environment (O'Neill et al., 1986). In many cases, this requires that we use not a single scale for all organisms, but rather selected scales in which to examine some subset of the biological environment that adequately represents the activities at that hierarchical level *(hierarchy theory)*. This is just as true for human activities as it is for other aspects of our environment. Recognizing how people act and react in their own neighborhood doesn't necessarily indicate how they react in their city, county, or nation. Both time and space become variably important at different scales. Even directionality might very well affect how we as humans interact in space. If, for example, we are going someplace that we don't want to go to versus going someplace where we do want to go influences how we perceive time and space and may have an impact on how we interact with our environment. You might notice, for example, that you tend to move faster going home for a refreshing drink and a relaxing chair than you do going to work or school. Attempts to use fractals to represent geographic space have proven less than totally effective because many of the properties of a wide range of geographic phenomena are not "strictly repeated across multiple spatial or temporal scales" (UCGIS, 1999).

The scale problem permeates nearly every aspect of GIS—from cognition and recognition to representation and modeling; from observation to communication. Among the more important future themes for research on scale are the following:

1. Role of scale in decision making, especially the identification of critical scales in decision making.

2. Evaluation of the appropriate scales for data analysis, especially where integrating multiple datasets. This includes the avoidance of the **ecological fallacy** (the problem of making disaggregate decisions from aggregate data) and problems associated with the **modifiable area unit problem** (the problem of obtaining different results from different area unit definitions) (Wong and Amrhein, 1996).

3. Examination of the cognitive nature of scale. The next section in this text, on visualization, presents an excellent example of a project that examines scale explicitly using visualization as the primary vehicle.

4. Identification of appropriate methods of multiscale analysis. Currently, few methods for analysis are not scale dependent.

5. Development of appropriate, intelligent methods of generalization for representing cartographic documents at different scales.

GEOGRAPHIC VISUALIZATION

The United States National Science Foundation Panel on Graphics, Image Processing, and Workstations defined **visualization** as "a method of computing ... a tool both for interpreting image data fed into a computer, and for generating images from complex multi-dimensional data sets." It went on to explain that its goal was to "leverage existing scientific methods by providing new insight through visual methods" (McCormick et al., 1987, pg. 1). Only five years later, MacEachren et al. (1992, p. 101) expanded that view by indicating that, beyond a method of computing, it was also "first and foremost an act of cognition, a human ability to develop mental representations that allow us to identify patterns and create or impose order." Given that expanded definition, one can then define geographic visualization to encompass spatial hypothesis formulation, identification of spatial patterning, the construction of knowledge from pattern, and the facilitation of decision making (Buckley et al., 2000).

The traditional methods of representing and visualizing geographic space most often fall into some form of cartography. Although the map is still among the most powerful methods of representing geographic phenomena, even its digital version has serious limitations. We have seen, for example, the addition of new tools to allow the cartographer and GIS professionals not only to display multiple datasets on a single perspective map view but also to show these as flythroughs, simulating the movement of an aircraft around topographic or even nontopographic statistical surfaces. Such visualizations assist the GIS professional in identifying patterns that may not be readily visible in a static orthogonal map. Moreover, sampling of datasets at different levels of spatial or temporal aggregation may lead to answers to questions of appropriateness of scale representation itself.

One example of using visualization to determine the appropriateness of scale for landform analysis involved employing numerous two-dimensional and three-dimensional visualization techniques for the same topographic surface to determine the scale dependency of slope over the surface (accessed [September 12, 2002] http://www.le.ac.uk/jwo/research/conferences/ibg99/). The premise is that the properties of a landscape are strongly associated with the scale at which it is observed. Within a fractal-generated (scale-independent) landscape, the author employed colored elevation, shaded relief, draped slope, distant three-dimensional view, foreground and background emphasizing view, and near view only (Wood 1998, 1999: Figure 17.1). None of the two-dimensional views was adequate; the authors suggest that the flythrough of a three-dimensional visualization provides the best view for visualization because it immerses the viewer in the landscape at a variety of scales. This seemed to provide the most information for the interpreter to evaluate the landscape.

The previous example of geographic visualization illustrated the strong relationship of scale to visualization. It also demonstrated the role of dynamic display in capturing the viewer's attention and taking advantage of the human capacity to incorporate changing amounts of space and time data simultaneously. Other nondynamic methods have proven to be equally effective, such as that developed by Hargrove et al. (2002) (Figure 17.2) at Oak Ridge National

(a) Coloured elevation   (b) Shaded relief   (c) Draped slope

(d) Distant 3d view   (e) Foreground and background   (f) Near view

Figure 17.1   Different views used to visualize a geomorphological surface. The surface is actually a simulation so that it can be experimentally controlled. (Wood: 1999) Used with Permission.

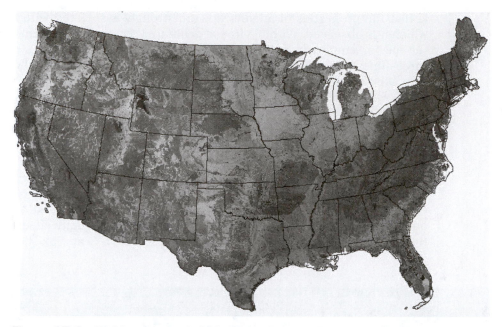

Figure 17.2   Multivariate statistical approach to characterizing ecoregions. The fall line / Atlantic flatwoods, the coastal plain, the Piedmont, the Arkansas Blacklands and the Ozarks are readily separable using this visualization technique. http://gis.esri.com/ library/userconf/proc98/PROCEED/TO350/PAP333/P333.HTM

Laboratories. They suggest the interaction of spatial statistical techniques with the visualization tools of digital mapping to examine the many ways that biogeographic regions can effectively be developed. The research in geographic visualization is still in its infancy. Yet it is likely to become one of the more exciting in the GIScience discipline because it incorporates technology, psychology, programming, theory development, and many other fields of knowledge. For the researcher who yearns for challenges and who likes to integrate seemingly disparate fields of study, this is an ideal subject for investigation.

Myriad approaches to geographic visualization exist—far too many to list here or to discuss in detail. The topic, however, is among the emerging themes recommended by the University Consortium for Geographic Information Science (http://www.ucgis.org) during its 2001 meetings. A white paper on the subject (Buckley et al., 2001) provides an extensive overview of the subject, including a detailed list of URLs for specific, well-equipped laboratories that focus on the study of geographic visualization. The white paper is also available on the UCGIS web site.

DATA MINING AND KNOWLEDGE DISCOVERY

Among the many uses of visualization is that of **data mining** and **knowledge discovery** (KD). Fayyad et al. (1996, pg. 37) define data mining and knowledge discovery as "the non-trivial process of identifying valid, novel, potentially useful and ultimately understandable patterns in data." Thus, in geographic data mining and knowledge discovery, we attempt not only to identify heretofore unseen spatial patterns of phenomena through unique and innovative approaches of geographic display, integration, and analysis but also to find ways to exploit them (see Chapter 16). These same techniques might allow us to envision previously unseen aggregations of patterns that themselves constitute new knowledge.

In earlier chapters, we saw that it was generally better to know beforehand what you were looking for before you even began building a geographic database. In most GIS applications this is true, but imagine if we were able to analyze vast quantities of geospatial data that are collected on a regular database with the sole purpose of finding out what we could learn from it. This is the spatial equivalent of decoding thousands of e-mail messages, telephone conversations, and telegraphs to tease out whether or not there was a terrorist threat that we could identify. Recent events, most notably the toppling of the World Trade Towers in 2001, suggest that, in addition to many other aspatial approaches to intelligence gathering and analysis, there is likely to be an increasing role in the mining of geospatial data for discovering knowledge and predicting the actions of terrorist groups, rogue governments, or individuals.

Other applications of knowledge discovery and data mining might be to examine the vast amounts of crime data that have yet to be mapped to determine whether trends in crime are predictable at different scales, or to study the millions of pieces of health-related data to determine disease patterns and suggest possible solutions. Global climate modeling, earthquake predictions, global economic market analysis, and many more could conceivably benefit from geospatial data mining and knowledge discovery. There is a need for new tools (Buttenfield et al., 2000) to allow us to effectively identify, categorize, and characterize the latent patterns hidden among the monstrous volumes of data that we collect on a regular basis.

As suggested, KD and data mining are nontrivial, usually because of the shear volume of the dataset, although the integrative or fused nature of the dataset may also result in a level of complexity that challenges existing techniques for analyzing data. This extends to inductive approaches to data analysis—often mistaken for knowledge discovery. Although some form of inductive logic is likely to be applied to many KD tasks, the ideal methodology would be one of **abductive logic** involving a simultaneous process of uncovering some structure within the data and an hypothesis to explain it (Buttenfield et al., 2000). Three important aspects of KD and data mining are that the structure uncovered through the process should be valid, novel, and useful (that is, useful in the context in which the search for knowledge was conducted). Although many patterns have been discovered through a wide range of GIS and non-GIS related approaches, the number of successful KD applications has been limited, primarily because the pattern's relevance and the cause are extremely difficult to prove. The subset of successful geographic applications of knowledge discovery and data mining is even smaller.

The first major effort to mine geographic data was the geographic analysis machine developed in the early 1990s (Openshaw et al., 1990). This approach was a computationally exhaustive, brute force method of analysis—a successful approach when the dataset is small. Given that the assumption in data mining of massive datasets, such as methodology would probably be ineffective. Instead, we need to develop methods that are computationally efficient, often employing highly selective sampling techniques that allow for the smallest loss of data possible. The development of such sampling methods might prove an important GIS research topic in itself.

One advantage of KD and geographic data mining is that if we can identify verifiable, useful patterns of geographic data, these may well lead to a new revolution by providing an answer to the following question: "What are the future problems we have not been able to answer in the past?" Some of these new approaches may come from improved automated methods of map interpretation and information extraction, such as that employed by Malerba et al. (2001). Similarly, Gopal, Liu, and Woodcock (2001) have shown how neural networks can be combined with visualization techniques to extract information from remotely sensed imagery.

Many of the geographic problems of knowledge discovery stem from the complexity and multidimensionality of the attributes. To extract knowledge from these, Eklund, Kirkby, and Salim (1998) and Lees and Ritman (1991) have shown that inductive applications of artificial neural networks and decision trees might prove useful for mapping soils and vegetation in complex environments. The temporal domain was added in research by Mesrobian et al. (1996) to query, explore, and visualize meteorological data by extracting unusual air pressure and wind patterns over time. This approach was extended to include interaction, flow, and movement in human systems. Because the environmental structure often imposes constrictions and rules on the flows and interactions of humans, their interactions with environment may often provide insights into the spatial structure of the system itself. In most cases, research examining these interactions looks primarily at direct and proximal effects. As the interactions become more subtle and more complex, the influences may become buried. Some researchers are now exploring the use of visualization and trajectory techniques for examining the massive datasets that are becoming more common through real-time monitoring systems, intelligent transportation systems, and other "position-aware" technologies such as cellular telephones and wireless Internet clients (Buttenfield et al., 2000; Marble et al., 1997). As new devices and new datasets become available, additional techniques need to be developed to exploit them to acquire space-time data.

ERROR AND UNCERTAINTY (ESPECIALLY ERROR PROPAGATION)

Among the early focal points of GIScience research was the analysis of error and uncertainty in geographic databases. These two concepts are related and are differentiated by Goodchild et al. (1994) who clearly indicate that uncertainty is a relative term, whereas error is a quantifiable difference from some real-world value. Whether your interest is in error analysis or in uncertainty analysis, the topic is a difficult and complex one. Despite decades of research on the topic, we still know little about error and uncertainty, particularly as they propagate through the GIS during modeling. Burrough and McDonnell (1998) have provided a comprehensive review of what has already been done on the subject. Their discussion shows the complexity of modeling uncertainty and error in a modeling context yet provides a readable discussion of some of the approaches that have been taken. They also indicate that there much remains to be learned.

The UCGIS (www.ucgis.org) members provide a useful list of appropriate research topics for the future. Among the more important topics are those providing detailed understanding of the mechanics of how uncertainty arises in geographic data. Once identified, it is also important to develop techniques for quantifying, reducing, and representing geographic data uncertainty and for predicting its propogation of this uncertainty through GIS-based data analyses.

INSTITUTIONAL, CULTURAL, AND SOCIOLOGICAL IMPLICATIONS OF GIS

Although we typically treat GIS and GIScience as somehow isolated from the world in which we live, or monolithically as a method of modeling and manipulating that world, it is important to recognize that GIS itself is part of our world as well. Questions of access to GIS are becoming as important as access to legal services, transportation, or any other commodity. There is a need to examine the possible impact of differential access to GIS on different societal groups, just as there is a need to examine the impact of differential access to a good attorney in capital murder cases on death sentences. Those who have access to and knowledge of GIS technology often have huge advantages over those who don't for a variety of human activities. As an example, imagine the advantage a real estate firm that uses GIS on a regular basis might have over other firms that do not.

GIS-related technologies may also impact our personal freedoms and privacy. When you or someone you know goes shopping on-line, check to see how often the vendor will make suggestions for things that you should buy. This is a direct result of data gathering, both actively and passively, through your interactions with the computer. It is not a large leap to add geospatial data to this mix to evaluate your probable personal income, size of family, and a vast array of additional pieces of information through extrapolation of readily available census data. These suggest a growing need for research on the societal aspects of GIS in all its many aspects. Who is using GIS? What factors cause some to use it and others not to? How is GIS being used in ways that might contribute to improvements in society? The list is nearly endless.

The University Consortium for Geographic Information Science (www.ucgis.org) recommends that future research related to society include the following topical groups of questions:

1. Capabilities and limitations of the current state of the art in GIS software with regard to the cultures and social groups that might benefit from an empowering technology. Are the GIS technologies being employed altering the nature of the geographic space to which they are applied?

2. Questions of organizational implementation of GIS, particularly regarding how adoption of the technology is affecting how organizations operate, its effectiveness in performing necessary tasks, who is or is not adopting GIS and why. It is also important to know how the adoption of GIS affects

the organizational structure (i.e., is it increasing the economic and governmental power of the organization?).

3. Legal and ethical questions such as what are the ethical implications of using GIS? Is there a need for a code of ethics for employing GIS? Who should have access to geographic data? How do we handle sensitive data? What are the possible impacts of GIS as a surveillance and/or regulatory tool?

APPLICATIONS

In its early years, management of natural resources drove the development of GIS technology. Most of the applications are still devoted to natural resources, but this is often still as much a function of necessity as that of the availability of data. Although we constantly collect georeferenced physical data through field surveys, remote sensing, GPS, and a host of other methods, cultural, social, and economic data are collected regularly. We have already seen that vast amounts of population data are collected through national censuses. Many of us provide economic data every time we go to a store and our purchases are scanned into a computer. Or some of us are polled over the telephone. More recently, we have begun to experience polls through the web sites we visit and even some we don't. These, together with emerging data collection technology—especially technology that collects data on individual people's activity—suggest a whole suite of new applications. This circumstance is especially true if some of the previous research themes prove successful.

Given the terrorist attacks of September 11, 2001, all aspects of domestic security are going to comprise an increasing amount of GIS applications. Although domestic security is not necessarily limited to acts of terrorism, the identification of threats, whether domestic or foreign, is likely to be an increasing component of security (Figure 17.3). Because such threats require analysis of vast amounts of interacting data, and because those conspiring to commit such acts are doing everything in their power to conceal their intentions from their potential victims, these data volumes and interactions will require improvements in data mining and knowledge discovery techniques.

Another set of applications likely to increasingly involve GIS is the prediction of large-scale global climate change and the impacts of those changes on the world's people. This includes predictions of flooding potential along susceptible coastlines, identification of potential food shortages, even preparation of contingency plans for responding to all types of natural disasters at a continental scale. Although the technology for this approach exists, detailed knowledge of the geography does not. For us to be successful, it is important to improve our basic understanding of the geography behind GIS.

Associated with predictive modeling is the need for dynamic GIS modeling. Although some technologies exist or are being developed for dynamic GIS modeling, much remains to be done before it becomes readily available. Still, the rapid increase in computer speed is likely to make existing methods of dynamic GIS modeling more available. Dynamic GIS modeling of urban expansion and its associated impact on the transportation net is likely to become more important

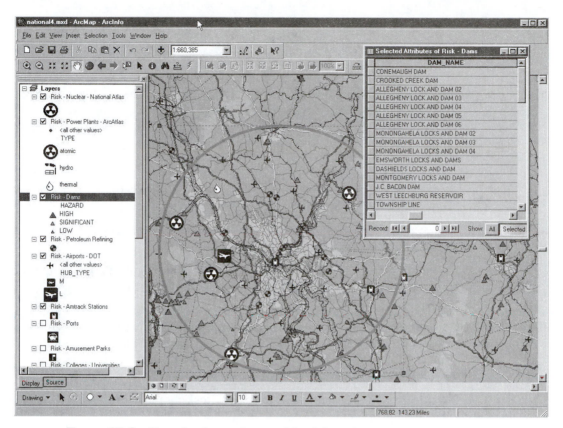

Figure 17.3   Homeland security graphic. Selected attributes of risk. In this case the selected attributes are dams.

as urbanization continues its pace. Other forms of dynamic modeling will also involve examining dispersal of aerosols, especially those associated with transnational boundaries, predicting insect outbreaks, and a host of others. An extreme form of dynamic modeling already taking advantage of GIS, and likely to continue to do so, is that of battlefield simulation and other forms of simulation as well. As advanced visualization techniques and tools increase, it is highly likely that simulation and gaming will be used to evaluate a vast array of potential scenarios no matter what domain of knowledge is employed. This will also prove very useful for the development of truly prescriptive GIS models—another weakness of GIS as it is currently conceived.

Terms

abductive logic

data mining

ecological fallacy

geographic information
  science

hierarchy theory

knowledge discovery

modifiable area unit
  problem

ontologies

visualization

Review Questions

1. List and describe at least five major areas of research likely to impact GIScience and GIS technology in the next 10 years. 20 years.

2. What is GIScience? What is the difference between research with GIS technology and research on GIS technology?

3. Describe some of the differences and similarities between Cartography and Geographic Visualization as well as some of the similarities.

4. Define data mining and knowledge discovery. What's the difference between these topics and simple spatial cognition and spatial data representation? Are they related? Are they related to Geographic Visualization?

5. What does the term ontology mean in general? Describe what it means in the context of GIScience.

6. With all the satellite imagery, digital GIS datasets, and GPS locational data available, why do we need to study more about them?

7. The terms error and uncertainty seem to be pretty much the same. What is the fundamental difference between them? Why, with all the years of research examining these topics have we still not developed a systematic method of accounting for them? Where is error and uncertainty most problematic?

8. You learned about map scale in the earliest chapters of this book. Why is it such a problem in GISience?

9. Why do you think algorithm development is such a problem? Aren't there plenty of computer programmers out there that can solve these issues?

10. List and describe three areas where GIS applications are likely to become increasingly important. Explain your response.

11. Discuss how GIS exists in a sociological environment by describing the impact of society on the future direction of GIS technologies.

References

Buckley, A. R., M. Gahegan, and K. Clarke, 2000. Geographic Visualization. UCGIS White Paper, December 1 (http://www.ucgis.org).

Burrough, P. A., and R. McDonnell, 1998. *Principles of Geographical Information Systems.* London: Oxford University Press.

Buttenfield, B., M. Gahegan, H. Miller, and M. Yuan, 2000. Geospatial data mining and knowledge discovery. UCGIS White Paper, December 1 (http://www.ucgis.org).

Eklund, P.W., S.D. Kirkby, and A. Salim, 1998. "Data Mining and Soil Salinity Analysis," *International Journal of Geographical Information Science,* 12, 247–268.

Fayyad, U., G. Piatetsky-Shapiro, and P. Smyth, 1996. "From Data Mining to Knowledge Discovery in Databases," *AI Magazine,* fall, pp. 37–54.

Goodchild, M.F., B. Buttenfield, and J. Wood, 1994. "Introduction to Visualizing Data Quality," in H. M. Hearshaw and D. J. Unwin, Eds., *Visualization in Geographical Information Systems.* New York: John Wiley & Sons, pp. 141–149.

Gopal, S., W. Liu, and C. Woodcock, 2001. "Visualization Based on Fuzzy ARTMAP Neural Network for Mining Remotely Sensed Data," in H. J. Miller and J. Han, Eds., *Geographic Data Mining and Knowledge Discovery.* London: Taylor and Francis.

Hargrove, W.W., F.M. Hoffman, and P.M. Schwartz, 2002. "A Fractal Landscape Realizer for Generating Synthetic Maps." *Conservation Ecology,* 6(1):2. [on-line] URL: http://www.consecol.org/vol6/iss1/art2

Lees, B.G., and K. Ritman, 1991. "Decision-Tree and Rule-Induction Approach to Integration of Remotely Sensed and GIS Data in Mapping Vegetation in Disturbed or Hilly Environments." *Environmental Management,* 15, 823–831.

MacEachren, A., B. Buttenfield, J. Campbell, D. DiBiase, and M. Monmonier, 1992. "Visualization," in *Geography's Inner Worlds: Pervasive Themes in Contemporary American Geography.* Piscataway, NJ: Rutgers University Press, pp. 101–137.

Malerba, D., F. Esposito, A. Lanza, and F.A. Lisi, 2001. "Machine Learning for Information Extraction from Topographic Maps," in H.J. Miller and J. Han, Eds., *Geographic Data Mining and Knowledge Discovery.* London: Taylor and Francis.

Marble, D.F., Z. Gou, L. Liu, and J. Saunders, 1997. "Recent Advances in the Exploratory Analysis of Interregional Flows in Space and Time," in Z. Kemp, Ed., *Innovations in GIS 4.* London: Taylor and Francis, pp. 75–88.

McCormick, B., T. DeFant, and M. Brown, 1987. "Visualization." *Scientific Computing in Computer Graphics,* 21: i-E-8.

Mesrobian, E., R. Muntz, E. Shek, S. Nittel, M. La Rouche, M. Kriguer, C. Mechoso, J. Farrara, P. Stolorz, and H. Nakamura, 1996. "Mining Geophysical Data for Knowledge." *IEEE Expert,* 11(5):34–44.

O'Neill, R.V., D.L. DeAngelis, J.B. Waide, and T.F.H. Allen, 1986. *A Hierarchical Concept of Ecosystems.* Princeton, NJ: Princeton University Press.

Openshaw, S., A. Cross, and M. Charlton, 1990. "Building a Prototype Geographical Correlates Machine." *International Journal of Geographical Information Systems,* 4(4):297–312.

UCGIS, 1999. Scale. White paper available at http://www.UCGIS.org.

Wong, D., and C. Amrhein, Eds., 1996. "The Modifiable Areal Unit Problem." Special issue of *Geographical Systems,* 3:2–3.

Wood, J. 1998. "Modelling the Continuity of Surface Form Using Digital Elevation Models." In T. Poiker, and N. Chrisman, Eds., *Proceedings of the 8th International Symposium on Spatial Data Handling,* IGU, Vancouver, pp. 725–736.

Wood, J. 1999. "Visualization of Scale Dependencies in Digital Elevation Models." *Proceedings, International Cartographic Association Annual Meeting,* Ottawa. 98 99. Downloaded from (http://www.soi.city.ac.uk/~jwo/ica99/wood_paper322.doc.), retrieved November 16, 2003.

Software and Data Sources

This appendix is a sample of available GIS software and sources of data for use with the software. Although this list is not complete, it should be enough to get you started either in a learning setting or a commercial operation. Additional data sources and vendors can be found by searching through some of the new GIS, remote sensing, and surveying trade journals, as well as through your instructor.

GIS SOFTWARE VENDORS

Arc/Info, Arc/View, MapObjects
    Environmental Systems Research Institute
    380 New York Street
    Redlands, CA 92372
    (714) 793–2853

Spans GIS
    Intera TYDAC Technologies Inc.
    1500 Carling Avenue, Ottawa, Ontario K1Z 8R7,
    Canada
    (613) 722-7508

GeoMedia
    Integraph Corporation
    Huntsville, AL 35894
    (205) 730–2000

ERDAS IMAGINE
    ERDAS, Inc.
    2801 Buford Highway
    Atlanta, GA 30329
    (404) 248-9000

GRASS
GRASS Information Center
USACERL
Attn: CECER-ECA
P.O. Box 9005
Champaign, IL 61826-9005

Atlas GIS
Strategic Mapping Inc.
4030 Moorpark Avenue
San Jose, CA 95117-4103
(408) 985-7400

GisPlus
Caliper Corporation
1172 Beacon Street
Newton, MA 02161
(617) 527-4700

MapInfo
MapInfo Corporation
200 Broadway
Troy, NY 12180-3289
(518) 274-6000

IDRISI
The Clark Labs for Cartographic Technology and Geographic Analysis
Clark University
950 Main Street
Worcester, MA 01610
(508) 793-7526

MapGrafix
ComGraphix, Inc.
620 E Street
Clearwater, FL 34616
(813) 443-6807

OSU Map-for-the-PC
Duane Marble
Department of Geography
The Ohio State University
Columbus, OH 43210

U.S. DEPARTMENT OF AGRICULTURE (USDA)

Natural Resources Conservation Service (NRCS)

The following information is available from

National Cartographic and Geographic Information Systems Center
USDA-NRCS
P.O. Box 6567
Fort Worth, TX 76115
(817) 334-5559
Fax: (817) 334-5290

Soil Survey Geographic Database (SSURGO). This is the most detailed soil mapping for the United States, ranging in scale from 1:12,000 to 1:31,680. It duplicates the original soil survey maps found in county soil survey documents.

State Soil Geographic Database (STATSGO). Generalized SSURGO data at a scale of 1:250,000, which covers state, regional, and multicounty areas.

National Soil Geographic Database (NATSGO). Data based on generalized state-level soils maps and formed from the major land resource area (MLRA) and land resource region (LRR) boundaries. Digitized at 1:7,500,000.

U.S. DEPARTMENT OF COMMERCE

Bureau of the Census

Census data in the form of TIGER files are available from

Customer Services
Bureau of the Census
Washington Plaza, Room 326
Washington, D.C. 20233
(301) 763-4100
Fax: (301) 763-4794

National Oceanic and Atmospheric Administration (NOAA)

National Environmental Satellite, Data, and Information Service (NESDIS)

U.S. weather records, collected by the National Climatic Data Center (NCDC), and international weather records, collected by the World Data Center A (WDC-A), are available from:

NOAA
National Climatic Data Center
Climate Services Division
NOAA/NESDIS E/CC3
Federal Building
Asheville, NC 28801-2696
(704) 259-0682
Fax: (704) 259-2876

National Geophysical Data Center (NGDC)

Solid earth, solar-terrestrial physics, and snow and ice data from worldwide sources are available from

National Geophysical Data Center
Information Services Division
NOAA/NESDIS E/GC4
325 Broadway
Boulder, CO 80303-3328
(303) 497-6958
Fax: (303) 497-6513

National Oceanic Data Center (NODC)

Oceanographic station data, bathythermograph, current, biological, sea surface, and GEOSAT data are available from

National Oceanographic Data Center
User Services Branch
NOAA/NESDIS E/OC21
1825 Connecticut Avenue, NW
Washington, D.C. 20235
(202) 606-4549
Fax: (202) 606-4586

National Ocean Service (NOS)

Sailing charts, coast charts, and harbor charts are available from NOS. For information, contact

Coast and Geodetic Survey
National Ocean Service, NOAA
1315 East-West Highway, Station 8620
Silver Spring, MD 20910-3282
(301) 713-2780

U.S. DEPARTMENT OF THE INTERIOR

U.S. Geological Survey (USGS)

Digital line graphs, digital elevation models, side-looking airborne radar (SLAR), advanced very high resolution radiometer (AVHRR), digital orthophotoquad, land satellite (*LANDSAT*) data, *système probatoire d'observation de la terre* (SPOT) data, defense mapping agency (DMA) data, national wetlands inventory (NWI) data, and national biological service (NBS) data are all accessible from

Earth Science Information Center
U.S. Geological Survey
507 National Center
Reston, VA 22092
(703) 648-5920
Fax: (703) 648-5548
Toll free: 1-800-USA-MAPs

or

Sioux Falls ESIC
U.S. Geological Survey
EROS Data Center
Sioux Falls, SD 57189
(605) 594-6151
Fax: (605) 594-6589

CANADIAN DATA

National Digital Topographic Data are available from

> Topographic Surveys Division
> Surveys and Mapping Branch
> Energy, Mines and Resources Canada
> 615 Booth Street
> Ottawa, Ontario K1A 0E9, Canada
> (613) 992-0924

Data from the Canada Land Data System (including the Canada Geographic Information System) are available from

> Environmental Information Systems Division
> State of the Environment Reporting Branch
> Environment Canada
> Ottawa, Ontario K1A 0H3, Canada
> (613) 997-2510

Canadian soils information system data are available from

> CanSIS Project Leader
> Land Resource Research Centre
> Agriculture Canada, Research Branch
> K.W. Neatby Building
> Ottawa, Ontario K1A 0C6, Canada
> (613) 995-5011

Landsat and SPOT data in Canada are available from

> Canada Centre for Remote Sensing
> 2464 Sheffield Road
> Ottawa, Ontario K1A 0Y7, Canada
> (613) 952-2171

or

> Canada Centre for Remote Sensing
> Prince Albert Receiving Station
> Prince Albert, Saskatchewan S6V 5S7
> (306) 764-3602

Census data for Canada are available from

> User Summary Tapes
> Electronic Data Dissemination Division
> Statistics Canada
> 9th Floor, R.H. Coates Building
> Ottawa, Ontario K1A 0T6, Canada
> (613) 951-8200

Using the Wiley World Wide Web to Find Data and GIS Examples

In Appendix A, I described a number of locations for data available through the mail. Many of these sources of data are now accessible using the World Wide Web (WWW). In fact, even a cursory perusal of the net using the key words *geographic information system* will provide a wide array of data sources, listserves, interest groups, class materials, active GIS research agendas, GIS output, sources of GIS employment, software and hardware vendors, and the like. A major problem with the Web is that you can spend an enormous amount of time surfing before you find the object of your search. I have compiled a list of sources that have come to my attention while searching. My intent is to shorten the amount of time you spend surfing and increase your productivity. This list is available directly from the New Mexico State University geography department Web page. The universal resource locator (URL) is

http://www.nmsu.edu/~geoweb/gisrc.html

Among the more useful websites for finding and sharing data is the GIS Data Depot at http://www.gisdatadepot.com. They call themselves the ". . . largest geo-spatial data repository on the web!" The site provides sources for both free data and data for purchase. In many cases the data for purchase are bundled to reduce the cost. Additionally, data can be downloaded directly from the web in some cases, or even created as a custom CD.

Another recent source of GIS data, often for free or in-kind sharing is the geography network at http://geographynetwork.com. Ranging from academic sources, to professional data providers and from ready-to-use datasets to software, this growing site is sure to be an important resource for both teaching and research.

Abductive logic: The synthetic inference that allows for finding explanations for observations in an existing theory.

Absolute barriers: Barriers that prevent movement through them.

Absolute location: Location on the earth or on a map that has associated with it a specific set of locational coordinates.

Accessibility: A measure of arrangement that focuses on the connectedness of an area object to other area objects.

Address matching: The process of defining exact addresses and linking them to specific locations along linear objects.

Affective objective: One of two parts of the purpose of a map, this one relating to the overall appearance of the map. Controls how information contained in a map is to be portrayed.

Aggregation: One of a number of numerical processes that place individual data elements into classes.

Alber's equal area projections: Widely used equal area conic projections that contain two standard parallels. Good for midlatitude areas of greater east-west than north-south extent.

Alidade: A telescopic survey device that uses magnetic bearings for determining locations on the earth. It is commonly used with an associated traverse table or plane table, so scaled values can be recorded directly on paper or mylar.

Allocate: The process of selecting a portion of geographic space to satisfy a predefined set of locational criteria. This process is usually performed on a vector data structure for the purpose of selecting a portion of a network.

Alpha index: A measure of network complexity that examines network circuitry.

AM/FM: Automated mapping/facilities management. Computer assisted cartography, par-

ticularly applied to the display and subsequent analysis of facilities associated with the function of urban and rural areas.

Analytical paradigm: A conceptual model, sometimes called the holistic paradigm, where the map is viewed both as a means of graphic communication and as a means of numerical spatial analysis.

Angle of view: The angular distance above the ground from which a perspective map can be shown.

Angular conformity: The property of some map projections that retain all angular relations on the map.

Application developers: GIS professionals specializing in the development of specific databases, analytical functional capabilities, and appropriate graphical user interfaces to allow specific applications to be performed by non-GIS specialists.

Application requirements model: A database design model focusing on the application needs as a means of properly completing the design.

Applications development: The process employed by a group of GIS professionals to allow non-GIS users easy access to the technology for specific tasks.

Arbitrary buffers: Buffers whose distances are selected without regard for any known criteria.

Area cartograms: Cartograms that vary the size of the polygons on the basis of the statistical value represented, rather than on their actual areal dimensions.

Arrangements: 1. A number of measurements indicating how spatial data are organized through such attributes as nearness and connectivity, as well as others. 2. The spatial distribution of internal markings for cartographic symbols to allow visual separation.

Aspect: The azimuthal direction of surface features.

Association: Spatial relationship that exists between different elements of the earth that occur at the same locations. An example would be a particular vegetation type occurring on north-facing slopes.

Attribute error: Incorrect or missing attributes.

Attribute pseudo nodes: Spatial pseudo nodes that are a result of explicit attribute changes along a line.

Attributes: Nongraphic descriptors of point, line, and area entities in a GIS.

Audience: Map design consideration based on the background and map-reading skills of the potential map user.

Available data: Map design limitation based on the quality, quantity, and reliability of the data used for mapping.

Azimuthal equidistant projections: Types of azimuthal map projections in which the linear scale is uniform along the radiating straight lines through the center. Allows for the entire sphere to be shown on a single projection.

Azimuthal projections: A family of map projections resulting from conceptually transferring the earth's coordinates onto a flat surface placed perpendicular to the sphere.

Barrier: Some object whose attributes either stop or impede movement through the rest of the coverage.

Binary maps: Maps that contain only two possible values for the same characteristic or attribute. The attribute is either present or absent.

Biophysical mapping: A traditional analog technique used by environmental planners to combine relevant mapped biological and physical data into meaningful combinations for decision making.

Block codes: A compact method of storing raster data as multifaceted blocks of homogeneous grid cells.

Block function: GIS modeling function that uses a manipulation of the properties of a defined, nonoverlapping block of cells to produce an output block of cells showing the new values.

Boolean overlay: A type of overlay operation that relies on Boolean algebra.

Borders: Higher order groups of line objects that exemplify some functional or political demarcation from one region to another.

Bottom-up implementation: One of the major weaknesses of the classical (waterfall model) project life-cycle design methodology where final system testing is left until last, thus requiring large chunks of uninterrupted computer time. Because such large chunks of computer time are often hard to obtain, the development often falls behind schedule.

Buffering: The process of creating areas of calculated distance from a point, line, or area object.

CAD: See Computer-assisted drafting (CAD).

Cadastral: Having to do with cadastre, which involves interests in landownership and management.

Cartesian coordinate system: A mathematical construct defined by an origin and a unit distance in the X and Y direction from that origin.

Cartograms: Maps that emphasize the communication paradigm to a degree that often modifies actual geographic space in deference to the phenomena being displayed.

Cartographic database: A digital database developed from an existing cartographic document. This process results in the appearance of increased accuracy where it does not necessarily exist.

Cartographic modeling: The process of combining individual GIS analytics to create complex models used for decision making.

Cartographic process: The steps involved in producing a map, beginning with data collection and ending with the final map product.

Causative buffers: Buffers whose distances are selected on the basis of some physical phenomena that dictate their location.

Census: A method of collecting data that normally involves tabulating data about an entire population rather than a sample of the population.

Center of gravity: Center of a polygon based on the arrangements of objects within it rather than simply as a measure of the area geometry.

Central place: A node that exhibits a high degree of network linkage intensity.

Central point linear cartograms: Cartograms that show distances from a central point as concentric lines, thus requiring the underlying graphic objects to be modified to conform to those lines.

Centroid-of-cell method: A method of raster encoding where a grid cell is encoded based on whether the object in question is located at the exact centroid of the geographic space occupied by each cell.

Centroids: For graphical objects, the exact geometric center.

Chains: In the POLYVRT vector data model, collections of line segments that begin and end with specific nodes that indicate topological information such as to and from directions as well as left and right polygons.

Changes of dimensionality: The process of either increasing or decreasing the dimension of cartographic objects often on the basis of the scale at which the map is produced. For example, a line (1 dimension) at one scale may be represented as an area at a substantially larger scale.

Chi-square: A statistical test that examines the relationship between observed and expected distributions of objects.

Chroma: The perceived amount of white in a hue compared with a gray tone of the same value level.

Circuitry: The degree to which nodes are connected by circuits of alternative routes. Defined by the alpha index.

Circuits: Type of network that allows movement to and from the initial point.

Circular variance: The circular analog to standard deviation. A measure of the degree of variation from a mean resultant length.

Class interval selection: The process of using a variety of statistical techniques to group cartographic data into classes for cartographic display. This data aggregation results in loss of data for analysis but is essential for many map production purposes to allow readability of the map.

Classed choropleth map: The traditional method of value by area mapping that first aggregates the data into classes before the polygons are assigned color or shading patterns.

Classless choropleth mapping: A proposed mapping technique that assigns a color or shading pattern to each polygon on the basis of its unit value rather than grouping polygons into classes.

Closed cartographic form: Polygonal object whose entire extent is contained within a single map.

Clustered: Spatial arrangement of objects where they occur as groups located close to one another, leaving large empty spaces between. This type of arrangement indicates that the processes operating on the objects are different near the clusters rather than in the intervening space.

Clustering: The process that produces a spatial arrangement where some objects are located very close to one another whereas others are widely separated.

Color video: A passive remote-sensing device designed to produce color images of the ground surface and storing it on videotape for later retrieval and analysis.

Communication paradigm: A conceptual model where the map is viewed primarily as a means of graphic communication.

Communities: Higher order groups of objects that either occur in a particular pattern that differs from their surrounding objects or that are linked as a functional region.

Computer-assisted cartograpy (CAC) systems: A computerized system designed primarily to assist in the production of maps.

Computer-aided drafting (CAD): A set of computer programs designed to assist in the process of drafting. It is normally used for architectural purposes but can also be used for drafting maps and as an input to GIS.

Computer file structures: Any of a number of ways in which files are named and where they are placed logically for storage and retrieval.

Concavity: A measure of the degree to which an area object produces concave shapes along its perimeter.

Conditions of use: Map design consideration based on such factors as viewing distance and lighting that limit the readability of the document as it is used.

Conflation: The general process of integrating or merging map data from multiple sources. This process is usually employed to either match two or more maps along edge or to co-register maps. The process often requires some form of rubber sheeting.

Conformal: The property of angular relations maintained by conformal projections.

Conformality: See Conformal.

Conformal projection: A type of map projection that maintains all angular relations normally found on the reference globe.

Conformal stereographic projections: Azimuthal projections that exhibit symmetrical distortion around the center point. Useful when the shape of the area to be mapped is relatively compact.

Conical projections: A family of map projections resulting from conceptually transferring the earth's coordinates onto a cone.

Connectivity: The degree to which lines in a network are linked to one another.

Constrained math: Any of a number of mathematical functions applied to cartographic coverages to reclassify their attributes.

Contiguity: A measure of the degree of wholeness within a region or of the degree to which polygons are in contact with one another.

Contiguous regions: Regions that are defined on the basis both of category homogeneity and of all parts' being in direct contact with one another.

Continuous: Data that occur everywhere on the surface of the earth. An example would be temperature.

Continuous surface: Any statistical surface whose values occur at an infinite number of possible locations.

Contour interval: Class interval illustrating the difference in elevation between contour lines.

Contour line: An isarithm connecting points of identical elevation.

Control points: Any points on a cartographic document for which the geographic coordinates are well known and reasonably accurate. These points are often used during the process of coregistration of two or more coverages or for adjusting the locations of other objects on the same coverage through conflation.

Controls: The limitations on the map design process.

Conventions: For maps, agreed-upon rules that have been carefully selected and are generally accepted by the mapping community.

Convexity: A measure of the degree to which an area object produces convex shapes along its perimeter.

Coregistered: The result of the process of precisely locating the coordinates for two or more GIS coverages so that their spatial locations match each other.

Coverage: The common terminology signifying a single thematic map in ESRI's ArcInfo data model. The more generic terminology for thematic maps in a multilayered GIS is layer or theme.

Correlative predictive models: Any of a number of models that employ logical or statistical correlation to make predictions about the variables under consideration.

Cross-sectional profile: A method of visualizing statistical surfaces by examining one transect through the surface viewed from the side.

Crosswalk: The process of matching disparate categories between maps or GIS layers.

Cumulative distance: Usually associated with cost surfaces. The accumulation of cost in terms of time, energy, etc., as one travels from place to place.

Cylindrical projections: A family of map projections resulting from conceptually transferring the earth's coordinates onto a cylinder.

Dangling node: A node located at the end of an undershoot.

Dasymetric mapping: A variety of techniques designed to improve area homogeneity for choropleth maps.

Dasymetry: See Dasymetric mapping.

Data dictionary: Detailed description of the data contents of a database, with particular attention being paid to explanations of categories.

Data requirements model: A database design model focusing on the required data needs as a means of properly completing the design.

Database: A collection of many files collectively associated with a single general category.

Database management system: Any of a variety of computer organizational structures that allow search and retrieval of individual files or items in the database. These systems are generally of three primary types: hierarchical, network, and relational.

Database structures: A set of approaches to organizing large collections of computer files commonly relating to a single major subject. Database structures are designed to enable the user to store, edit, search, and retrieve files or individual pieces of data within those files.

Data mining: Any of a number of techniques designed to identify patterns from large, often data-rich environments.

Datum: A reliable starting point from which accurate measurements of the earth's size can be made.

Dead reckoning: A method of surveying that measures distance and direction from one point to the next.

Decision system matrix: An aid to determining system design requirements, especially for linking spatial information products to necessary data elements.

Deductive models: Cartographic models that move from a general goal to the selection of individual components needed to achieve it.

DEMs: See Digital elevation models.

Density: Measure of the number of objects per unit area.

Density of parts: A type of dasymetric mapping that uses improved quantitative information about subarea density on maps to calculate improved information about the remaining polygons.

Density zone outlining: A type of dasymetric mapping that creates neighborhoods of uniform density of cartographic objects that are then mapped as individual polygons. Also known as density of parts.

Descriptive: A type of cartographic model that describes an outcome based on a set of conditions.

Design creep: Process that occurs when the design process exceeds the organizational learning curve. Usually results in a system that is overdesigned and offers more than the client needs.

DD: Decimal degrees method of designation for geocoded objects.

Differential rectification: Process used to adjust aerial photographs for planimetric errors due to changes in aircraft altitude and elevational features on the ground.

Diffusion: A spatiotemporal process wherein objects move from one area to another through time, or wherein additional objects appear where they had not previously existed.

Digital elevation models: Digital model of landform data represented as point elevation values. Also called digital terrain models (DTM).

Digital line graphs: Digital representations of the graphics contained on USGS topographic maps. These do not include topographic data as one might find in a digital elevation model.

Digital orthophotoquads: Digital version of aerial photographs that are constructed to eliminate image displacement due to changes in aircraft tilt and topographic relief.

Digital photography: A passive remote-sensing device that produces still digital images similar to that from video, but as single frames.

Digitizers: Electronic devices designed to transfer analog cartographic data to digital form.

Direct files: Within an indexed file structure a condition in which the data items themselves provide the primary order of the files.

Directed network: Type of network that places restrictions on the direction of movement (e.g., a one-way street).

Directional filter: A high-pass filter that enhances linear objects that lie in a particular direction (e.g., NE-SW).

Dirichlet diagrams: See Thiessen polygons.

Discrete: Data that occur only in selected places on the earth's surface. An example is human population.

Discrete altitude matrix: A lattice of point values that represents elevation, used to model topography in a GIS.

Discrete surface: Any statistical surface whose values are limited to selected locations.

Dispersed: A spatial arrangement that exhibits the greatest possible distance between objects in a confined space.

Dispersion: A measure of arrangement that focuses on the relationship between one area and its neighbors, especially regarding the average distances and average density of patches in the map.

DMS: Degrees, minutes, seconds method of designation for geocoded objects.

Dominant type method: A method of raster encoding where a grid cell is encoded based on whether the object in question occupies greater than 50% of the geographic space occupied by each cell.

Dot distribution map: See Dot mapping.

Dot mapping: The cartographic technique of representing one or more objects as points on a map, and where the exact geographical locations are not precisely recorded.

Doughnut buffer: A series of buffers of varying distance produced one inside the other.

Drift: The random, spatially correlated elevational component for a Kriging model. The general trend in elevational values.

Dumpy level: A telescopic survey device that has an elongated base with a movable joint that allows vertical movement so that elevations can be measured easily in steep terrain.

Easting: On some grid systems, a measurement of distance east of a preselected standard starting meridian.

Edge enhancement: A type of filter, often called a high-pass filter, that enhances values that change rapidly from place to place so that these changes can easily be observed.

Edge matching: The process of aligning the edges of two or more cartographic documents.

Edginess: Measure of the amount of edge in a polygonal cartographic form.

Ellipsoid: Hypothetical, nonspherical shape of the earth that resembles a flattening of the earth at the poles due to rotation. Basis for datums.

Encode: To place analog graphic data into a form that is compatible with computer cartographic data structures.

Entities: Points, lines, and polygons as they are represented by computerized cartographic data structures.

Entity-attribute agreement error: Related to attribute error. A mismatch between entity objects and the attributes assigned to them.

Entity error: Error of position for cartographic objects.

Equal area projections: A group of map projections that maintain the property of equal area for cartographic objects. These projections are useful for small-scale general reference and instructional maps.

Equidistant projections: A group of map projections that maintain the property of equal distance along straight lines radiating through a central starting point from which the projection was made.

Equivalent projections: See Equal area projections.

Euclidean distance: Distance measured simply as a function of Euclidean geometric space.

Euler function: Mathematical measure of spatial integrity that compares the number of regions to the number of perforations within all the regions.

Euler number: Number resulting from the implementation of the Euler function.

Exclusionary variables: Coverage variables whose importance outweighs the results of normal Boolean overlay operations for the purposes of decision making.

Explanatory predictive models: Any of a number of GIS models designed to explain or confirm predictions or hypotheses based on maps.

Extended neighborhoods: Regions that extend beyond those immediately adjacent to the focal point of analysis.

Extrapolation: The process of numerically predicting missing values by using existing values that occur on only one side of the point in question.

False eastings: On some coordinate systems, arbitrary, large values given to the Y axis of the false origin to allow only positive distance measured east of that point.

False northings: On some coordinate systems, arbitrary, large values given to the X axis of the false origin to allow only positive distance measured north of that point.

False origins: Arbitrary starting points on a rectangular coordinate system designed to allow only positive distances east and north.

Federated databases: Databases whose component parts are derived from disparate sources, each with its own data quality standards.

Filter: In raster GIS and digital remote sensing, a matrix of numbers used to modify the grid cell or pixel values of the original through a variety of mathematical procedures.

First normal form: Based on the theory of normal forms, the first requirement that all tables must contain rows and columns and that the column values cannot contain repeating groups of data.

Fishnet maps: A term often applied to wireframe diagrams but representing cartographic statistical surfaces.

Flowchart: A graphical device that illustrates the exact coverages and data elements, the operations to be applied to each, and the order in which the operations are applied to produce a cartographic model.

Focal function: GIS modeling function that uses properties of an immediate neighborhood for characterizing the target cell.

Foreign key: In relational database management systems, a column in a secondary table linked by a primary key in the primary table that is being used to join two tables.

Fragmented regions: Regions based on category homogeneity but with parts spatially separate from each other.

Freeman-Hoffman chain codes: Compact raster data models that use eight unique directional vectors to indicate the directional orientation or change of linear features.

Free sampling: In the join-count statistic, the method of testing that assumes that we can determine the expected frequency of within and between category joins based on theory or known patterns.

Friction surface: Some assigned attribute on a surface representation of a portion of the earth that acts to impede movement.

Functional distance: Distance measured as a function of difficulty rather than as a simple geometric distance.

Fuzzy tolerance: A user-defined distance of error to allow for minute digitizing mistakes.

Gamma index: A measure of complexity of a network that compares the number of links on a given network to the maximum possible number of links.

GBF/DIME: Geographic base file/dual independent map encoding system. A topological vector data model created by the U.S. Bureau of the Census and based on graph theory.

General reference map: A map, usually at small scale, whose primary objective is to show locations of different features on a single document.

Geocoding: The process inputting spatial data into a GIS database by assigning geographical coordinates to each point, line, and area entity.

Geodatabase: The GIS data model specific to ESRI's ArcGIS software where all GIS data are stored in relational database management systems tables.

Geodetic framework: A carefully measured system of ground-based coordinates designed to assure accurate locations for cartographic documents.

Geodetic reference system (grs80): The reference spheroid used for the NAD83 data.

Geographic data measurement: The processes and data levels that combine to characterize data observed on the surface of the earth.

Geographic database: A digital database developed from field-based spatial data. This process results in decreased accuracy and increased generalization because of the sampling necessary to produce it.

Geographic information science: The science underlying the principles of geographic information systems and their technology.

Geographic visualization (GVIS): A spatial approach to scientific visualization whereby the cartographic output is designed to elicit a response from the map reader that results in the formulation of new scientific hypotheses.

Geoid: Generalized physical model of the earth based on compilations of all variations from the ellipsoid.

GIS systems analysts: GIS professionals specializing in the proper design of the overall GIS operations.

Global positioning system (GPS): A satellite-based device that records locational (X,Y,Z) and ancillary data for portions of the earth.

Global view: Organizational view of a GIS database requiring the integration of multiple local views.

Gnomonic projections: Azimuthal projections where all great circles are represented as straight lines. Useful for marine navigation.

GPS: See Global positioning system (GPS).

Graph theory: A mathematical theory that examines the relationships among linear graphs.

Graphic data structure: A method of storage of analog graphical data into a computer that enables the user to reconstruct a close approximation of the analog graphic through some output procedure.

Graphic scale: One method of representing map scale where a preselected, real-world distance is shown as a bar or line on the map document. Sometimes called a bar scale.

Graphicacy: The level of understanding of graphic devices of communication, especially maps, charts, and diagrams.

Gravity model: A measure of the interaction of nodes based on their distance and some functional measure of their individual importance.

Grid: Term normally given to represent the collective set of grid cells that make up a thematic map in raster GIS.

Grid cells: Raster data structures in which the geographic space is quantized into rectangular shapes of equal size and shape.

GRID/LUNR/MAGI: Raster GIS data model where each grid cell is referenced or addressed individually and organized in columnar fashion.

Grid system: A system of horizontal and vertical lines on a globe or projected map that allow for locating objects in geographic space and making measurements among objects.

Ground control points (GCPs): Points of known geographic location used to register satellite imagery and other coverage data to the geodetic framework.

GRS80: Standard geographic reference system based on NAD83.

Heterogeneity: The amount of diversity within a selected region. Opposite of homogeneity.

Hierarchical data structures: Computer database structures employing parent child or one-to-many relationship that requires direct linkages among items for a search to be successful.

Hierarchical structure: The graphic appearance of some cartographic objects appearing to be more important than others, resulting from one or more cartographic techniques to produce this result.

Hierarchy theory: A dialect of general systems theory having emerged as part of the general science of complexity. It focuses on levels of organization and issues of scale, especially as they deal with biological systems.

High-pass filter: See Edge enhancement.

Holistic paradigm: See Analytical paradigm.

Homogeneity: The degree to which attributes in a region are similar. If all attributes are identical within a selected region, they are said to be homogeneous.

Hot-link: A link between two applications such that changes in one affect the other. For example, a document might be hot linked to a map so that when the embedded link is activated the document appears.

Hot-spots: Geographic locations of high levels of activity. An example might be a location where an unusually high level of crime activity occurs.

Hue: The color of a cartographic symbol based on the wavelength of electromagnetic radiation reaching the viewer.

Hybrid systems: GIS systems whose entity and attribute tables are separate and are linked through a series of pointers and identification codes.

Identity overlay: The use of tabulations of data from two or more coverages to make decisions about how attribute values are to be combined. This technique can be employed to enhance a number of other methods of overlay by providing a decision tool for their use.

IMGRID: Raster GIS data model where grid cells are referenced as part of a two-dimensional array and where the thematic attributes are coded in Boolean fashion (i.e., 1 and 0 only).

Immediate neighborhoods: Regions limited to those immediately adjacent to the focal point of analysis.

Impedance value: A user-specified numerical value used to simulate the effects of barriers or friction surfaces. These are also employed in the same way for network modeling where the impedance values indicate the degree to which a network allows travel.

Incorrect attribute values: Attribute error resulting from the assignment of incorrect or nonexistent attribute values to entities.

Incremental distance: The simple addition of distance at each successive step.

Indexed files: Computer file structures that allow faster search than ordered sequential because each entry is assigned an index location through the use of software pointers.

Inductive models: A method of cartographic model that begins with individual observations (data elements) and proceeds to develop general patterns.

Information theory: A field of study that deals with information content. It is used in GIS primarily as an estimate of the amount of sampling necessary to encode a graphic object.

Innovation diffusion: A set of geographic models that examine the movement of ideas, innovations, or strategies through geographic space.

Inscribe: To draw within so as to meet yet not cut the boundaries. For example, a circle is inscribed in a polygon, when it touches each side of the polygon.

Institutional design: Design issues concerning the role of institutional considerations such as training and personnel, data security, organizational functioning, and many others.

Institutional setting: The size, type, organizational hierarchy, goals, and overall objectives of each organization likely to adopt GIS.

Integrated systems: GIS systems whose entity and attribute tables are shared.

Integrated terrain unit (ITU): A unit of geographic space defined by the explicit collection of various attributes at one time, usually from aerial photography.

Interactions: A measure of arrangement that focuses on the proximity, sizes, and amount of edge between neighboring areas, especially regarding the edges of these areas.

International date line: Meridian drawn at 180 degrees east and west of the prime meridian.

Interpolation: A process of predicting unknown elevational values by using known values occurring on multiple locations around the unknown value.

Interval: Comparative data with a relatively high degree of accuracy, but with an arbitrary starting point (e.g., degrees Centigrade).

Intervisibility: Analytical technique that allows for the determination of visibility from one object to another and back.

Inverse map projection: The process of converting from two-dimensional (projected) map coordinates to geographical coordinates.

Inverted files: Within an indexed file structure a condition in which the data items are organized by a second or topic file that provides the primary order of the files. In such a system, the topic or inverted file is searched rather than the data items themselves.

Irregular lattice: A series of point locations whose interspatial distances are not identical.

Isarithm: General term for a line drawn on a map to connect points of known continuous statistical surface values.

Isarithmic map: The result of the process of isarithmic mapping.

Isarithmic mapping: The process of using line symbols to estimate and display continuous statistical surfaces.

Island pseudo nodes: Spatial pseudo nodes that result when a single line connects with itself.

Isolated State Model: An early model to explain the arrangement of agricultural activities on the basis of their distance to a single market.

Isolation: A measure of arrangement that focuses on the distance between objects in geographic space.

Isoline: A line connecting points of known or predicted equal statistical surface value.

Isometric map: A map composed of isolines whose known Z values are sampled at point locations.

Isoplethic map: A map composed of isolines whose known Z values are recorded for polygonal areas rather than at specific point locations.

Isotropic surface: A measure of simple Euclidean distance from some central point outward throughout a coverage where no obstructions or frictional changes exist.

ITUs: Integrated terrain units.

Knowledge discovery: Any of a number of techniques to identify pattern in seemingly unpatterned collections of data.

Kriging: An exact interpolation routine that depends on the probabilistic nature of surface changes with distance.

Lag: On a semivariogram, the distance between sample locations, plotted on the horizontal axis.

Lambert's conformal conic projections: Conformal map projections that have concentric parallels and equally spaced, straight meridians meeting the parallels at right angles. Provide good directional and shape relationships for east-west midlatitudinal zones.

Lambert's equal area projections: Azimuthal equivalent projections with symmetrical distortion around a central point. Useful for areas that have nearly equal east-west and north-south dimensions.

Land information systems: Subset of geographic information systems that pay special attention to land-related data.

Latitude: Angular measurement north and south from the equator.

LCGU: Least common geographic unit. A computer graphics construct that provides topological information for vector overlay operations to determine the ability or lack of ability of polygons to be further divided.

Least convex hull: Straight-line segments connecting all outside points within a distribution in such a fashion that the smallest possible polygon is produced.

Least-cost distance: A distance measure based on the minimization of friction or cost for a single path.

Least-cost surface: A distance measure based on the minimization of friction or cost for an entire surface.

Legibility: The ability of a cartographic object to be visually observed and recognized by the map user.

Limiting variables: Variables that result in the absence of one or more attributes because of how they interact with one another. A method of dasymetric mapping.

Line dissolve: In vector GIS, the process of eliminating line segments between polygons so that they become larger polygons with identical attributes.

Line-in-polygon: A method of overlay that combines a coverage with linear objects with a polygon coverage to determine the numbers and extents of linear objects that fall within selected polygons.

Line intersect methods: A group of techniques for analyzing the spatial distribution of linear objects by drawing one or more sample lines at random and noting where they intersect the coverage lines.

Linear interpolation: A type of interpolation linearity: measure of the ability of a digitizer to be within a specified distance or tolerance of the correct value as the puck is moved over large distances.

Linearity: The property of having a single dimension.

Local operator models: Methods of interpolation that rely heavily on existing elevational data that are located within close proximity to the points being predicted.

Local views: Individual users' views of a GIS database.

Locate: Determining the best location for some form of geographic activity, most often used for economic activities.

Locational information: Any information about the absolute or relative coordinates of points, lines, or areas. This information is often used to reclassify coverage values.

Location allocation: A group of models designed to determine the best locations for activities in geographic space and to assign a portion of geographic space to existing facilities based on demand and location.

Longitude: Angular measurement east and west from the prime meridian.

Low-pass filter: See Smoothing.

Mandated buffer: Buffers whose distances are dictated by legal mandate (e.g., frontage along homes dictated by zoning ordinances).

Map Analysis Package (MAP): Raster GIS data model where grid cells are referenced as part of a two-dimensional array and where the thematic attributes for each coverage are referenced by separate number codes or labels, thus allowing a range of values for each category in a single theme.

Map legend: The portion of the map document that describes the symbols. Conceptually, the map legend ties the entities and attributes together in an analog map document.

Map projections: Any of a number of approaches to transfer the spherical earth onto a two-dimensional surface. Each projection is an approximation and imposes its own limitations on the utility of the map.

Marble model: A spiral-structured GIS design model that uses an increasing level of detail to assure design flexibility.

Mathematical overlay: A method of overlay, usually associated with raster GIS, that uses mathematical and algebraic expressions to create new coverage attribute variables.

Mean center: Center of a group of point objects located in space derived by calculating the average X and Y coordinate distances.

Mean resultant length: The result of averaging the resultant length by the number of observed component vectors.

Measurable buffers: Buffers whose distances are selected based on exact measurements of existing phenomena as they occur in space.

Mercator projections: Conformal map projections introduced specifically for nautical navigation. It maintains the property that rhumb lines always appear as straight lines.

Meridians: Lines of longitude drawn north and south on the globe and converging at the poles.

Metadata: Data about data. An overall description of the contents of a database.

Method of ordinates: A simplified method for calculating volumes.

Missing attributes: Attribute error resulting from a failure to assign labels to point, line, or polygonal features or to grid cells.

Missing labels: Entity-attribute agreement error resulting from failure to label a polygon during the input process.

Mixed pixels: See Mixels.

Mixels: Mixed pixels-pixels whose spatial extent is coincident with more than one category of object on the earth. Mixels often result in category confusion during the classification process.

Model verification: The process of comparing expected results to observed results of a cartographic model to ensure it is performing as it should.

MrSID: A pyramid based method of spatial data compression.

Mythical-man month: The erroneous concept that if a problem arises, the simple addition of more workers to the problem will result in its solution. This idea ignores the necessary learning curve for people asked to perform tasks with which they are unfamiliar.

Nation building: The process of creating or restoring the social, cultural, economic, and political mechanisms responsible for a viable nation to survive.

Nearest neighbor analysis: A statistical test to compare the distance between each point object and its nearest neighbor to an average "between-neighbor" distance.

Neighborhood functions: GIS analytical functions that operate on regions of the database within proximity of some starting point or grid cell.

Network complexity: Overall network patterning that combines the number of links and the degree of connectivity.

Network systems: Computer database structures employing a series of software pointers from one data item to another. Unlike hierarchical data structures, network systems are not restricted to paths up and down hierarchical pathways.

Networks: 1. Two or more interconnected computers often used to allow communications between them. 2. Higher order linear objects. Interconnected lines (arcs, chains, strings, etc.) defining the boundaries of polygons or a linear object that allows movement (e.g., road network).

Node: In vector data structures, a point that acts as the intersection of two or more lines (links) and explicitly identifies either the be-

ginning or ending of each line to which it is attached.

Nominal: Level of data measurement that is noncomparative, usually representing a description or name.

Nonfree sampling: In the join-count statistic, the sampling procedure that compares the number of joins of a random pattern of joins to those measured.

Nonlinear interpolation: A method of interpolation that accounts for the nonlinear nature of elevational change with distance.

Normal forms: In relational database management systems, a theory of table design that specifies what types of values columns may contain and how the table columns are to be dependent on the primary key.

North American datum (NAD27): Accepted U.S. datum as of 1927 based on a flatness ratio of the earth of $1/294.9787$.

North American datum (NAD83): Datum established for the United States as of 1983, based on estimates of the center of the earth rather than surface measurements.

Northing: On some grid systems, a measurement of distance north of a preselected standard starting parallel.

Nugget variance: Spatially uncorrelated noise associated with a Kriging model.

Object-oriented database management systems: GIS systems based on object-oriented programming methods and demonstrating object inheritance.

Ontologies: An explicit formal specification of how to represent the objects, concepts, and other entities that are assumed to exist in some area of interest and the relationships that hold among them.

Ordered sequential files: Computer file structures that are ordered based on some form of alphanumeric scheme much like alphabetizing a mailing list. This file structure makes it more difficult for data input, but allows searches and retrieval with greater speed than with simple lists.

Ordinal: Ranked data (e.g., good, better, best) that are comparable only within a given spectrum.

Organizational diagram: Hierarchical diagram showing the structure of an organization from management to workers.

Orientation: 1. The azimuthal directions in which linear objects are placed. For area objects this can also be applied to the major axis of the object. 2. The azimuthal directions of internal symbol markings used to separate one category from another.

Orthogonal: The normal point of view for viewing a cartographic document directly from above.

Orthographic projections: Azimuthal projections that look like perspective views of the globe at a distance. Useful for illustrations where the sphericity of the globe must be maintained.

Orthomorphic projections: See Conformal projection.

Orthophotoquads: Aerial photographs that are constructed to eliminate image displacement due to changes in aircraft tilt and topographic relief.

Overlay: The operation of comparing variables among multiple themes or layers.

Overshoot: An arc that extends beyond the arc with which it was meant to connect.

Parallels: Lines of latitude drawn east and west around the globe parallel to the equator.

Passive remote sensing: A method of remote sensing that uses ambient energy from the object being sensed, rather than sending out a signal for later sensing.

Passive simplification: The process of cartographic simplification that does not require the explicit intervention of a human operator.

Pattern: The regular arrangement of cartographic objects or internal symbol markings.

Pattern characterization: Any of a number of techniques for quantitatively describing patterns.

Pattern comparison: Any of a number of techniques for quantitatively describing the similarity or dissimilarity between two or among many sets of patterns.

Pattern detection: Any of a number of techniques for identifying the existence of patterns in seemingly randomly distributions of spatial objects.

Pattern exploitation: If spatial patterns have been identified, this allows the user to take advantage of such identified patterns. One example of a pattern that might be exploited is a set of well-ordered streets that will allow increased traffic flow.

Percent occurrence: A method of raster encoding where a grid cell is encoded so that it includes all categories falling within the geographic space occupied by each cell as well as the percentage of each cell it occupies.

Perforated regions: Regions that demonstrate category homogeneity, but within which other dissimilar polygonal forms exist (e.g., islands within a lake).

Perimeter-area ratio: Method of measuring polygons that compares the perimeter of each polygon to its area as a ratio.

Pixels: In remote sensing, picture elements, each of which contains a preselected amount of geographic space and a set of electromagnetic measurements assigned to each. Pixels most often resemble grid cells when displayed at small scale.

Plane coordinates: See Rectangular coordinates.

Plane table and alidade: A common traditional field device used for surveying and consisting of a table upon which a draft survey document would be placed, a telescopic device for determining line of site at a distance, and a tripod upon which they both rest.

Planimeters: Mechanical devices designed to measure lengths and areas on analog maps.

Point-in-polygon: The process of overlaying a point coverage with a polygon coverage to determine which polygons contain which points.

Polygon: A multifaceted vector graphic figure that represents an area.

Polynomials: Mathematical equations with two or more terms, used for approximating surface trends. The more complex the polynomial, the more complex the surface estimate.

POLYVRT: Polygon converter model. A vector data model that uses explicit topological data stored separately for each type of entity.

Position measures: Analytical techniques that characterize cartographic objects on the basis of their absolute positions or of their positions and arrangements relative to other cartographic objects.

Precision agriculture: An application area of GIS, GPS, and remote sensing to support agriculture.

Predictive models: A group of cartographic models (either descriptive or prescriptive) that rely on the prediction of outcomes given a set of conditions.

Prescribe: To recommend the best solution.

Prescriptive: A type of cartographic model that determines the best possible solution from a set of conditions.

Presence/absence method: A method of raster encoding where a grid cell is encoded based on whether the object in question is either present or absent from that portion of geographic space occupied by each cell.

Primary data acquisition: The process of acquiring data directly from the earth either in-situ or using remote-sensing devices.

Primary key: A set of attributes in a relational database that are designed as the primary search criteria for the database. For example, a database of students may have as a primary key the student name.

Prime meridian: Arbitrary starting point for lines of longitude located at Greenwich, England.

Principal scale: A given representative fraction for a reference globe derived by dividing the earth's radius by the radius of the globe.

Project life cycle: A highly structured model developed to ensure proper software development that proceeds from module testing to subsystem testing and finally system testing.

Projection family: A group of map projections based on a conceptual model of transferring the spherical earth onto a cylinder, a cone, or a flat surface. The conceptual model is visualized as if the spherical coordinates are physically transferred onto these individual surfaces by projecting a light source located at the center of the sphere toward the projected surface.

Proximal region: The area of influence defined for each point object when subjected to Thiessen polygonal analysis.

Pseudo nodes: A node where two, and only two, arcs intersect, or where a single arc connects with itself.

Public land survey system (PLSS) grid: Common grid system used in the United States to divide the land into square mile sections, each equal to 640 acres.

Purpose: A primary limitation on the design of a cartographic document relating to what the map is to be used for.

Quadrat analysis: A standard method of testing point distribution patterns by comparing the numbers of objects from one subarea (quadrat) to another.

Quadrats: Uniform subareas designed for sampling point objects.

Quadtrees: Compact data structures that quantize geographic space into variably sized quarters, each of which exhibits attribute homogeneity.

Quantize: The process of dividing data into quanta or packets. In GIS the process is most commonly associated with the encoding of geographic space into some form of raster data structure.

Radiotelemetry: A method of tracking large animals with the use of radio collars that transmit a signal that can be received either on the ground or through low-flying aircraft.

Random: Spatial arrangement of objects where their locations relative to one another are unpredictable. This type of arrangement is indicative of processes that operate through pure chance occurrences.

Random noise: The nonspatially correlated error or residual component for a Kriging model.

Random walk: A line-intersect method for analyzing the distribution of linear objects by drawing a zigzag path and recording its points of intersection with lines on the coverage.

Range: On a semivariogram, the critical value where the variance levels off or stays flat.

Range graded: The result of a cartographic classing technique that groups ranges of numerical values into single classes.

Raster: A form of GIS graphic data structure that quantizes space into a series of uniformly shaped cells.

Raster chain codes: A compact method of storing raster data as chains or grid cells bordering homogeneous polygonal areas.

Raster database structures: Any of a number of methods of quantizing geographic space into discrete uniform chunks.

Ratio: The highest level of data measurement that includes an absolute starting point and allows ratios of values to be produced (e.g., salaries).

Ray: A line based on optical geometry representing the uppermost bound of visible objects from a viewer location.

Ray tracing: Analysis of locations on a map that are visible from a viewer location by using a series of rays.

Reality: A map design constraint resulting from the often irregular shapes of real geographic objects.

Rectangular coordinates: Coordinate system based on a two-dimensional planar surface, using X and Y coordinate locations from an origin or starting point. The origin is normally based on an arbitrary Cartesian coordinate system. This is the common coordinate system used by digitizer tablets.

Reductions: In mapping, the idea that the cartographic document is normally smaller than the area it represents.

Reference globe: A hypothetical reduction of the earth and its coordinates mapped onto a sphere that is reduced in size to that chosen for the scale of the flat map being produced.

Region growing: In remote sensing the procedure that is most often used in unsupervised classification as it attempts to define groups of pixels (regions) with similar spectral signatures.

Regions: Areas of relative uniformity of content within a coverage.

Registration points: Digitizer locations that locate the corners of the map document so that the user can digitize different parts of the document at different sessions.

Regular: Spatial arrangement of objects where all objects are the same distance away from their immediate neighbors. Such an arrangement implies some form of process that places these objects this way.

Regular grid: A series of locational lines whose interspatial distances are identical.

Regular lattice: A series of point locations whose interspatial distances are identical.

Related variables: Variables from two coverages whose interactions can be defined by mathematical equations. A method of dasymetric mapping.

Relational database structures: Computer database structures employing an ordered set of attribute values or records known as tuples grouped into two-dimensional tables called relations.

Relational join: In relational database management systems, the process of creating functional links between two or more tables sharing some common attribute.

Relations: Two-dimensional tables of relational database attribute records.

Relative barriers: Barriers that, like friction surfaces but at discrete locations, act to impede movement.

Relative location: Location determined in relation to a second object rather than by its own coordinates.

Remote sensing: The observation of objects or groups of objects, normally at a distance, most often with the use of some form of mechanical or electronic device. The data can either produce an image or be stored for later retrieval.

Repeatability: Deals with how close the digitizer will read to an original value if it is recorded more than once.

Representational fraction (RF): One method of representing map scale where both map units and earth units are represented by the same measurement units and are shown as a fraction (e.g., 1:24,000).

Resolution: 1. The amount of earth surface represented by a single grid cell in a raster GIS. 2. The ability of a digitizer to record increments in space. The smaller the units it can handle the better its resolution.

Resultant length: The length of the resultant vector calculated through the use of the Pythagorean theorem on component vectors.

Root mean square error (RMS): The distance between the input (source) location of a GCP and the retransformed location for the same GCP. Measure used to assess the accuracy of geometric correction

Rose diagram: A circular graph that displays the orientation of linear objects or phenomena by starting at the center of a circle and drawing each observation as a single line outward from that point.

Rotation: The process of moving part or all of a cartographic object in some azimuthal direction around some focal point.

Rough surface: A statistical surface whose values change rapidly and often unpredictably with changes in horizontal distance.

Routed line cartograms: Cartograms that show the order of objects located along a line but do not show the actual distances between them.

Routing: The process of using networks to define travel paths from one place to another.

Roving window neighborhood functions: A group of analytical functions that employ a moving filter to reclassify the values of the coverage.

Rubber sheeting: A method to adjust coverage features in a nonuniform manner. Used to rectify objects within a coverage.

Rules-of-combination overlay: A method of overlay that allows the user to define which categories will be combined, which will take precedence, and which may be ignored.

Run-length codes: A compact method of storing raster data as strings of homogeneous grid cells.

Runs test: A simple statistical technique for analyzing the pattern of sequences of data such as those derived through line intersect methods.

Saddle-point problem: For interpolation, the problem that arises when one pair of diagonally opposite Z values forming the corners of a rectangle is located above and the second pair below the value the algorithm is attempting to solve.

Sampled area: The portion of a larger area that is to be sampled to draw inferences about the whole.

Sampling frame: A complete list of all members of a population to be sampled, or the entire area within which a sample is to be selected.

Scale: In mapping, the size relationship or ratio between the map document and the portion of the earth it represents.

Scale change: The process of changing the size of part or all of a cartographic object.

Scale-dependent error: Spatial data error that is primarily a function of the scale of the input map document.

Scale factor (SF): On a reference globe, the actual scale divided by the principal scale. By definition, the scale factor for a reference globe should be 1.0 because the actual scale and the principal scale should be the same everywhere.

Scan lines: Lines of pixels resulting from the process of scanning used by the remote-sensing device.

Scanning radiometers: Passive remote-sensing devices that receive electromagnetic radiation in groups of wavelengths.

Secondary data acquisition: The acquisition of data not collected through direct observation of the subject objects themselves.

Second normal form: Based on the theory of normal forms, the second requirement that every column that is not part of the primary

key must be completely dependent on the primary key.

Selective overlay: See Rules-of-combination overlay.

Semivariance: One half the square of the standard deviation between each elevational value and its neighbors.

Semivariogram: A graphical device used in the process of Kriging that compares the variance of the difference in value of elevational points to their distances.

Shape measures: Analytical techniques that characterize cartographic objects based on comparisons of their own dimensional attributes (e.g., perimeter area ratios), comparisons to known geometric shapes (e.g., circles), or to alternative geometry's (e.g., fractal geometry).

Shortest path: A coverage created by determining the shortest Euclidean distance from one place to another. This can be determined for a surface or for networks.

Shortest path surface: A coverage created by determining the distance from a point, line, or area to all other locations on the coverage. This coverage includes the shortest path as part of the result.

Side-looking airborne radar (SLAR): A system mounted at the base of an aircraft that sends out and receives radar signals in a sweeping motion perpendicular to the flight path of the aircraft.

Simple list: The simplest computer file structure that stores data as it is input but does not organize it in any sequence or order. This type of file structure allows very easy input but makes searches and retrieval operations very difficult.

Sinuosity: The relationship between straight-line distance and actual distance of linear objects.

Sinuous: Wandering wildly in a lateral direction along its length.

Size measures: Analytical techniques that characterize cartographic objects based on their absolute sizes or comparisons of their sizes with other cartographic objects.

Skew: Measure of the squareness of the results of digitizing four corners of a document on a digitizing table.

Slicing: A group of GIS functions that selectively change the class interval for statistical surfaces.

Sliver polygons: Small polygons, often without attributes, that result either from digitizing the same line twice or following the overlay of two or more coverages.

Slope: Measure of the amount of rise divided by the amount of horizontal distance traveled.

Smooth surface: A statistical surface whose values change at a relatively constant rate with changes in horizontal distance.

Smoothing: A type of filter, often called a low-pass filter, that reduces the value of extraordinarily high cell or pixel values through some process of averaging.

Software design: The structured process of transforming software functional requirements to specific code.

Software engineering: A subfield of computer science specializing in the systematic approaches to software code generation.

Spacing: The attribute of spatial objects that shows their distances from one another.

Spaghetti model: The simplest vector data structure representing a one-for-one translation of the graphic image. Often used as a data structure for vector input.

Spatial: Anything dealing with the concept of space. In the geographic context, primarily dealing with the distribution of things on the surface of the earth.

Spatial arrangement: The placement, ordering, concentration, connectedness, or dispersion of multiple objects within a confined geographic space.

Spatial information product (SIP): The desired output from a set of GIS analysis operations.

Spatial integrity: The degree of perforation of a perforated region.

Spatial pseudo nodes: Pseudo nodes that represent island polygons or where some attribute along a line changes attributes. Both of these are acceptable pseudo nodes.

Spatial surrogates: Spatially explicit data used to replace data that do not exhibit spatial properties.

Spatially correlated: Sets of objects that are spatially associated may also show statistically significant relationships that may indicate some connection between the processes acting on both objects or even the processes acting on one set of objects influencing those of another set of objects in the same area.

Splines: Mathematical methods of smoothing linear cartographic objects. Often used in interpolation methods.

Stability: The property of a digitizer that deals with the propensity of the device to change the readings it provides as it warms up.

Standard parallels: Lines on a reference globe that maintain a scale factor of 1.0.

Static neighborhood functions: Analytical functions that modify the attributes of neighboring cells without the use of a moving filter.

Statistical surface: Any ordinal, interval, or ratio data that can be represented as and operated on as a surface.

Step rate: The incremental distance covered by an output device during the process of drawing lines. The smaller the step rate, the finer the resolution of the output document.

Stereographic projections: See Conformal stereographic projections.

Structure: See Drift.

Substantive objective: One of two parts of the purpose of a map, this one relating to the information the map must include. Controls what is to be mapped.

Supervised classification: In remote sensing the process of classifying digital data with the interaction of a user. The user normally selects pixels that are of known category and uses these to train the software about the electromagnetic properties likely to be associated with each category.

Supplier for GIS (GIS supplier): Third-party vendor for GIS software, and sometimes hardware, used by the GIS community.

Surface-fitting models: Interpolation methods that attempt to fit the existing data points into a mathematical surface equation.

System design: The process whereby the output from system analysis is transformed from technology independent statements of user requirements to the system that is going to be used to implement the GIS.

System operators: GIS professionals responsible for ensuring that the system functions correctly and for maintaining the hardware and software in a condition that allows the users to perform their tasks.

System sponsor: The organizational leadership responsible for making funding and logistical decisions and for selecting priorities for resource allocation within the organization.

System users: GIS professionals that interact most closely with the GIS through data input, storage, editing, and performing the actual analytical functions necessary to provide answers to queries.

Target population: That portion of a whole population that is likely to contain attributes for which a sample is being taken.

Targeted analysis: An analysis that is directed toward reclassifying a target cell or target groups of cells as a function of neighboring values.

Technical design: GIS design issues that deal with system requirements, hardware and software needs, and other noninstitutional issues.

Technical limits: A set of limitations on map design based on the physical limitations of the equipment to produce readable results.

Template: A user-selected area on a map that is employed as the outline for subsequent coverages to compare their attributes.

Tessellation: Any of a number of subdivisions of a two-dimensional plane (or three-dimensional volume) into polygonal tiles (polyhedral blocks) that completely cover a plane or volume. Planes or volumes can be either divided into regular or irregular tessellations.

Texture: The visual appearance of a symbol resulting from the spacing and placement of internal markings.

Thematic maps: Maps whose primary purpose is to display the locations of a single attribute or the relationships among several selected attributes.

Theodolite: A modern optical survey device that includes digital readouts and improved accuracy over its predecessors. This device is still limited in its ability to provide survey data for large areas.

Thiessen polygons: A method of creating polygons or proximal regions around point objects by defining them mathematically, dividing the space between each point, and connecting these distances with straight lines.

Third normal form: Based on the theory of normal forms, the requirement that every nonprimary key column must be nontransitively dependent on the primary key.

Tick marks: Digitizer locations of known latitude and longitude, used for the process of geocoding.

Tiling: Process of separating large databases into predefined subsections for archival purposes.

Too many labels: Entity-attribute agreement error resulting from placing more than one label point within a single polygon during input.

Topological models: A vector data structure that incorporates explicit spatial information about the relative locations of objects in the database. These are necessary to allow advanced analysis in a GIS.

Topologically Integrated Geographic Encoding and Referencing System (TIGER): A topological vector data model created by the U.S. Bureau of the Census as an improvement over GBF/DIME. Points, lines, and areas are explicitly addressed, allowing direct retrieval of census block data, and real-world objects are portrayed in their true geographic shape.

Topology: The mathematically explicit rules defining the linkages of geographical elements.

Total analysis of neighborhood: A function that analyzes the entire neighborhood and returns new values for that entire neighborhood.

Translation: The process of moving part or all of a cartographic object to a new location in Cartesian space.

Transverse projections: A type of Mercator projection that does not exhibit the property that all rhumb lines are straight. As a result, scale exaggeration increases away from the standard meridian, limiting the usefulness of this type of projection to a small zone along the central meridian.

Trapezoidal rule: A methodology for calculating the centroid of a highly irregular polygonal shape.

Trend surface analysis: An interpolation routine that simplifies the surface representation to allow visualization of general or overall changes in elevation value with distance.

Triangulated irregular network (TIN): A vector data model that uses triangular facets as a means of explicitly storing surface information.

Triangulation: A method of surveying that measures distance and direction from a baseline outward to the selected points.

Trilateration: A method of surveying that measures both from and to a baseline and each point to be located.

Tuples: Individual records (rows) in a relational database structure.

Undershoot: An arc that does not extend far enough to connect with another arc. Sometimes called dangling arcs or just dangles.

Undirected networks: Type of network that does not place restrictions on the direction of movement.

Uniform: A distribution of objects that exhibits the same density in one small subarea as that in each additional subarea.

Unit value: The actual numerical value assigned to individual polygons in a choropleth map.

Universal polar stereographic (UPS) grid: A grid system used to cover the polar areas of the globe. Each circular polar zone is divided in half at 0 and 180 degrees meridian.

Universal Transverse Mercator (UTM): An international plane (rectangular) coordinate system developed by the U.S. Army. The UTM divides the world into 60 zones of 6 degrees longitude. Each zone extends 3 degrees east and west from its central meridian and are numbered consecutively west to east from the 180-degree meridian. Transverse Mercator projections may then be applied to each zone.

Unsupervised classification: In remote sensing the process of classifying digital data without the interaction of a user. The process, most often statistical, randomly selects pixels and attempts to find spectral similarities between the selected pixels and the remainder of the pixels in the image.

Value: The lightness or darkness of a cartographic object compared to a standard black and white.

Value-by-area mapping: A method of mapping whereby each polygon is assigned a specific ordinal, interval, or ratio attribute value.

Variance-mean ratio (VMR): An index showing the relationship between the frequency of subarea variability to the average number of points per quadrat.

Vector: A graphic data structure that represents the points, lines, and areas of geographical space by exact X and Y coordinates.

Vector resultant: The result of combining the magnitudes and directions of component vectors.

Verbal scale: One way of representing map scale where both map units and earth units are stated as text (e.g., one inch equals 100 feet).

Vertical exaggeration: A technique used for production of cross-sectional profiles that increases the vertical dimension for easier viewing of surface changes.

View integration: The process of satisfying multiple user needs for a GIS database and its applications by incorporating each local view into the global view.

Viewing azimuth: The geographic direction from which a perspective map can be displayed so as to observe its different sides.

Viewing distance: One of a number of factors that can be modified to effectively display perspective maps.

Viewshed: A polygonal map resulting from analysis of topographic surfaces that portrays all locations visible from a preselected viewpoint.

Viewshed analysis: The techniques employed to produce surfaces that portray all locations visible from one or more viewing locations.

Visibility analysis: See Viewshed analysis.

Virtual reality markup language (VRML): A computer language that provides a three-dimensional environment for traditional Internet browsers, resulting in a simple form of virtual reality available over the Internet.

Visual contrast: The relationship between cartographic symbols and the underlying map background, often responsible for legibility.

Voronoi diagrams: See Thiessen polygons.

Waterfall model: See Project life cycle.

Weighted mean center: Center of a group of point objects located in space and modified by a weighting factor applied to each point. The weight is representative of some physical, economic, or cultural factor.

Weighting methods: Nonlinear interpolation methods that weight the predicted elevation values by the distance to its nearest neighbor elevation values.

Weird polygons: Graphical artifacts resembling real polygons but lacking nodes. Usually the result of digitizing points for polygonal objects out of sequence.

WGS84: Modified version of GRS80, developed for the United States military in 1984.

Wire-frame diagrams: A perspective view of a set of statistical data displayed as a skeleton form.

Witness trees: Trees left behind by surveyors during pioneer periods in the United States. These trees were left as witnesses for the metes and bounds survey method and have also been used more recently as indicators of previous forest vegetation.

World geodetic system (WGS84): The reference ellipsoid used by NASA/GSFC. It is the Defense Mapping Agency (DMA) World Geodetic System 1984 (WGS84) ellipsoid model.

Zonal function: GIS modeling function that uses properties of a defined region of cells (in-cluding fragmented regions) to characterize the target cell.

Zones: On the UTM grid system the earth is divided into columns 6 degrees of longitude wide. Each of these is called a zone and is numbered from 1 to 60 eastward beginning at the 180th meridian.

Zoom transfer scope: A mechanical, optical device designed for accurately transferring information from aerial photographs to projected map documents.

Chapter 2
Page 22 (left and right): Courtesy Del Webb Corporation. Page 35: Courtesy NASA. Page 36: Courtesy NAVSTAR/UNAVCO.

Chapter 5
Page 115: Courtesy Calcomp, Inc. Page 117: Courtesy Agfa Corp. Page 118: Photo courtesy The International Boundary and Water Commission. Graphics courtesy Dona Ana County, New Mexico.

Chapter 7
Page 178: Courtesy GeoHealth, Inc. Pages 180 and 183: Courtesy Michael N. DeMers.

Chapter 10
Page 243: Courtesy USGS.

Chapter 14
Page 372: Courtesy Topozone.

Chapter 16
Page 406: Courtesy Department of Geography, New Mexico State University. Page 409: Courtesy Michael N. DeMers.

Chapter 17
Page 430: Courtesy Environmental Systems Research Institute.